Advances in the Geological Storage of Carbon Dioxide

NATO Science Series

A Series presenting the results of scientific meetings supported under the NATO Science Programme.

The Series is published by IOS Press, Amsterdam, and Springer in conjunction with the NATO Public Diplomacy Division

Sub-Series

I. Life and Behavioural Sciences	IOS Press
II. Mathematics, Physics and Chemistry	Springer
III. Computer and Systems Science	IOS Press
IV. Earth and Environmental Sciences	Springer

The NATO Science Series continues the series of books published formerly as the NATO ASI Series.

The NATO Science Programme offers support for collaboration in civil science between scientists of countries of the Euro-Atlantic Partnership Council. The types of scientific meeting generally supported are "Advanced Study Institutes" and "Advanced Research Workshops", and the NATO Science Series collects together the results of these meetings. The meetings are co-organized by scientists from NATO countries and scientists from NATO's Partner countries – countries of the CIS and Central and Eastern Europe.

Advanced Study Institutes are high-level tutorial courses offering in-depth study of latest advances in a field.
Advanced Research Workshops are expert meetings aimed at critical assessment of a field, and identification of directions for future action.

As a consequence of the restructuring of the NATO Science Programme in 1999, the NATO Science Series was re-organized to the four sub-series noted above. Please consult the following web sites for information on previous volumes published in the Series.

http://www.nato.int/science
http://www.springer.com
http://www.iospress.nl

Series II: Mathematics, Physics and Chemistry – Vol. 65

Advances in the Geological Storage of Carbon Dioxide

International Approaches to Reduce Anthropogenic Greenhouse Gas Emissions

edited by

S. Lombardi
Universita di Roma "La Sapienza", Rome, Italy

L.K. Altunina
Siberian Branch-Institute of Pertoleum Chemistry
Russian Academy of Sciences,
Tomsk, Russia

and

S.E. Beaubien
Universita di Roma "La Sapienza", Rome, Italy

Springer

Published in cooperation with NATO Public Diplomacy Division

Proceedings of the NATO Advanced Research Workshop on
Advances in CO_2 Geological Sequestration in Eastern and Western
European, Countries
Tomsk, Russia
November 15-18, 2004

A C.I.P. Catalogue record for this book is available from the Library of Congress

ISBN-10 1-4020-4469-0 (HB)
ISBN-13 978-1-4020-4469-4 (HB)
ISBN-10 1-4020-4471-2 (e-book)
ISBN-13 978-1-4020-4471-7 (e-book)

Published by Springer,
P.O. Box 17, 3300 AA Dordrecht, The Netherlands.

www.springer.com

Printed on acid-free paper

Printed in the Netherlands.

Contents

Contributing Authors ix

Preface xiii

Acknowledgements xv

PART I Anthropogenic Greenhouse Gases in the Atmosphere **1**

Study of Long-Term Variations of CO_2 and CO Concentrations in the Ground Atmospheric Layer near the City of Tomsk (Western Siberia) 3
B.D. Belan, A.V. Fofonov, T.K. Sklyadneva and G.N. Tolmachev

Dynamics of the Vertical Distribution of CO_2 and CO Concentrations over Western Siberia (1997-2003). 11
M.Yu. Arshinov, B.D. Belan, G. Inoue, T. Machida, M.V. Panchenko and A.V. Fofonov

Carbon Balance and the Emission of Greenhouse Gases in Boreal Forests and Bogs of Siberia 17
E.A. Vaganov, S.P. Efremov and A.A. Onuchin

The Interaction of CO_2 Between the Atmosphere and Surface Waters of Lake Baikal and the Influence of Water Composition 35

V.M. Domysheva, M.V. Sakirko, M.V. Panchenko and D.A. Pestunov

Remote Sensing and GIS for Spatial Analysis of Anthropogenic Carbon Oxide Emissions 47
Yu.M. Polishchuk and O.S. Tokareva

The Sources of Carcinogenic PAH Emission in Aluminium Production using Soderberg Cells 57
L.I.Kurteeva, S.V.Morozov and A.G.Anshits

PART II Permafrost CO_2 Storage **67**

Distribution of Permafrost in Russia 69
V.P. Melnikov and D.S. Drozdov

Characteristics of Permafrost in Siberia 81
A.D. Duchkov

Possibilities of SO_2 Storage in Geological Strata of Permafrost Terrain 93
A.G. Anshits, N.P. Kirik and B.V. Shibistov

Cryogels – A Promising Material for Underground Works in Permafrost 103
L.K. Altunina, V.A. Kuvshinov and S.N. Dolgikh

Subsurface Carbon Dioxide Storage Through Clathrate Hydrate Formation 111
P. Jadhawar, A.H. Mohammadi, J. Yang, B. Tohidi

PART III Natural Analogues of CO_2 Storage **127**

What Can We Learn from Natural Analogues? 129
J. M. Pearce

Near-Surface Gas Geochemistry Techniques to Assess and Monitor CO_2 Geological Sequestration Sites 141
S. Lombardi, A. Annunziatellis, S.E. Beaubien and G. Ciotoli

Geochemical Interactions between CO_2, Pore-Waters and Reservoir Rocks 157

I. Czernichowski-Lauriol, C. Rochelle, I. Gaus, M. Azaroual, J. Pearce, and P. Durst

Study of Natural CO_2 Emissions in Different Italian Geological Scenarios 175
N. Voltattorni, G. Caramanna, D. Cinti, G. Galli, L. Pizzino and F. Quattrocchi

Natural Leakage of Helium from Italian Sedimentary Basins of the Adriatic Structural Margin 191
G. Ciotoli, S. Lombardi and F. Zarlenga

Tectonically Controlled Methane Escape in Lake Baikal 203
J. Klerkx, M. De Batist, J. Poort, R. Hus, P. Van Rensbergen, O. Khlystov and N. Granin

PART IV Active CO_2 Injection Sites **221**

The IEA Weyburn CO_2 Monitoring and Storage Project 223
J.B. Riding

Assessment of the Long-Term Fate of CO_2 Injected into the Weyburn Field 231
M.J. Stenhouse, W. Zhou and R. Arthur

Strontium Isotope ($^{87}Sr/^{86}Sr$) Chemistry in Produced Oil Field Waters: The IEA CO_2 Monitoring and Storage Project 243
F. Quattrocchi, M. Barbieri, R. Bencini, D. Cinti, K. Durocher, G. Galli, L. Pizzino, M. Shevalier and N. Voltattorni

Optimization of CO_2 Injection for Sequestration / Enhanced Oil Recovery and Current Status in Canada 261
T. Babadagli

The Use of CO_2 and Combustion Gases for Enhanced Oil Recovery in Russia 271
V.A. Kuvshinov

Controls of CO_2 Filtration in Heterogeneous Reservoirs with Foam-Emulsion Systems 277
A.G. Telin

State of CO_2 Capture and Subsurface Storage Activities in Germany 289
B.M. Krooss and F. May

Geophysical Monitoring of the CO_2 Plume at Sleipner, North Sea 303
A. Chadwick, R. Arts , O. Eiken , P. Williamson and G. Williams

4-D Seismics, Gas-Hydrate Detection and Overpressure Prediction as a Combined Methodology for Application to CO_2 Sequestration 315
S. Persoglia, J.M. Carcione, G. Rossi And D. Gei

PART V The Way Forward **325**

The Role of Fossil Fuels in the 21st Century 327
I. Lakatos and J. Lakatos-Szabó

Stakeholder Acceptance and Understanding of CO_2 Geological Storage 345
S. Vercelli and R. Tambelli

CO_2GeoNet – An EC-funded "Network of Excellence" to Study the Geological Storage of CO_2 359
N. Riley

Contributing Authors

Below is an listing of all participating authors, organised by country and institution. Delegates to the workshop are highlighted in bold.

Belgium

International Bureau for Environmental Studies - Brussels, Belgium
- ***J. Klerkx,*** *S.E. Kolobova*

University of Ghent - Renard Centre of Marine Geology, Ghent, Belgium
- *M. De Batist, J. Poort, R. Hus, P. Van Rensbergen*

Canada

Alberta Research Council - 3608-33rd Street, NW; Calgary, AB, T2L 2A6, Canada
- *K. Durocher*

University of Calgary - Department of Geology & Geophysics; 2500 University Dr. NW; Calgary, AB, T2N 1N4, Canada
- ***M. Shevalier***

University of Alberta - Civil and Environmental Engineering; School of Petroleum and Mining; 3-112 Markin/CNRL – NREF; Edmonton, Alberta, T6G 2W2, Canada
- ***T. Babadagli***

France

French Geological Survey (BRGM) - 3 avenue Claude Guillemin, BP 6009; 45060 Orleans Cedex 2, France
- *M. Azaroual,* ***I. Czernichowski-Lauriol,*** *P. Durst and I. Gaus*

Germany

Bundesanstalt für Geowissenschaften und Rohstoffe (BGR) - Stilleweg 2 D-30655 Hannover, Germay
- *F. May*

RWTH-Aachen University - Institute of Geology and Geochemistry of Petroleum and Coal; Lochnerstr. 4-20; D-52056 Aachen, Germany
- ***B.M. Krooss***

Holland

Netherlands Institute of Applied Geoscience (TNO) - National Geological Survey; Kriekenpitplein 18, PO Box 80015; 3508 TA Utrecht, Holland
- *R. Arts*

Hungary

University of Miskolc - Research Institute of Applied Chemistry; 3515 Miskolc-Egyetemváros POB 2; Hungary
- ***I. Lakatos*** *and J. Lakatos-Szabó*

Italy

ENEA-Cr- Casaccia- Prot/Prev, Italy
- ***F. Zarlenga***

IGAG-CNR - Università di Roma "La Sapienza"; Rome, Italy
- *M. Barbieri*

Istituto Nazionale di Oceanografia e di Geofisica (OGS) – Trieste; Italy
- *J.M. Carcione, D. Gei,* ***S. Persoglia*** *and G. Rossi*

Istituto Nazionale di Geofisica e Vulcanologia (INGV) - Rome, Italy
- *R. Bencini, G. Caramanna, D. Cinti, G. Galli, L. Pizzino,* ***F. Quattrocchi*** *and* ***N. Voltattorni***

Università di Roma "La Sapienza" - Dipartimento di Psicologia Dinamica e Clinica; Rome, Italy
- *R. Tambelli and* ***S. Vercelli***

Università di Roma "La Sapienza" - Dipartimento di Scienze della Terra; Piazzale Aldo Moro, 5; 00185 Roma, Italy
- *A. Annunziatellis, S.E. Beaubien, G. Ciotoli and* ***S. Lombardi***

Japan

National Institute for Environmental Studies - 16-2 Onogawa; 305-8506, Tsukuba, Japan
- *G. Inoue and T. Machida*

Norway

Statoil Research Centre - Rotvoll, N-7005; Trondheim, Norway
- *O. Eiken*

Russia

Atmospheric Optics Institute - Siberian Branch of the Russian Academy of Sciences (SB RAS); 1 Akademicheskii Ave.; Tomsk 634055, Russia
- ***M.Yu. Arshinov, B.D. Belan****, A.V. Fofonov,* ***M.V. Panchenko****, D.A. Pestunov, T.K. Sklyadneva and G.N. Tolmachev*

Chemistry and Chemical Technology Institute - Siberian Branch of the Russian Academy of Sciences; K.Marx Str. 42; Krasnoyarsk 660049, Russia
- ***A.G. Anshits, N.P. Kirik*** *and* ***L.I. Kurteeva***

Earth Cryosphere Institute - Siberian Branch of the Russian Academy of Sciences (SB RAS); P/O 1230; Tyumen, Russia
- ***D.S. Drozdov*** *and V.P. Melnikov*

Geophysics Institute - Siberian Branch of the Russian Academy of Sciences (SB RAS); 3, pr. Koptyug; Novosibirsk, 630090, Russia
- ***A.D. Duchkov***

Krasnoyarsk Regional Geological Association - 62 K.Marx St. Krasnoyarsk, 660049, Russia
- *B.V. Shibistov*

Limnological Institute - Siberian Branch of the Russian Academy of Sciences (SB RAS); Irkutsk, Russia
- *V.M. Domysheva, N. Granin, O. Khlystov and M.V. Sakirko*

Mechanics Institute - Ufa Research Center of Russia; Revolutionary Str., 96/2; Ufa 450078, Russia
- ***A.G. Telin***

N.N.Vorozhtsov Novosibirsk Institute of Organic Chemistry - Siberian Branch of the Russian Academy of Sciences (SB RAS); Lavrentyev Ave. 9; Novosibirsk 630090, Russia
- *S.V. Morozov*

Institute of Petroleum Chemistry - Siberian Branch of the Russian Academy of Sciences (SB RAS); 3 Academichesky Ave.; Tomsk 634021, Russia
- ***L.K. Altunina, V.A. Kuvshinov, Yu.M. Polishchuk*** *and O.S. Tokareva*

V.N. Sukachev Forestry Institute - Siberian Branch of the Russian Academy of Sciences (SB RAS); Krasnoyarsk, Russia
- *S.P. Efremov, A.A. Onuchin and* ***E.A. Vaganov***

"Yakutniproalmaz" Institute - Stock Company "ALROSA"; Mirny, Yakuti, Russia
- S.N. Dolgikh

United Kingdom

British Geological Survey (BGS) - Keyworth, Nottingham NG12 5GG; United Kingdom
- *J. M. Pearce, C. Rochelle,* ***J.B. Riding****, A. Chadwick, P. Williamson and G. Williams,* ***N. Riley***

Heriot-Watt University - Centre for Gas Hydrate Research; Institute Of Petroleum Engineering; Edinburgh, United Kingdom
- *P. Jadhawar,* ***A.H. Mohammadi****, B. Tohidi and J. Yang*

United States of America

Monitor Scientific LLC - Suite #555, 3900 S. Wadsworth Blvd.; Denver CO 80235, U.S.A.
- *R. Arthur,* ***M.J. Stenhouse*** *and W. Zhou*

Preface

As is now generally accepted mankind's burning of fossil fuels has resulted in the mass transfer of greenhouse gases to the atmosphere, a modification of the delicately-balanced global carbon cycle, and a measurable change in world-wide temperatures and climate. Although not the most powerful greenhouse gas, carbon dioxide (CO_2) drives climate change due to the enormous volumes of this gas pumped into the atmosphere every day. Produced in almost equal parts by the transportation, industrial and energy-generating sectors, atmospheric CO_2 concentrations have increased by about 50% over the last 300 years, and according to some sources are predicted to increase by up to 200% over pre-industrial levels during the next 100 years. If we are to reverse this trend, in order to prevent significant environmental change in the future, action must be taken immediately. While reduced use of fossil fuels (through conservation, increased efficiency and expanded use of renewable energy sources) must be our ultimate goal, short to medium term solutions are needed which can make an impact today.

Various types of CO_2 storage techniques have been proposed to fill this need, with the injection of this gas into deep geological reservoirs being one of the most promising. For example this approach has the potential to become a closed loop system, whereby underground energy resources are brought to surface, their energy extracted (via burning or hydrogen extraction), and the resulting by-products returned to the subsurface. In fact the re-injection of waste CO_2 even has the potential to recover energy reserves that would otherwise be trapped forever, such as is seen in CO_2 Enhanced Oil, Gas or Coal Bed Methane Recovery, while at the same time isolating the CO_2 deep underground.

The present volume organises contributions given at a NATO Advanced Research Workshop entitled "*Advances in CO_2 Geological Sequestration in Eastern and Western European Countries*", held in Tomsk, Russia from November 15-18, 2004. This workshop was a unique opportunity for researchers from all over the world to share experiences and to describe innovative approaches to this highly promising environmental technology. The book is divided into 5 parts. Part 1 provides a background for the discussion by describing various examples of how human activities are modifying the atmosphere in industrially-active areas in Siberia. Part 2 outlines the innovative idea of using deep permafrost layers as either an impermeable boundary, below which CO_2 could be injected, or as a cooling source for the formation CO_2 clathrates. Part 3 gives numerous examples of recent studies conducted on naturally-occurring sealed or leaking CO_2 reservoirs, sites which have the potential to help us understand what geological characteristics can result in the long-term isolation of CO_2, what are the effects on the ecosystem should CO_2 leak to surface, and what are the most promising technologies for the monitoring of an engineered CO_2 geological storage site. Part 4, which outlines various industrial-scale applications of CO_2 geological storage, shows that this approach is technically practical, economically feasible and, to date, is very safe. Finally Part 5 gives us a view of the future, showing how energy uses are predicted to change in the future, how the public must be involved in any future decisions regarding climate change abatement and how Europe has decided to move forward in its research on geological storage of CO_2.

Acknowledgements

The organizers gratefully acknowledge the financial support of the Scientific Affairs Division of NATO, which made this workshop possible.

A sincere thank you is also extended to all the staff at the Petroleum Chemistry Institute in Tomsk for all their hard work in providing excellent facilities and logistical support, and to Prof. Jan Klerkx of The International Bureau for Envirnmental Studies (IBES) for his contribution to the successful organisation of the event. Finally the attendance of many members of the EC-funded Network of Excellence "CO2GeoNet" contributed to the successful outcome of the Workshop.

PART I

ANTHROPOGENIC GREENHOUSE GASES IN THE ATMOSPHERE

Sources, Concentrations and Processes

STUDY OF LONG-TERM VARIATIONS OF CO_2 AND CO CONCENTRATIONS IN THE GROUND ATMOSPHERIC LAYER NEAR THE CITY OF TOMSK (WESTERN SIBERIA)

B.D. Belan, A.V. Fofonov, T.K. Sklyadneva and G.N. Tolmachev
Institute of Atmospheric Optics SB RAS 1, Akademicheskii Ave., 634055, Tomsk, Russia

Abstract: Continuous measurements of carbon dioxide near Tomsk started in 1992 while CO monitoring began in 1998. Both analyzers operate at the TOR Station of the Institute of Atmospheric Optics. All readouts are recorded every hour all year round. In this paper we present some results of these measurements.

Key words: surface long-term measurements, carbon dioxide, carbon monoxide, air mass transport.

1. INTRODUCTION

The TOR automated monitoring station was designed and put into operation within the framework of the Tropospheric Ozone Research Project of the EUROTRAC Programme (Belan et al., 1992; Arshinov et al., 1994). This station allows for the continuous monitoring of the main greenhouse gases, aerosol content, and standard meteorological parameters. Real-time data of these observations are presented at the web page of the Institute of Atmospheric optics (http://meteo.iao.ru).

Continuous measurements of carbon dioxide at the TOR-Station were initiated in December 1992. Until 1995 a GIAM CO_2 gas-analyser (Russian manufactured) was used, which was subsequently replaced with a NDIR Vaisala GMM-12 gas-sensor. Monitoring of CO was started in 1998 using an electrochemical sensor. Prior to analysis an air sample is pumped through tubing for 10 minutes for ventilation and calibration.

S. Lombardi et al. (eds.),
Advances in the Geological Storage of Carbon Dioxide, 3–9.

Although the measurement site is located in a boreal forest area near the city of Tomsk, this location cannot be considered as a background observational point because it is sometimes influenced by industrial pollution sources from the city (Fig. 1). However, the analysis of long-term measurements reveals local, regional and even global variations of these atmospheric parameters.

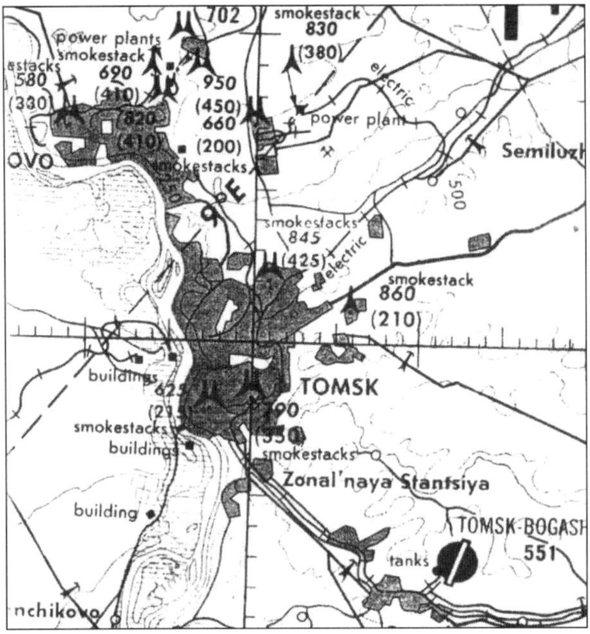

Figure 1. Arrangement of the measurement site relative to Tomsk.

2. RESULTS

This paper presents the results of the last decade of observations.

It can be seen from Figures 2 and 3 that long-term behaviour of CO_2 and CO exhibit opposite tendencies. The concentration of carbon dioxide increases from year to year while carbon monoxide concentrations generally decrease.

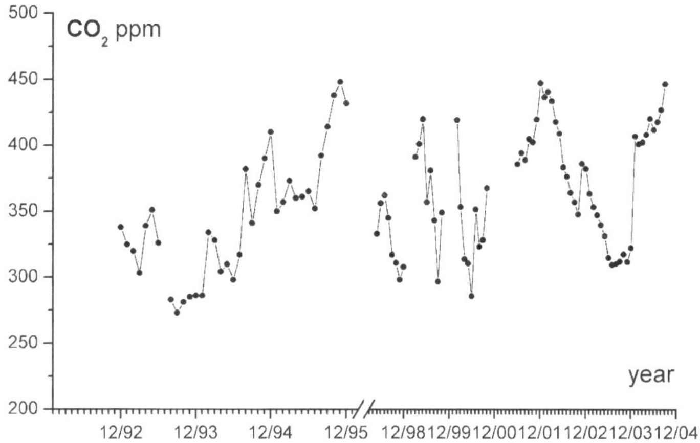

Figure 2. Long-term variations of carbon dioxide concentrations. (1992-2004).

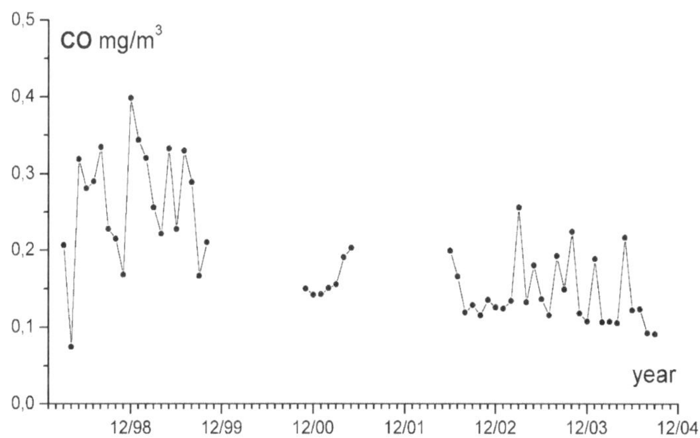

Figure 3. Long-term variations of carbon monoxide concentrations. (1998-2004).

Figures 4 and 5 show that gases of the carbon cycle have similar seasonal variations, with concentration maximums during the cold period of year and minimums during the warm period. Such behaviour reflects the normal annual cycle of vegetation activity in the region. Analysis of the seasonal variations of carbon oxides also allows one to conclude that the strength of the carbon sink exceeds the source during a four month period (from April to July). This can mean that in spite of the fact that Siberia has vast areas of forestry, on the whole the region is an additional source of one of the main greenhouse gases.

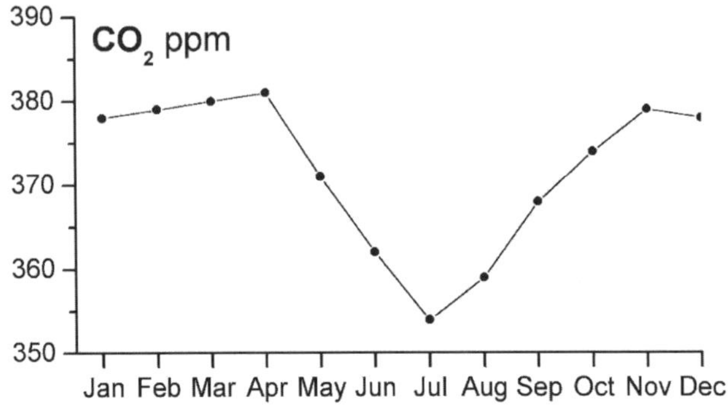

Figure 4. Seasonal variations of carbon dioxide concentrations near Tomsk.

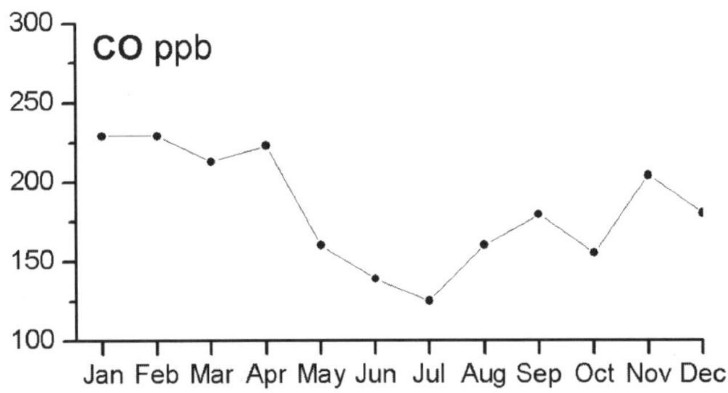

Figure 5. Seasonal variations of carbon monoxide concentrations near Tomsk.

The terrestrial surface is a natural source of nitrogen monoxide. Vegetation generates nitrous oxide, and then N_2O transforms to NO (Crutren, 1974). It can be seen from Figure 6 that nitrogen monoxide concentrations are higher during the warm period of year when the ground surface is not covered by snow. During the winter season nitrogen monoxide concentrations decrease significantly. Nitrogen dioxide is formed from nitrogen monoxide as result of interaction with ozone or hydroxyl radicals, and thus NO_2 has a seasonal variation which is similar to that of NO. That said, nitrogen dioxide concentrations do not decrease as significantly as does nitrogen monoxide during the cold period.

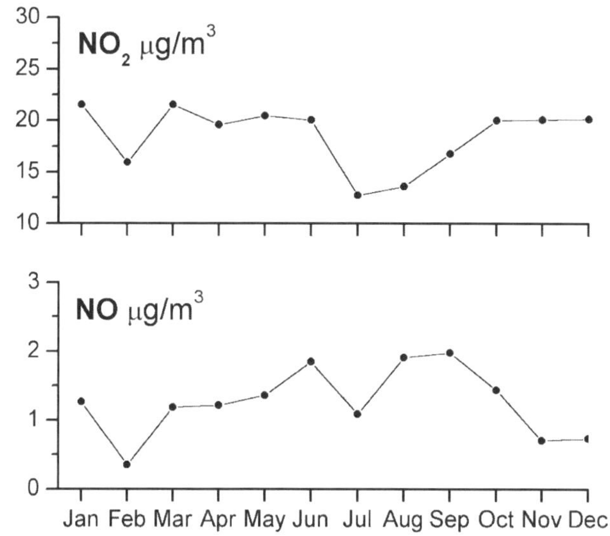

Figure 6. Seasonal variations of nitrogen dioxide and nitrogen monoxide near Tomsk.

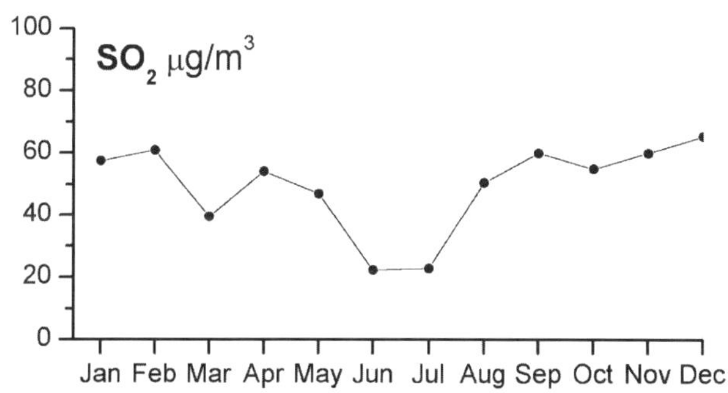

Figure 7. Seasonal variations of sulphur dioxide concentrations near Tomsk.

The seasonal variations of sulphur dioxide concentrations are shown in Figure 7. It is clear that SO_2 has a primary minimum during the summer and two secondary minima in March and October. Since the main natural source of sulphur dioxide in the atmosphere is hydrogen sulphide generated from the ground surface this kind of seasonal behaviour is typical for this region, although it should be remembered that the burning of coal, natural gas, etc.

in western Siberia is also a comparable source of SO_2. During the summer, when the mixing layer reaches maximum altitudes and turbulence is well developed, sulphur dioxide is redistributed in a larger volume, so its surface concentration decreases under the condition of a stable source. In general during the winter, when both mixing layer height and the inversion layer are lower, accumulation of impurities occurs in smaller volumes and the surface concentration of SO_2 increases. Most likely the minima in March and October are caused by wet deposition, but this assumption should be verified by measurements of precipitation acidity.

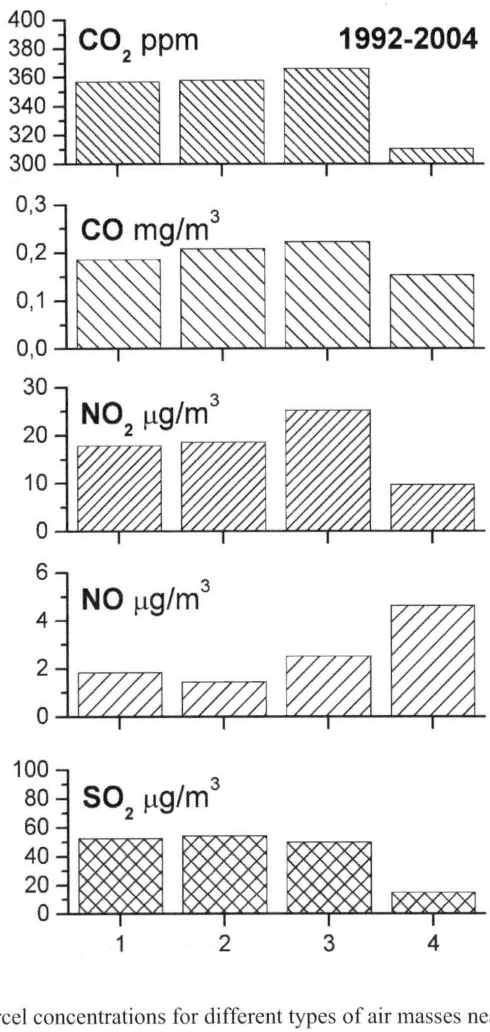

Figure 8. Gas parcel concentrations for different types of air masses near Tomsk: 1- arctic, 2 – temperate, 3 – subtropical, 4 – tropical.

The average concentration of trace gases observed in different air masses is shown in Figure 8. It can be seen from this figure that for CO_2, CO, NO_2, and NO the warmer the air mass the higher the observed concentration, except for the tropical air mass (which can be explained by poor statistics for this unit in the Siberian region).

3. CONCLUSIONS

In summary long-term tendencies of seasonal gas variations in the surface atmospheric layer reflect the influence of both regional sources and sinks as well as of long-range transport of background air mass values.

REFERENCES

Belan B.D., Kolesnikov L.A, Luk'yanov O.Yu., Mikushev M.K., Sem'yanova O.I., Tolmachev G.N., and Usmanova V.M., 1992, Variations of the ozone concentration in the ground air layer. *Atmospheric and Oceanic Optics*. **5**, (6), pp.400-403.

Arshinov M.Yu., Belan B.D., Zuev V.V., Zuev V.E., Kovalevskii V.K., Ligotskii A.V., Meleshkin V.E., Panchenko M.V., Pokrovskii E.V., Rogov A.N., Simonenkov D.V., and Tolmachev G.N., 1994, TOR-station for monitoring of atmospheric parameters. *Atmospheric and Oceanic Optics*. **7**, (8), pp.580-584.

Crutren P.J., 1974, A review of upper atmospheric photochemistry. *Can. J. Chem.* **52**, (8), pp.1569-1581.

DYNAMICS OF THE VERTICAL DISTRIBUTION OF CO_2 AND CO CONCENTRATIONS OVER WESTERN SIBERIA (1997-2003)

M.Yu. Arshinov[1], B.D. Belan[1], G. Inoue[2], T. Machida[2], M.V. Panchenko[1] and A.V. Fofonov[1]

[1]*Institute of Atmospheric Optics SB RAS, 1, Akademicheskii Ave., 634055, Tomsk, Russia;*
[2]*National Institute for Environmental Studies, 16-2 Onogawa, 305-8506, Tsukuba, Japan*

Abstract: In this paper we present some results of airborne measurements of main greenhouse gases over the southern part of West Siberia, which were started in 1997 within the framework of a joint Japanese-Russian Project on the study of greenhouse gases in Siberian ecosystems.

Key words: greenhouse gases, carbon oxides, vertical distribution

1. INTRODUCTION

Airborne measurements of greenhouse gases over the southern part of western Siberia were started in 1997 within the framework of a joint Japanese-Russian Project on the study of greenhouse gases in Siberian ecosystems (Belan et al., 2000). The study region and the flight routes are shown in Figure 1. The measurement procedure consists of flask sampling of ambient air at various elevations using an Antonov-30 aircraft laboratory. After the flight all flask samples were analysed at the National Institute of Environmental Studies (Japan). Measurements were carried out in the 0 to 7 km atmospheric layer at the end of each month, all year round, and as such a seven-year database now exists which includes CO_2, CO, CH_4, H_2 and N_2O concentration distributions. A NDIR gas-analyser (LI-6262) mounted on board the aircraft was used to perform continuous CO_2 measurements as a control of the flask sampling measurements.

S. Lombardi et al. (eds.),
Advances in the Geological Storage of Carbon Dioxide, 11–16.

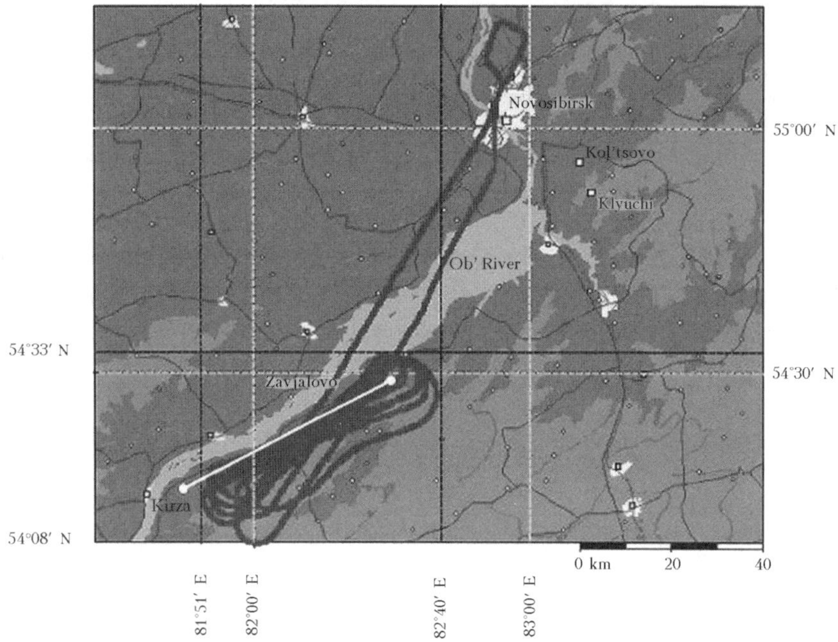

Figure 1. Region of airborne surveys.

2. RESULTS

Long-term variations in CO_2 and CO mixing ratios are shown in Figure2. This figure shows that carbon dioxide concentrations had a positive trend during the study period, which is a global characteristic feature. At the same time carbon monoxide concentrations tended to decrease.

In spite of the fact that a summer minimum and a winter maximum is typical for both gases there is one important difference in there variations. It is clear that the higher mixing ratio values of the carbon oxides observed in the lower troposphere during the winter are the result of anthropogenic emissions and vegetation respiration. Also the presence of frequent temperature inversions, which are typical for Siberia in the winter, lead to the accumulation of CO and CO_2 in the atmospheric boundary layer and higher mixing ratios in the lower troposphere as compared to the free troposphere. Differences are due to the fact that carbon monoxide concentrations are higher during the whole year at lower altitudes, while CO_2 concentrations in the summer are significantly lower compared to the free troposphere, as seen in Figure 3.

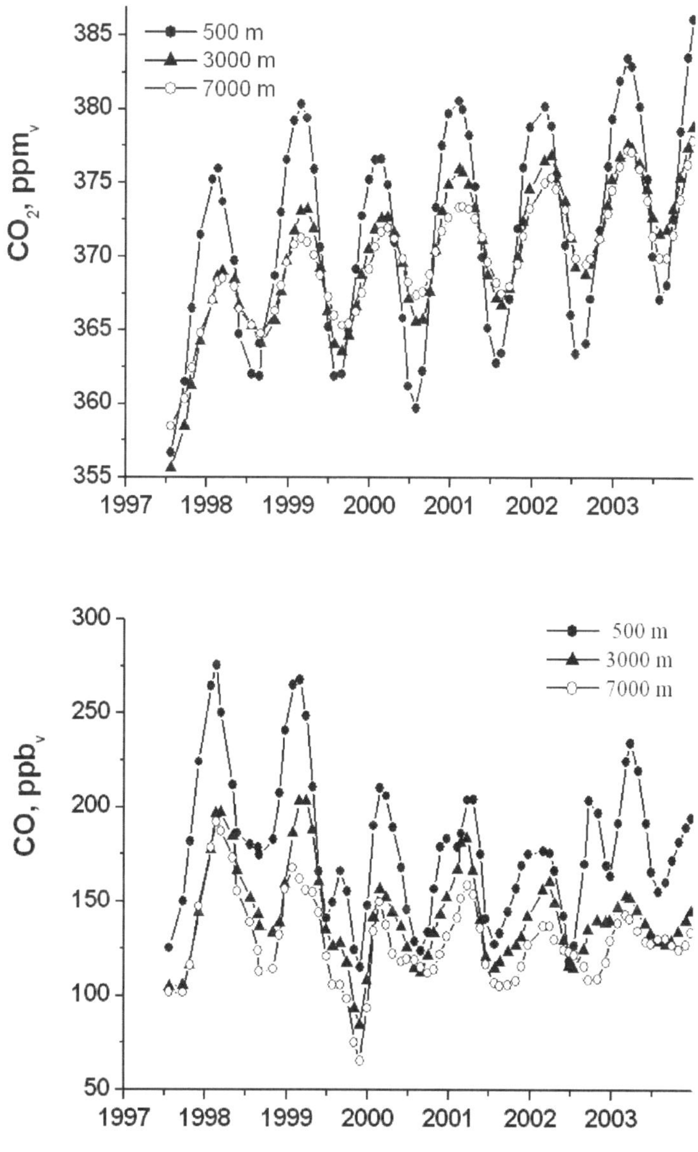

Figure 2. Long-term variations in CO_2 and CO mixing ratios.

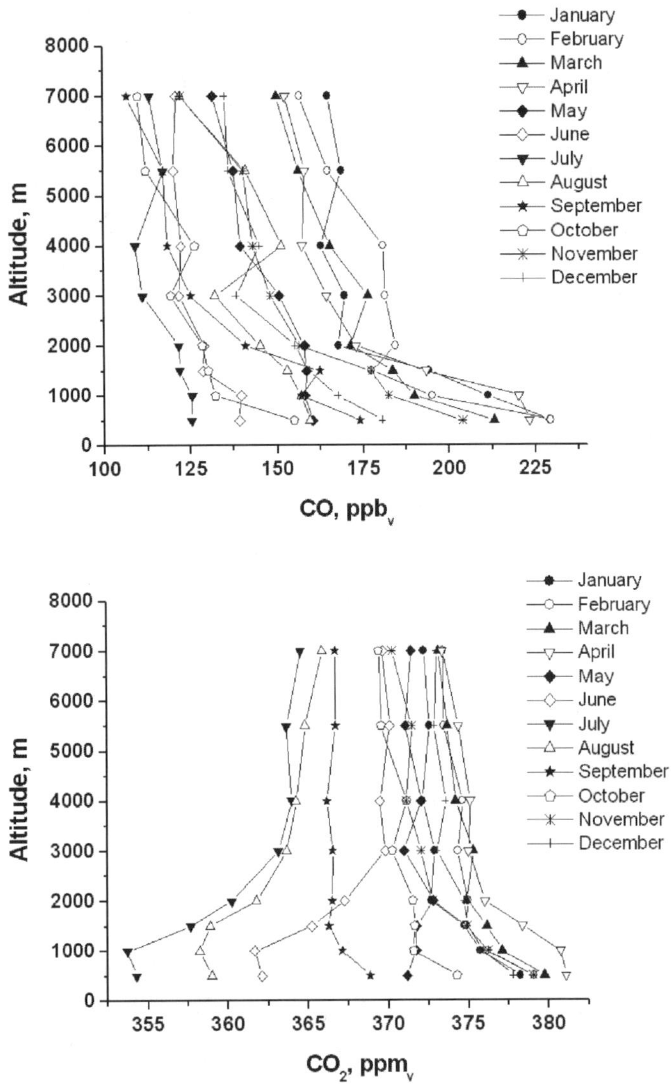

Figure 3. Averaged vertical distributions of CO_2 and CO mixing ratios.

During the summer photosynthesis leads to the decrease of CO_2 mixing ratios in the surface atmospheric layer, and well-developed vertical mixing results in distribution homogeneity in the boundary layer. As for CO, the difference in its mixing ratio at higher and lower altitudes becomes smaller during the summer, that is it does not vary with altitude.

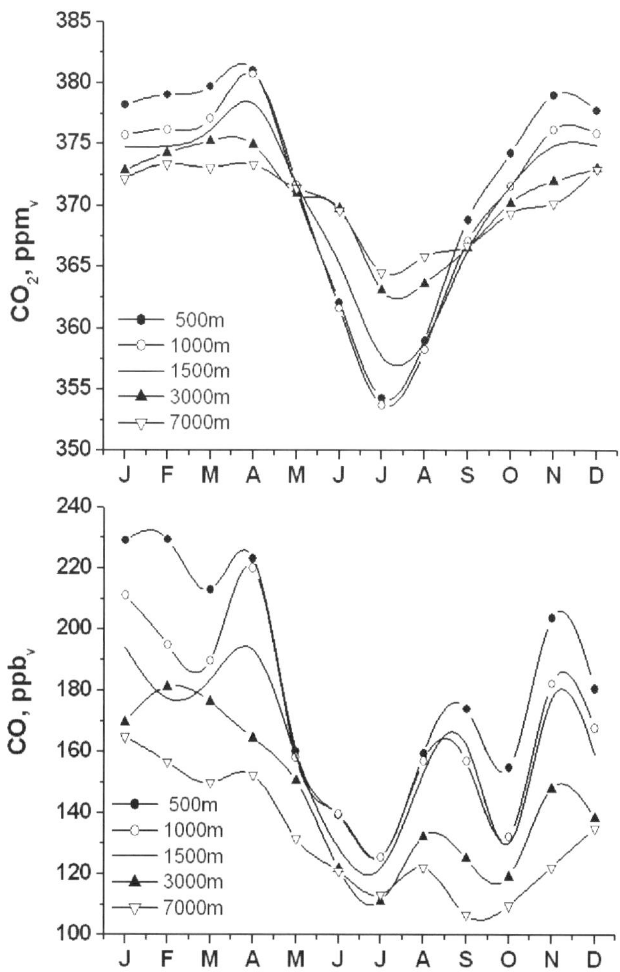

Figure 4. Averaged seasonal behaviour of CO_2 and CO mixing ratios.

The vertical CO_2 distributions observed in September and May, when CO_2 concentrations are practically invariant with altitude, represent two opposite intermediate processes. In May photosynthesis is increasing in strength while in September it is weakening. Averaged seasonal variations are shown in Figures 4 and 5. Large amplitudes are observed in the atmospheric boundary layer.

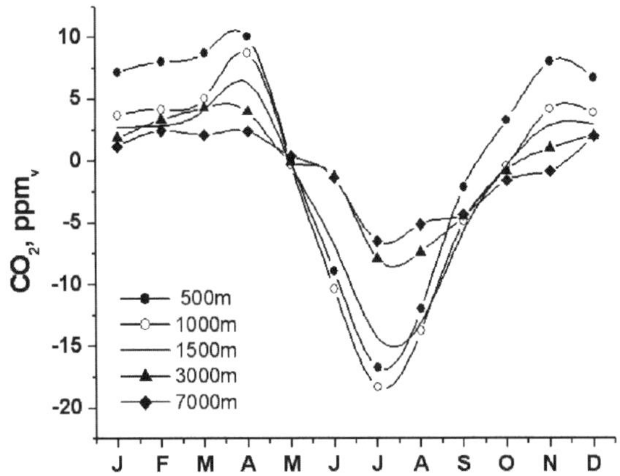

Figure 5. Seasonal amplitudes of CO_2 and CO mixing ratios.

3. CONCLUSIONS

The study has shown that long-term variations of carbon dioxide in the free troposphere have a positive trend during all seasons. At the same time there is no such tendency for the boundary atmospheric layer where CO_2 concentrations remain constant from year to year during summertime, that leads to increase of seasonal amplitude. So, we can draw the conclusion that Siberian forests are an important CO_2 sink. However, the question of what leads to the global increase of carbon dioxide still needs to be addressed. At present it is generally recognized that this is a result of direct industrial and transportation emissions. But analysis of variations of CO_2 and CO mixing ratios in the lower troposphere suggests that anthropogenic activity can affect the biosphere, which leads to changes in vegetation activity and, consequently, plants respiration behaviour; this can result in global increase of carbon dioxide being a very powerful source of carbon on a global scale.

REFERENCES

Belan, B.D., Inoue, G., Machida, T., Nadeev, A.I., Panchenko, M.V., Plotnikov, A.P., 2000, Variations of atmospheric greenhouse gases over Siberia. *Proceeding of the 8th Symposium on the Joint Siberian Permafrost Studies between Japan and Russia.* National Institute for Environmental Studies, Tsukuba, Japan, 2000, pp. 275-279.

CARBON BALANCE AND THE EMISSION OF GREENHOUSE GASES IN BOREAL FORESTS AND BOGS OF SIBERIA

E.A. Vaganov, S.P. Efremov and A.A. Onuchin
V.N. Sukachev Institute of Forest SB RAS, Krasnoyarsk, Russia

Abstract: Zonal patterns of above-ground phytomass dynamics and carbon storage in above-ground vegetation, phytodetritus and humus were revealed based on the study of the carbon balance and its components in forest ecosystems of the Yenisei meridian transect. Results indicate that the carbon storage ratio in different plant layers is determined by climatic regimes. For example pine stands were used to calculate the full carbon budget using data on its fluxes and storage in different biogeocenosis components. Biological productivity indices and the carbon pool of hydromorphic complexes are highly dependant on the mineral nutrition regime and morphological characteristics of the stands. Experimental study results show the importance of forest and bog ecosystems as carbon cycle regulators is determined by the complex interaction of zonal-climatic and forest conditions as well as by forest vegetation characteristics (which depend on varying carbon balance structure and energy- mass exchange processes).

Key words: greenhouse gases; forest ecosystems; post- fire carbon emissions; phytomass of forest bogs.

1. INTRODUCTION

Forests and bogs are dominant ecosystem types in middle and high latitudes of the northern hemisphere. The present paper details the results of studies into the role of Siberian forests and bogs in the global carbon cycle and outlines some trends for further studies. Carbon stocks and fluxes formed by Siberian forests and bogs make a significant contribution to the global system, based in part on its great size (approx. 10 million km^2).

S. Lombardi et al. (eds.),
Advances in the Geological Storage of Carbon Dioxide, 17–34.

The carbon cycle is fundamental in forest ecological and climatological studies due to the uptake and release of various natural and anthropogenic greenhouse gases during the life cycle of plants (the so-called temporary carbon depots). Siberian forests and bogs have several peculiarities with respect to their participation in the global carbon cycle: 1) tree stands typical of Siberia (with 5 coniferous species as dominants) as well as forest-bog and bog ecosystems provide long- term carbon conservation; 2) most Siberian forests are located in the permafrost zone, which has low productivity (about 100 m^3 per ha, or 1.27 m^3 per ha / year) but extra potential for carbon sequestration in soil; 3) a large portion of Siberian forests periodically suffer wild fires (variable in different taiga subzones) (Furyaev et al., 2001) that greatly shift the balance towards emission of C containing gases to the atmosphere.

2. STUDY METHODS

The information outlined above explains the great interest of the scientific community in studying the carbon cycle of Siberian forest and bog areas. At present more than 20 joint international projects are aimed at examining this problem. The methods used in these projects can be divided into direct and indirect, ground-based and remote-sensing, and combination techniques (i.e. methods based on direct measurements combined with simple models which compute the difficult or non-measured characteristics of the cycle).

Direct methods physically measure carbon cycle parameters, such as carbon content measurements within the forest litter, in different soil horizons, in gases below and above ground or in flowing water which removes carbon compounds from the ecosystem.

Methods for measuring respiration (stem and soil respiration, etc.) as well as the eddy-covariance method used for measuring carbon dioxide fluxes along the vertical profile of ecosystems are also direct ones. The eddy-covariance method requires special towers which are less than or much higher than the tree canopy and which are equipped with a series of gas analyzers located along its length (Shibistova et al., 2002a,b,c; Schulze et al., 1999). Vertical concentration profiles of carbon dioxide and other carbon containing greenhouse gases given in papers by Levin and Lloyd are more spatially integrating (Levin et al., 2002; Lloyd et al., 2002), and these measurements are used as a base for computing exchangeable carbon within the biogeochemical cycles of vast forest areas (up to hundreds of thousands of square kilometers).

Flux measurements of carbon-containing gases made using tall (up to 300 m) towers have a special importance in this group of methods (Bakwin et al., 1998), as the surface atmospheric layer (c. 0-200 m) is characterized by eddies induced by the friction of air flowing against earth surface roughness. However, the dynamics of the overlying layer (200-2000 m) is determined by heavy convective mixing during the day and by layer degradation or the absence of turbulence during the night. This interval is known as the mixing layer, and a wide range of diurnal carbon signals within the surface layer are smoothed and weakened in the mixing layer. Therefore, the mixing layer is an efficient integrator both of diurnal carbon cycles and of small-scale ecosystem heterogeneities. It enables not only the precise measurement of gas / aerosol compositions and their vertical profile dynamics but also to compute exchange fluxes of carbon containing gases which are integrated for the tower footprint area (about 40,000 km^2).

Some meteorological instruments also host partially-automated systems for measuring the absolute concentrations of CO, CO_2, O_2, N_2, CH_4 N_2O and the stable isotope ratios of $^{13}C/^{12}C$ in CO_2, CH_4 and CO, $^{18}O/^{16}O$ in CO_2 and $^{15}N/^{14}N$ in N_2O. Isotope ratios can be used to differentiate carbon sink and release processes. For example CO has information on anthropogenic emissions since the incomplete burning of fossil fuels is one of its main sources.

Concentration measurements of CO_2 fractions having different C and O isotopic signatures make it possible to separate oceanic and terrestrial carbon fluxes (Keeling et al., 1996) due to the strong contrast in the number of heavy carbon isotopes in CO_2 adsorbed during terrestrial C_3 photosynthesis compared to oceanic photosynthesis. Measurements of $^{18}O/^{16}O$ ratios in atmospheric CO_2 also help assess respiration intensity of terrestrial ecosystems (Mortazavi and Chanton, 2002). Modeling has shown that respiration and evaporation are the main factors responsible for spatial and seasonal changes of $C^{16}O^{18}O$ content in the atmosphere (Ciais et al., 1997; Peylin, 1999). Observations of methane will help to better understand the relationships between climate and ground vegetation since trends in methane (and its isotope $^{14}CH_4$) indicate changes in the respiration of peat bogs and permafrost soils.

Construction of the tall tower in the middle of the taiga forest region (Zotino, Krasnoyarsk krai) will provide measurements of energy- and mass exchange characteristics in real time, with a high temporal resolution and an assessment of the impact of meteorological conditions on exchange processes.

Studying carbon cycle parameters on the scale of Siberia requires the use of remote sensing methods owing to its great size and the limited access of many of its regions. For example satellite images can be used to determine species composition, the age of tree stands, phytomass stocks and other large scale forest ecosystem characteristics.

Computation of the total carbon budget of forests using periodic forest inventory data is one of the most developed approaches for understanding the role of Russian forests in the global carbon cycle This can be considered a combined approach, as forest inventory data is not enough. Instead, theoretical estimates of some carbon budget parameters should be introduced and models should be applied which take into account the age and species composition of the forest ecosystems in different physical and geographical regions. The total carbon budget of Russian forests was calculated in a cooperative project involving Russian forest research institutions and the Institute of Applied System Analysis (Austria). In this work a special database was established which includes a geographic information system (GIS) model. The multi-layered GIS contains land status data (landscapes, soils, vegetation, forests, land use etc.) and attributive data (phytomass, heterotrophic respiration, disturbed lands and forests etc.) which is based on focused studies and forest inventory reports as well as some auxiliary models (Shvidenko et al., 2003; Nilsson et al., 2000; Shvidenko and Nilsson, 2003).

Even the most thorough measurements and assessments of carbon cycle parameters will show only a small part of its value if they are not coordinated with other studies within an integrated description of the functioning ecosystems. Reducing together the minimum set of both direct and indirect (forest inventory or remote sensing) measurements of carbon cycle parameters needed for a relatively complete description is possible only within integrated models due to the different spatial scales of the various methods. An adequate model of ecosystem functioning not only links measurement data obtained by different methods but it is also able to assess the processes which has the most errors. For example the inaccuracy of assessing carbon emissions caused by forest fires is one of the key factors resulting in imprecise descriptions of the carbon cycle in Siberian forests (Vaganov et al., 1998; Arbatskaya, 1998; Conard and Ivanova, 1997; Conard et al., 2002). Great variability in the area of Siberian forest burned annually (from 1 to 10 million ha) greatly affects the balance between carbon accumulation and emission (Isaev et al., 1998).

3. RESULTS AND DISCUSSION

3.1 Spatial variability of carbon stocks in forest ecosystems along the Yenisei meridian

During several years researchers at the Forestry Institute SB RAS carried out detailed studies of the structure of the Yenisei transect forest ecosystems in order to assess their diversity and composition as well as various dynamic components, such as carbon, nitrogen and water turnovers (Forest ecosystems, 2002). Availability of all boreal forest subzones in the transect area makes it possible to link the stock and annual phytomass productivity indices, the rate of organic decomposition, heterotrophic respiration, detritus stocks, frequency and intensity of wild fires, as well as potential assessments of carbon emission during and after forest fires as a result of the main climatic factor - temperature. Latitudinal regularities in the forests of the Yenisei meridian can be considered as a spatial analog of changing ground vegetation structure due to large climate changes, such as the shifting of forest zones during warming (Furyaev et al., 2001; Tchebakova et al., 1994).

Tree stand phytomass and ground vegetation stocks show appropriate changes along the meridian and agree well with average annual temperature changes and growing season length (Forest ecosystems, 2002). The growing stock increase in forests of middle and southern taiga, as compared to the northern taiga, parallels the decrease of ground vegetation phytomass like mosses, lichens, grasses and small shrubs. In the more northern ecosystems the contribution of ground vegetation to the formation of annual ecosystem production is commensurable or exceeds wood growing stock contribution (Knorre, 2003).

The organic matter (OM) carbon pool in the soil cover (i.e. the sòil layer equivalent to the organogenic and mineral horizon thickness, 0-50 cm depth) of forest tundra and boreal forests of the Yenisei meridian is 114 t ha^{-1}, which is divided between forest tundra (17%) and northern (23%), middle (39%) and southern (21%) taiga. Stable humus consists of 60.5 t ha^{-1}, 33.4 t C ha^{-1}, which is concentrated in mobile organic matter and 19.8 t C ha^{-1} is in phytodetritus on the soil surface and below (Fig. 1).

The upper 20-cm soil layer is characterized by a maximum accumulation of both mobile and stable carbon. The light mineralised OM fraction makes up 40% of the carbon stock in this part of the soil profile. With regards to carbon accumulation in the litter the fraction share increases to 50% of total carbon stock. In terms of structure, 19.8 t C ha^{-1} is phytodetritus carbon which takes an active part in exchange processes between phytomass, soil humus and the atmosphere while 21.3 t C ha^{-1} consists of soil organic matter involved in biological turnover and possibly rapid overturn. It is this latter which would be used during changing environment conditions and which controls the ratio of mineralisation and humification processes during decomposition. The ratio of carbon stocks in phytodetritus and in mobile soil OM (as in the most labile reserve during changing ecological states) decreases 5 times from forest tundra to southern boreal forests, the ratio of $C_{phytomass}$: $C_{phytodetritus}$ stock increases 6 times in the latitudinal range and C_{humus} stock in the mineral soil thickness (0-50 cm) varies from 72 to 121 t ha^{-1}.

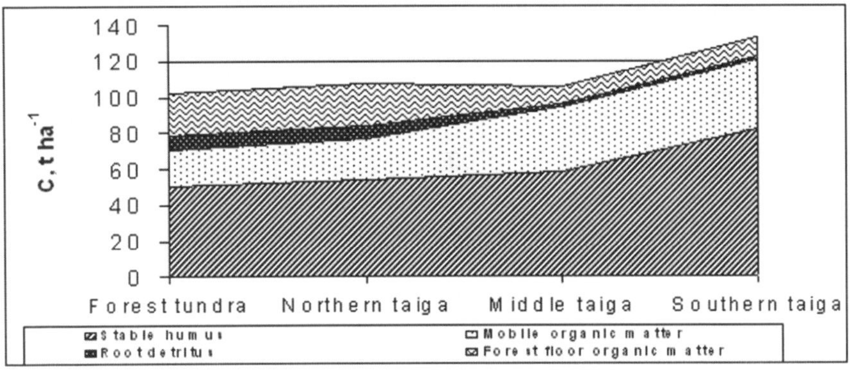

Figure 1. Carbon stocks in soil organic matter components.

Maximum phytodetritus stock in coniferous stands of forest tundra and northern taiga is clearly followed in latitudinal diapason. Existing permafrost, the high location of frost in the soil profile, its slow and shallow retreat during vegetation, low temperatures and high soil moisture contents determine the formation of specific ground vegetation. This vegetation usually consists of thick, often continuous coverage consisting of green mosses (of the genera *Pleurozium*, *Hylocomium*, *Politrichum*) as well as bushy lichens (of the genera *Cladonia*, *Cladina*, *Cetraria*). The slow decomposition of these plants due to numerous stable compounds (lignocellulose complex, cellulose etc.) results in the accumulation of variously decomposed moss and lichen remains on the soil surface. Their stocks are more than ten times that of tree leaf-fall remains. Soil mass cryoturbation (frost heave, frost cracks etc.) result in an increase of detritus in the upper mineral soil depth.

Phytodetritus stock increases towards the south, from middle taiga, are related to changes in the structure of the forest-forming tree species, to forest productivity increases and, as a result, to annual leaf fall amount. In addition, the southern taiga forests are characterized by increased dead wood (deadfall, stumps).

In ecosystems of all forest-forming tree species the distribution of phytodetritus components, up to class V of stand age, is: litter > root detritus > stem wood. Prevalence of over mature stands results in detritus accumulating due to slowly decomposed fallen stems. The latter clearly shows itself in southern taiga larch forests, pine forests and spruce forests.

Overall carbon accumulation in forest tundra and taiga ecosystems of the Yenisei meridian (in blocks "vegetation" and "soil") equals about 16,000 billion t (156 t C ha^{-1} of forested area). Approximately 26% consists of above and below ground phytomass, while the rest is accumulated in the organic matter of the upper 50 cm of the soil (39% in stable humus, 13% in mobile humus and 22% in phytodetritus). Studies in other countries show comparable values: forest ecosystems in the Netherlands are characterized by a similar ratio with 113; 59 and 19 t C ha^{-1} in humus, phytomass and dead plant residues, respectively (Nabuurs and Mohren, 1993, 1995); in Finland 34 t C ha^{-1} was found in phytomass and 72 t C ha^{-1} in the 0-17 cm soil interval (Kauppi et al., 1997); while in the USA Birdsey (1992) found 59% in soil, 31% in the phytomass and 9% in plant residues on soil surface.

Carbon stocks in both blocks increase in the meridian direction. The redistribution of pools between them varies along the transition from forest tundra to the northern taiga, such that the C_{soil} / $C_{phytomass}$ ratio diminishes from 4.8 to 2.9 and remains equal in ecosystems of middle and southern taiga (2.6 and 2.4, respectively). As Utkin et al. (2001) found, the C_{soil} / $C_{phytomass}$ ratio for subzones of western and eastern Siberian macro-regions equals 3.3. From the estimate of Shvidenko and Nilsson (2003) it follows that this ratio in the forested Russian area equals 6.8, 2.4 and 2.8 in forest tundra and northern taiga forests, middle and southern taiga, respectively.

Intensity of mineralised carbon flux to the atmosphere is practically 90%, determined by phytodetritus decomposition. Between 68% (in southern taiga) and 87% (in forest tundra) of its emission is formed via litter decomposition. Root detritus contribution does not exceed 11-17%, while stem fall increases from forest tundra and northern taiga towards middle and southern taiga (from 2-3 to 14-17%). Soil humus mineralisation in ecosystems varies from 3 to 13% in the southern taiga and from 2 to 6% in northern forests. The 2 - 5% of detritus being decomposed are humificated in southern taiga forests and 0.1-0.2% of it in forest tundra and in larch forests of the northern taiga.

Experimentally determined phytodetritus decomposing rates, averaged for forest zones and subzones of the Yenisei meridian, are given in Table 1. For the period having an average daily air temperature above 5°C the phytodetritus decomposition rate increases 3-fold (from 0.4 to 1.1 mg C g^{-1} day^{-1}) in the direction from forest tundra to southern taiga.

Table 1. Indices of phytodetritus decomposition rate.

Zone and subzone	Forest litter		Root detritus	
	k, $year^{-1}$	$T_{½}$, years	k, $year^{-1}$	$T_{½}$, years
Forest tundra, n = 41	0.035 ± 0.002	20	0.036 ± 0.003	17
Northern taiga, n = 35	0.046 ± 0.004	15	0.048 ± 0.007	14
Middle taiga, n = 49	0.089 ± 0.007	8	0.072 ± 0.007	10
Southern taiga, n = 44	0.211 ± 0.024	3	0.085 ± 0.011	8

Young forest ecosystems (tree stands of classes I-II in age) of the main Siberian forest-forming species, the southern taiga pine forests and birch stands and larch ecosystems of the northern taiga are a sink for atmospheric carbon. Prevailing over mature tree stands of southern taiga larch forests in the forested area of the subzone are considered as negative balance ecosystems with regards to carbon production and destruction processes (thus a C source to the atmosphere). Forest tundra ecosystems are considered at equilibrium.

The carbon sink (net ecosystem production) of young forest ecosystems consists of 50 to 70% of carbon net primary production (NPP) and is formed via the accumulation of phytomass (73-78% of NEP value), and conservation in phytodetritus (6-20%) and soil humus (5-18%). Pine and birch forests of the southern taiga are a sink for 16% of the photosynthetic assimilation of atmospheric carbon in the net primary production. NEP of larch forests of the northern taiga make up 30% of net primary production. The major part of the sink consists of ground vegetation and accumulation in the phytodetritus and only 9% on tree stand biomass.

3.2 Direct measurements of carbon dioxide fluxes by eddy- covariance methods

The assessment of spatial and temporal distributions of carbon sinks and sources in terrestrial ecosystems can be obtained based on data of regular vertical soundings of the atmosphere (Tans et al., 1996; Gloor et al., 2000). Measurements of vertical concentration profiles of carbon dioxide, methane and stable isotopes make it possible to characterize seasonal cycles and gas exchange intensity between the earth surface and atmosphere at a regional scale (Levin et al., 2002; Lloyd et al., 2002).

Since the eddy covariance method registers integrated CO_2 fluxes between the ecosystem and atmosphere, it can be difficult to assess separate carbon exchange constituents because of the difficulty of interpreting the balance between photosynthetic assimilation and the release of carbon dioxide during the respiration process (Grace et al., 1996). As such gas exchange chambers are used as an alternative approach which can give a quantitative estimation of CO_2 fluxes in separate ecosystem components (photosynthesis, soil respiration, tree stems, crown) as well as to register spatial and temporal variability.

Continuous research has been conducted on the carbon balance of a forest ecosystem in the southern subzone of middle taiga (60°45′N, 89°23′E) since 1998. The measurement system was installed in an even-aged (200-year old) pine tree stand of V bonitet class, growing on sandy soils. Mean tree height is 22 m, basal area is 30 m/ha, leaf area index is 1.5 of the projective cover, and crown closure is 0.7 (Wirth et al., 1999). Lichens pp. *Cladina spp. Cladonia spp.* dominate the ground vegetation. The grass-shrub layer mainly consists of *Vaccinium vitis-idaea* (L). Avror., while the presence of green mosses and *Vaccinium myrtillus* (L). is minor.

Photosynthetic activity was observed in the ecosystem from early May to late September. During the long autumn-winter period the tree stand was transformed into a weak CO_2 source to the atmosphere (average CO_2 flux equaled 0.05 mol C m^2 day^{-1} - Shibistova et al., 2002). On the whole, during the vegetation season the studied ecosystem was considered as a CO_2 "sink" and the accumulating activity (up to 0.4 mol $C \cdot m^{-2} \cdot day^{-1}$), in spite of a lesser leaf area index, was rather high and quite comparable with values characteristic of European and Canadian boreal forests (Schulze et al., 1999; Goulden et al., 1997; Jarvis et al., 1997).

The gross primary production (GPP) value of the studied tree stand was much less compared to European forests. For example, GPP values of the order of 100, 85 and 126 mol C m^2 $year^{-1}$ were observed for forest ecosystems in Sweden, Finland and the European part of Russia, respectively (Valentini et al., 2000; Milyukova et al., 2002). Nevertheless this index, varying in experiments from 46 to 53 mol C m^2 $year^{-1}$ (Shibistova et al., 2002; Lloyd et al., 2002), was similar to the 60 ±10 mol C m^2 $year^{-1}$ value obtained for Canadian forests (Ryan et al., 1997). It should be noted that the coefficient of GPP "expenditures" for autotrophic respiration (φ = 0.64) also agreed with computations given for coniferous forests of Canada (Baldocchi et al., 1997; Ryan et al., 1997).

Analysis of integrated CO_2 fluxes has shown a great variability in net gas ecosystem exchange (NEE) values, showing the difference between values of gross primary production and total constituent respiration during the vegetation season. These fluctuations reflect a complex effect of abiotic factors (photosynthetic active radiation, deficiency of aqueous tension and air temperature) on photosynthetic activity (Shibistova et al., 2002; Lloyd et al., 2002), and moreover they depend on the start time of photo-assimilating stand activity (Suni et al., 2003). The integrated value of annual net gas exchange of the ecosystem varied from 13 to 15 mol C m^2 $year^{-1}$.

The net gas exchange of forest ecosystems is considered to be primarily determined by respiration intensity (Valentini et al., 2000). From this point of view the CO_2 emission value from soil, which makes up to 60-80% of the total ecosystem respiration (Shibistova et al., 2002 a,b,c; Kelliher et al., 1999; Hollinger et al., 1998:), is a key component of carbon cycle.

Thus the direct measuring of carbon dioxide fluxes by the eddy covariance method has shown that the studied tree stand is a large accumulator of atmospheric CO_2. Carbon accumulation, on average, consisted of 1.8 t year $^{-1}$ per hectare of forest area for the studied period.

3.3 Total carbon budget of Russian forests

Two main fluxes, net primary production (with positive accumulation) and heterotrophic respiration (with negative carbon emission), are decisive in the carbon budget of Russia (Nilsson et al., 2003; Shvidenko et al., 2003). Net production results from green phytomass (~49%), above ground tree mass (~26%), and underground processes (~25%). Heterotrophic respiration is mainly determined by ground vegetation respiration, with 15% caused by above and below ground detritus decomposition. Carbon emission fluxes by wild fires are 37%, due to ecosystem disturbances, while biotic factors (most significantly insect reproduction) contribute the same amount. Carbon accumulation in the hydro- and lithosphere contributes no more than 2.2% of the net primary production, which means that over approximately 40-45 years the annual volume of organic matter assimilated by vegetation is essentially removed from the biological cycle. This is not so large a value relative to the inherent time scale of the biological cycle, however it is significant over geological time. Summarizing the various pools and fluxes shows the net biome production (absorbing carbon from the atmosphere) of Russian forests to be at a rate of 7.6% of the net primary production.

From flux assessments it follows that for the studied period of time Russian forested areas have absorbed about 240 Tg (million ton) of carbon per year from the atmosphere. It is now known that the “controlled” (by forestry measures) part of the carbon budget is limited by disturbances and other possible transformations of non-wooded lands to forested ones (forest regeneration and forest propagation). Better integrated and temporally averaged values were obtained when assessing the total carbon budget parameters for the other time periods. Thus for 38 years (1961-1997) Russian forested areas absorbed, on average, 430+/-70 million tons of carbon per year from the atmosphere, the third part of which was determined by changing ground vegetation. Due to disturbances in forest ecosystems the total flux varied from 90 to 400 Tg of carbon per year. Net biome production averaged in 5-year periods shows less variations: from 240 to 320 million t of carbon per year.

Assessment of the total carbon budget (Nilsson et al., 2003; Shvidenko et al., 2003) shows that: 1) the Russian forests are presently a huge reservoir of additional carbon sequestration (about 200-600 million t C / yr); 2) the assessed potential carbon sequestration may be obtained only for vast areas and using sustainable forest management; 3) realizing sustainable forest management requires a new state policy in managing the forestry sector, however this does not conflict with traditional forestry activities aimed at high forest productivity, improved forest protection and maintaining forest biodiversity.

3.4 Wildfires and carbon emissions – the main source of uncertainties in estimating parameters of carbon cycle of Siberian forest ecosystems

Official statistics show that between 20,000 and 40,000 fires occur annually in Russia, affecting an area of 2 to 3 million ha of forest and other lands (Davidenko et al. 2003). The fires are detected and controlled only in the so-called "protected forests" and on pasturelands. However, the application of space-borne sensors, such as the NOAA/AVHRR Terra/Aqua/MODIS, ENVISAT/MERIS and Terra/ASTER, have improved considerably the detection of active fires along with a better estimation of burned areas and large-scale impacts (Sukhinin et al., 2004a).

For example, before the 1980s it was believed that, on average, fires annually burned 1.5 million ha in the boreal forests within the former Soviet Union. Recent investigations based on satellite imagery revealed that the magnitude of the fires has been greatly underestimated, and now it is believed that boreal zone fires annually destroy an average of 8 million ha, with considerable annual fluctuation (Conard et al., 2002). For example, in 1987 satellite image evaluation revealed a total area burned in boreal forests and other lands in the East-Asian regions of Russia of about 14 million ha (Cahoon et al., 1994).

The Krasnoyarsk satellite receiving stations at the V.N. Sukachev Institute for Forest and the Russian Ministry of Emergency are now capable of downloading and processing both NOAA/AVHRR and Terra/MODIS data. They cover the whole Asian part of Russia, which consists of approximately one billion ha of vegetated land between the Urals to the West and Sakhalin Island in the Far East. The surveyed area includes all vegetation types (forest, tundra, steppe, etc.)(Soja et al., 2004).

Dynamics of the total area burned and number of fires in the Asian part of Russia, based on satellite monitoring data for 1996-2003, indicates an increase in both values (Sukhinin et al., 2004b). The fire seasons of 2002 and 2003 were especially catastrophic in Eastern Siberia (Sukhinin et al., 2003, 2004b) (Table2).

Table 2. Observed and remote sensing data on forest fires in 2002 and 2003 in Eastern Siberia.

Year	Governmental agencies reports based on ground and aircraft observations			Satellite-based data (NOAA/AVHRR)		
	Fire number mentioned in reports	Total area of burn scars, (ha)	Forest burned area, (ha)	Number of surveyed fires	Total area of burn scars, (ha)	Forest burned area, (ha)
2002	35 000	1 834 000	1 200 000	10 300	11 767 000	No data
2003	28 000	2 654 000	2 074 000	15 440	21 527 000	18 862 000

The differences between ground / aircraft-based observations versus satellite monitoring of forest fires are obvious. Data from the Aerial Fire Suppression (Avialesookhrana) Institute also do not provide the full overview of the situation.

Considering the large discrepancies between different satellite datasets on the one hand and conventionally collected fire data on the other, the question of absolute accuracy of satellite data seems to be of minor concern. Instead it is most important to close the extremely large gap between the statistics of the forestry services and the remote sensing institutions. For example, the satellite-derived burned forest area provides a total area affected by fire which is almost ten times higher than the assessments by aerial observations for the same region. According to the results of aerial surveys carried out in Buryatia and Chita in September 2003, it was decided that about 20% burned in Irkutsk and Buryatia area and that at least 10% of the Chita area was affected by crown fires because of extremely dry weather (Goldammer et al., 2004).

Rough assessments of carbon release to the atmosphere from fires of various intensities and duration based on 2003 satellite data and some empirical assumptions (Soja et al., 2004; Goldammer et al., 2004; Kasischke et al., 2005) gave a value about 46-100 Tg as a prompt release, and 190-400 Tg as post-fire release. This scenario of 2003 carbon emissions represents a crude assessment of the effect of fire on carbon release and probable patterns of carbon decrease dynamics (i.e. its sequestration). The most important source of errors in estimating fire emissions is the incomplete research into forest fuels and their full combustion depending on weather and environmental conditions.

3.5 The role of forest- bog and bog ecosystems in carbon accumulating in Siberian area

In spite of researcher efforts to make known the contribution of Siberian bogs to the global organic carbon cycle, this problem remains as urgent as before. Presently only the process of intensive accumulation and study of new data can be stated (P'yavchenko, 1985; Peat resources of RSFSR, 1991; Vompersky, 1994; Vompersky et al., 1999; Efremov et al., 1994; Efremova et al., 1997, 1998; Makhov et al., 1994; Panikov et al., 1993; Naumov et al., 1994; Naumov, 1997; etc.).

The main barriers to an objective assessment of this contribution are: a) absence of total and reliably differentiated account of boggy and peaty areas; b) geographically uneven and quantitatively insufficient measurements of peat deposits; c) fragmentary information on the group and fractional structure of organic matter of peat deposits and the transformation of carbon forms in these peat deposits related to environmental dynamics (e.g. T, moisture, redox potential); d) the weak experimental basis for the current retrospective assessments and prediction of the depositional effect of bogs and boggy forests of Siberia.

According to the forest mapping made by Korotkov (1994) two forest oblasts make up part of the western Siberian continental sector: the Altai-Sayan mountains and the western Siberian plains. Each is differentiated into forest provinces (FP), of which the first oblast covers 7 FP, and the second covers 3 FP.

The correlation of areas, peat stocks and deposited carbon for peaty bogs of the region are given in Table 3. One can see that plain areas, and taiga forest zones in them with a specific index distribution in subzones, have a decisive importance. Moreover, a rather contrasting increase of total peat and organic matter stocks from the north to the south at the more "smoothed" differentiation of bogs area is observed here.

Deposited peat stocks converted to bone-dry organic matter (BDM) within the BVB reach almost 18 billion tons, which can be subdivided into oligotrophic (50.2%), mesotrophic 18.1% and eutrophic (31.7%) peats. Uptake and long-term isolation of atmospheric carbon by BVB equals 9.3 billion tons (oligotrophic - 49.4%, mesotrophic - 18.2%, eutrophic - 32.4% of organic carbon). Based on calculations the biosphere contribution of BVB to the C- depositing indices of forest- bog- peat complexes of the western Siberian plain may be presented in generalized form as follows: in area – 6.7%, in deposited ACB – 16.5%, in organic carbon of peat deposits – 17.2%.

Table 3. Correlation of areas, peat stocks and deposited carbon in peaty bogs of western Siberian forest zones.

Ecoregions	% of total bog area	% of total peat stock	% of total carbon stock
Zauralsk- Yeniseisk FP			
pretundra forests and open forests:	15,3	3,6	3,5
Tundra	4,8	0,1	0,05
Forest tundra	10,5	3,5	3,5
Zauralsk- Yeniseisk FP			
taiga forests:	77,8	90,4	90,6
northern taiga	22,1	18,7	18,6
middle taiga	24,8	28,1	28,1
southern taiga	30,9	43,6	43,9
Plain steppe and forest- steppe	4,6	3,6	3,6
Total in plain area	97,7	97,6	97,7
Mountain and pre- mountain area	2,3	2,4	2,3
Total in western Siberian macro-region	100	100	100

Progressing climate warming will undoubtedly result in the heating of peat deposits, the fall of soil water level and the increased aeration of bog ecosystems, which will in turn increase microbiological and fermentative soil activity. This will result in accelerated decomposition of the organic matter in peat and in their humification. Polysaccharides will be transformed first, however eventually all groups and organic fractions will be affected, including even more stable components like lignin, humic coal and wax resins. The earlier stated proportions will be shifted in the group and fraction composition of humus, and it will be stabilized at another level which corresponds to the new oxidative- regenerative situation.

4. CONCLUSIONS

It is recommended that future studies focus on the following items: i) elucidation of the reasons for the discrepancy in assessing carbon pools and fluxes determined by various methods for some regions or smaller areas; ii) climate change impact on the dynamics of pools and fluxes, at different temporal and spatial resolutions; iii) the more precise assessment of carbon emissions as a result of forest area disturbance caused by cuts, wild fires, land use, outbreaks of mass insect reproduction, man-made pollution; iv) elaboration of a system of carbon cycle models at different spatial levels; v) study of models for assessing scenarios of carbon emission accumulation via different climate changes; vi) elaboration of ecological- economical models of the carbon budget for forest and bog areas.

ACKNOWLEDGEMENTS

The authors thank the researchers of the V.N.Sukachev Institute of Forest SB RAS, including Vedrova E.F., Verkhovets S.V., Efremova T.T., Kruglov V.B., Sukhinin A.I., Shibistova O.B., for presented materials and their help in preparing this paper.

REFERENCES

Arbatskaya, M.K., 1998, A long- term variability of climate, tree growth and fire occurrence in taiga forests of Middle Siberia, PhD thesis abstract, Krasnoyarsk, 21. (in Russia)

Bakwin, P. S., Tans, P. P., Hurst, D. F., and Zhao, C., 1998, Measurements of carbon dioxide on very tall towers: Results of the NOAACMDL program, *Tellus* **50B:**401-415.

Baldocchi, D., Vogel, C.A., and Hall, B., 1997, Seasonal variations of carbon dioxide excange rates above and below a boreal jack pine forest, *Agric. For. Meteorol.* **83**:147-170.

Birdsey, R.A., 1992, Carbon storage and accumulation in United States forest ecosystems, USDA Forest Service. General Technical Report WO-59. 51 p.

Cahoon, D.R., Stocks, B.J., Levine, J.S., Cofer, W.R., and Pierson, J.M., 1994, Satellite analysis of the severe 1987 forest fires in northern China and southeastern Siberia., *J. Geophys. Res.* **99** (D9):18627-18638.

Ciais, P., Denning, A.S., Tans, P.P., Berry, J.A., Randall, D.A., Collatz, G.J., Sellers, P.J., White, J.W.C., Trolier, M., Meijer, H.J., Francey, R.J., Monfray, P. and Heimann, M., 1997, A three dimensional synthesis study of δ18O in atmospheric CO2. Part 1: Surface fluxes, *Journal of Geophysical Research* **102**:5857-5872.

Conard, S.G., Ivanova, G.A., 1997, Wildfire in Russian boreal forests – potential impacts of fire regime characteristics on emissions and global carbon balance estimates, *Enviromental Pollution.* **98(**N3): 305-313

Conard, S.G., Sukhinin, A.I., Stocks, B.J., Cahoon, D.R., Davidenko, E.P., Ivanova, G.A., 2002, Determining effects of area burned and fire severity on carbon cycling and emissions in Siberia, *Climatic Change* **55**(1-2): 197-211.

Davidenko, E.P., Furyaev, V.V., Sukhinin, A.I., and Goldammer, J.G., 2003, Fire management needs in Russia's boreal forest zone, Paper presented at the *3rd International Wildland Fire Congress,* Sydney, Australia, 3-6 October 2003..

Efremov, S.P., Efremova, T.T., Melentieva, N.V., 1994, Carbon stocks in bog ecosystems, Carbon in forest and bog ecosystems of Russia, *Krasnoyarsk* 128-140. (in Russia)

Efremova, T.T., Efremov, S.P., Melentieva, N.V., 1998, Water resources of bogs of Russia and assessment of their chemical composition, *Geographia and prirodnye resursy (Geography and natural resources)* **2**:79-84. (in Russia)

Efremova, T.t., Efremov, S.P., Melentieva, N.V., 1997, Stocks and content of carbon compounds in bog ecosystems of Russia, *Pochvovedenie (Soil science)* **12**:1470-1477. (in Russia)

Forest ecosystems of the Yenisei meridian, 2002, Pleshikov, F.I., (ed) Novosibirsk, Publ. House of SB RAS,: 356 p. (in Russia)

Furyaev, V.V., Vaganov, E.A., Tchebakova, N.M., Valendik, E.N., 2001, Effects of Fire and Climate on Successions and Structural Changes in the Siberian Boreal Forests, *Eurasian J. For. Res.* **N2**: 1-15.

Gloor, M., Fan, S.-M., Pacala, S. W., and Sarmiento, J. L., 2000, Optimal sampling of the atmosphere for purpose inverse modelling: a model study, *Global Biogeochem. Cycles* **14**:407–428.

Goldammer, J.G., Sukhinin, A.I.,Chisar, I., 2004, Existing fire situation in Russian Federation: conclusions for extending international cooperation in UNO frames and global programs in monitoring and assessing fires. In: *Forest fire control at the ecoregional level,* World Bank, Publ House "Alex", Moscow,: 26-66. (in Russia)

Goulden, M.L., Daube, B.C., Fan, S.-M., Sutton, D.J., Bazzaz, Munger, J.W., Wofsy, S.C., 1997, Physiological responses of a black spruce forest to weather, *JGR-Atmospheres* **102(D24)**:28987-28996.

Grace, J., Malhi, Y., Lloyd, J., McIntyre, J., Miranda, A.C., Meir, P., Miranda, H., 1996, The use of eddy covariance to infer the carbon balance of Brazilian rain forests. *Global Change Biol.* **2**:209-218.

Hollinger, D.Y., Kelliher, F.M., Schulze, E.-D., Bauer, G., 1998, Forest-atmosphere carbon dioxide excahnge in Eastern Siberia. *Agric. For. Meteorol.* **90:**291-306.

Isaev, A.S., Korovin, G.N., Zavarzin, G.A., 1998, Carbon turnover on the Russian area, Moscow, RAS, C. 63-95. (in Russia)

Jarvis, P.G., Massheder, J.M., Hale, S.E., Moncrieff, J.B., Rayment, M., and Scott, S.L., 1997, *Journal of Geophysical Research.* **102**:28953-28986.

Jenkinson, D.S., Rainer, J.H., 1977, The turnover of soil organic matter in some of Rothamsted classical experiments, *Soil Sci.* **123**(5):298-305.

Kasischke, E. S., Hyer, E. J., Novelli, P. C., Bruhwiler, L. P., French, N. H. F., Sukhinin, A. I., Hewson, J. H., and Stocks, B. J., 2005, Influences of boreal fire emissions on Northern Hemisphere atmospheric carbon and carbon monoxide, *Global Biogeochem. Cycles* 19: GB1012, doi:10.10292004GB002300.

Kauppi, P.E., Posch, M., Hänninen, H.M., et al., 1997, Carbon reservoirs in peatlands and forests in boreal regions of Finland, *Silva Fennica.* **31**(1):13-25.

Keeling, R.F., Piper, S.C. and Heimann, M., 1996, Global and hemispheric CO_2 sinks deduced from measurements of atmospheric oxygen concentration, *Nature* **381**:218-221.

Kelliher, F.M., Lloyd, J., Arneth, A., et al., 1999, Carbon dioxide efflux density from the floor of a central Siberian forest, *Agric. Forest. Meteorol.* **94:**217-232.

Knorre, A.A., 2003, Use of registering structures at assessing annual production of components of phytocenoses in forest and bog ecosystems of Priyeniseisk Siberia, PhD thesis abstract. – Krasnoyarsk,: 21p. (in Russia)

Korotkov, I.A., 1994, Forest regioning of Russia and of republics of the former Soviet Union, Carbon in forest and bog ecosystems of Russia. Krasnoyarsk,: 170 p.

Levin, I., Ciais, P., Langenfelds, R., Schmidt, M., Ramonet, M., Sidorov, K., Tchebakova, N., Gloor, M., Heimann, M., Schulze, E.-D., Vygodskaya, N.N., Shibistova, O. and Lloyd, J., 2002, Three years of trace gas observations over the EuroSiberian domain derived from aircraft sampling a concerted action, *Tellus*, **54B(5):**696-712.

Lloyd, J., Langenfelds, R., Francey, R.J., Gloor, M., Tchebakova, N., Zolotukhine, D., Brandt, W., Werner, A., Jordan, A., Allison, C., Zrazhewske, V., Shibistova, O., and Schulze, E.-D., 2002, A trace-gas climatolog above Zotino, central Siberia, *Tellus*, **54B(5):**749-767.

Makhov, G.A., Bazhin, N.M., Efremova, T.T., 1994, Methane emission from bogs of the area between rivers Ob' and Tom', *Chemistry in the interest of stable development* **2**:619-622. (in Russia)

Milyukova, I., Kolle, O., Varlagin, A., Vygodskaya, N.N., Schulze, E.-D., and Lloyd, J., 2002, Carbon balance of a southern taiga spruce stand in European Russia. *Tellus*, **54B**(5):429-442

Mortazavi B., and Chanton J. P. 2002, Carbon isotopic discrimination and control of nighttime canopy 18O-CO2 in a pine forest in the southeastern United States, *Global Biogeochem. Cycles* **16**(1):10.10292000GB001390.

Nabuurs, G.J., and Mohren, G.M.J., 1993, Carbon in Dutch forest ecosystems, *Netherlands J. Agricultural Sci.* **41**:309-326.

Nabuurs, G.J., and Mohren, G.M.J., 1995, Modelling analysis of potential carbon sequestration in selected forest types, *Can. J. For. Res.*, **25**(7):1157-1172.

Naumov, A.V., Efremova T.T., Efremov, S.P., 1994, On emission of carbon dioxide and methane from bog soils of Southern Vasyuganie, *Sibirsky Ekologichesky jurnal (Siberian ecological journal),* **1**(3):269-274. (in Russia)

Naumov, A.V., 1997, Plant respiration and carbon dioxide emission in the bog ecosystem, *Sibirsky Ekologichesky jurnal (Siberian ecological journal)*, **4**:385-391.

Nilsson, S., Shvidenko, A.Z., Stolbovoi, et al., 2000, Full Carbon Account for Russia, IIASA, Laxenburg, Austria.

Panikov, N.S., Titlyanova, A.A., Paleyeva, M.V., Semenov, A.M., Mironycheva-Tokareva, N.P., Makarov, V.I., Dubinin, E.V., Efremov, S.P., 1993, Methane emission from bogs of West Siberian south, DAS, **330**(3):388-390. (in Russia)

Peat resources of RSFSR. 1991, Moscow: Torfgeologia (Peat geology), 100 p.

Peylin, P., 1999, ^{18}O isotopic composition in atmospheric CO_2: A new tracerto estimate photosynthesis at global scale. PhD Dissertation, Paris IVUniversity,. 221 pp.

P'yavchenko, N.I., 1985, Peat bogs, their natural and economical importance, Moscow, *Nauka (Science)* 152 p. (in Russia)

Ryan, M.G., Lavigne, L.B., and Gower, S.T., 1997, Annual carbon cost of autotrophic respiration in boreal forest ecosystem in relation to species and climate, *J. Gephys. Research* **102**:28871-28883.

Shibistova, O.B., Lloyd, D., Kolle, O., Arneth, A., Tchebakova, N.M., Zolotukhin, D.A., Zrazhevskaya, G.K., and Schulze, E.D., 2002a, Assessment of CO_2 accumulating by pine stand using the eddy covariance method, DAS, **3**:425-429. (in Russia)

Shibistova, O., Lloyd, J., Zrazhevskaya, G., Arneth, A., Kolle, O., Astrakhantceva, N., Shijneva, I., Knohl, A., and Schmerler, J., 2002b, Ecosystem respiration budget for a Pinus sylvestris stand in Central Siberia, *Tellus* **54B(5)**:568-589.

Shibistova, O., Lloyd, J., Evgrafova, S., Savushkina, N., Zrazhewskaya, G., Arneth, A., Knohl, A., Kolle, O., and Schulze, E.-D., 2002c, Seasonal and spatial variability in soil CO_2 efflux rates for a central Siberian Pinus sylvestris fores. *Tellus* **54B**(5):552-567.

Schulze, E.-D., Lloyd, J., Kelliher, F., Wirth, Ch., Rebmann, C., Luhker, B., Mund., M., Knohl., A., Milyukova., I., Schulze, W., Ziegler, W., Varlagin, A., Sogachev, A., Valentini, R., Dore, S., Grigoriev, S., Kolle, O., Panfyorov, M., Tchebakova, N., and Vygodskaya, N.N., 1999, Productivity of forests in the Eurosiberian boreal region and their potential to act as a carbon sink - a synthesis, *Global Change Biology* **5**(*6*):703-722.

Shvidenko A., Nilsson S. 2003 A synthesis of the impact of Russian forests on global carbon budget for 1961-1998, *Tellus* **55B**:391-415.

Shvidenko, A.Z., Vaganov, E.A., and Nilsson, S., 2003, Biospheric role of Russian forests at the start of the third millenium: carbon balance and Kyoto Protocol, *Sibirsky ekologichesky jurnal (Siberian ecological journal)* **6**:649-658. (in Russia)

Soja, A.J., Sukhinin, A.I., Cahoon, D.R., Shugart, H.H., and P.W., Stackhouse, 2004, AVHRR-derived fire frequency, distribution, and area burned in Siberia, *International Journal of Remote Sensing* **25(**10):1939-1960.

Soja, A.J., Cofer III, W.R., Shugart, H.H., Sukhinin, A.I., Stackhouse, Jr. P.W., McRae, D.J., and Conard, S.G., 2004, Estimating fire emissions and disparities in boreal Siberia (1998 through 2002), *Journal of Geophysical Research* **109**:(D14S06).

Sukhinin, A.I., Ivanov, V.V., Ponomarev, E.I., Slinkina, O.A., Cherepanov, A.V., Pavlichenko, E.A., Romasko, V.Y., and Miskiv, S.I., 2003, The 2002 Fire season in the Asian part of the Russian Federation: A view from space, *Int. Forest Fire News* **28**:18-28.

Sukhinin, A.I., French, N.H.F., Kasischke, E.S., Hewson, J. H., Soja, A.J., Csiszar, I.A., Hyer, E.J., Loboda, T., Conard, S.G., Romasko, V.I., Pavlichenko, E.A., Miskiv, S.I., Slinkina, O.A., 2004, Satellite-based mapping of fires in Russia: New products for fire management and carbon cycle studies, *J. Remote Sensing of Environment* **93**:546-564.

Sukhinin, A.I., 2004, Dynamics of the forest fire situation in the Asian part of Russia during the fire season of 2003 (April- September), *Int. Forest Fire News* **29**:113-118.

Suni T., Berninger, F., Vesala, T., Makkanen, T., Harri, P., Lindroth, A., Arneth, A., Shibistova, O., Lloyd, J., 2003, Air temperature triggers the recovery of evergreen boreal forest photosynthesis in spring, *Global Change Biology* **9**:1410–1426

Tans, P. P., Bakwin, P. S., and Guenther, D. W., 1996, A feasible global carbon cycle: a plan to decipher today's carbon cycle based on observations, *Global Change Biol.* 2, 309–318.

Tchebakova, N.M., Monserud, R.A., Nazimova, D.I., 1994, A Siberian Vegetation Model Based on Climatic parameters, *Can. J. For. Res.* **24**:1597-1607.

Utkin, A.I., Zamolodchikov, D.G., Chestnykh, O.V., Korovin, G.N., Zukkert, N.V., 2001, Forests of Russia as a reservoir of organic carbon of the biosphere, *Lesovedenie (Forest science)* **5**:8-23. (in Russia)

Vaganov, E.A., Furyaev, V.V., Sukhinin, A.I., 1998, Fires of Siberian taiga, *Priroda (Nature)* **7**:51-62. (in Russia)

Valentini, R., Metteucci, G., Dolman, A.J. et al., 2000, Respiration as the main determinant of carbon balance in European forests, *Nature* **404**:861-865.

Vompersky, S.E., Tsyganova, O.P., Kovalev, A.G., Glukhova, T.V., Valyaeva, N.A., 1999, Bog formation of the Russian area as a factor of fixing atmospheric carbon, carbon turnover on the area of Russia. Selected scientific proceedings on the problem "*Global evolution of the biosphere. Anthropogenic contribution*", Moscow, 124-145. (in Russia)

Vompersky, S.E., 1994, Role of bogs in carbon cycle, Readings to the memory of the academician V.N.Sukachev. XI: Biogeocenotic peculiarities of bogs and their efficient use. Moscow: *Nauka (Science)* **5-38**. (in Russia)

THE INTERACTION OF CO_2 BETWEEN THE ATMOSPHERE AND SURFACE WATERS OF LAKE BAIKAL AND THE INFLUENCE OF WATER COMPOSITION

V.M. Domysheva[1], M.V. Sakirko[1], M.V. Panchenko[2] and D.A. Pestunov[2]
[1]Limnological Institute SB RAS, Irkutsk; [2]Institute of Atmospheric Optics, SB RAS, Tomsk

Abstract: Investigations into the inventory of CO_2 in the hydrosphere are primarily aimed gas exchange between the atmosphere and oceans. However, lakes often are not considered as possible sources and sinks, even if they can be locally important for many geographical regions. Lake Baikal covers a tremendous area and has locally high biotic activity, and thus it can be expected to play an essential role in regional CO_2 dynamics. Measurements of atmospheric CO_2, dissolved gases and various aqueous chemical parameters were performed at the Limnological Institute SB RAS site near Bolshie Koty in 2003-2004

Key words: carbon dioxide, gas exchange, chemical composition of water, organic matter.

1. INTRODUCTION

The Kyoto Protocol on climate change occupies a particularly important place among international conventions and agreements. This Protocol, which will start in 2005, will for the first time apply economic market mechanisms to addressing intergovernmental and trans-boarder environmental problems. As such specialists involved in the investigation of CO_2 sources and sinks should now concentrate on regional scale processes.

Investigations of the hydrosphere CO_2 inventory are primarily focused on the study of gas exchange between the atmosphere and ocean. Lake systems, in contrast, are often not considered as possible sources and sinks even if

S. Lombardi et al. (eds.),
Advances in the Geological Storage of Carbon Dioxide, 35–45.

they can be important for many geographical regions. In this aspect, Baikal is a unique natural study site. Taking into account its tremendous area and biotic activity, one can expect that Baikal plays an essential regional role in the processes of CO_2 generation and consumption. Prior to the present study no other research group had monitored CO_2 (one of the most problematic greenhouse gases) in the atmosphere of the Baikal region. Measurements were performed at the Limnological Institute SB RAS site near Bolshie Koty in August–September 2003 and September 2004.

2. DIURNAL VARIABILITY IN THE WATER-TO-ATMOSPHERE GAS EXCHANGE PROCESS DURING DIFFERENT SEASONS

The main purpose of this chapter is to study the relationship between changes in the chemical composition of Lake Baikal surface water and the content of carbon dioxide in the near-water atmospheric layer. For this work the concentration of CO_2 in air and the pH values of the surface water were directly monitored in situ.

In order to decrease the effect of advective processes on the measurement results and subsequent estimation of the CO_2 fluxes from the water surface, a semi-immersed transparent chamber (Fig. 1) was used. Carbon dioxide concentration, temperature, relative humidity and wind velocity were recorded at 1 and 4 m above the water surface at a meteorological mast situated near the chamber. In situ observations were performed around the clock, while at the same time water samples were collected every 3 hours at a distance of 100 m from the coast for subsequent laboratory chemical analysis. The concentrations of dissolved O_2, bicarbonate, nitrate, phosphate and pH were measured in the upper water layer. The dissolved CO_2 concentration was then calculated from the pH and bicarbonate data. Results obtained during four measurement cycles are shown in Figures 2–5.

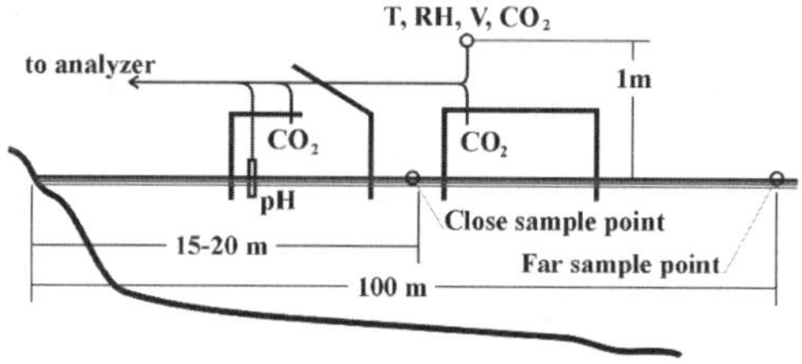

Figure 1. Diagram of the experiment.

As can be seen in all measurement series, the gas exchange process has a well-defined diurnal behaviour, which is correlated with the change of near-surface aqueous chemistry (pH). It can be seen that the rapid increase of water pH and simultaneous decrease of CO_2 concentration in the near-water layer of the atmosphere begins at sunrise. This is an illustration of the fact that the transfer of CO_2 from the atmosphere to the water surface begins practically simultaneously with the decrease of CO_2 content in the water. This process reverses direction shortly before sunset, and CO_2 begins to be transferred to the near-water layer of the atmosphere. The ratio between the durations of "positive" and "negative" stages of the process is different in different seasons, as well as the amplitude of the diurnal CO_2 fluctuations both in the atmosphere and in the measurement chamber, because these parameters are related to the duration of insolation.

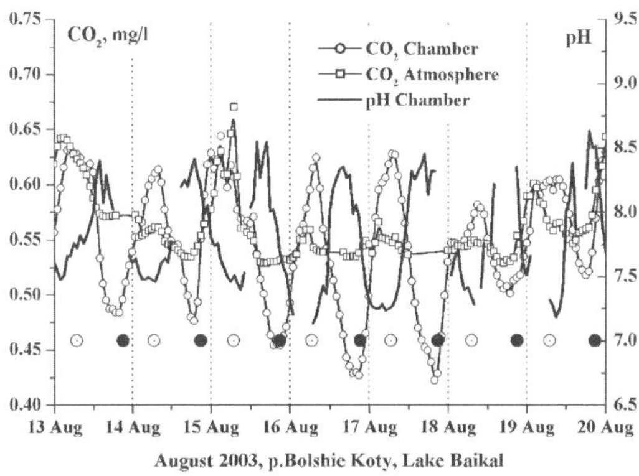

Figure 2. The results of measurements of CO_2 and pH of water in August 2003.

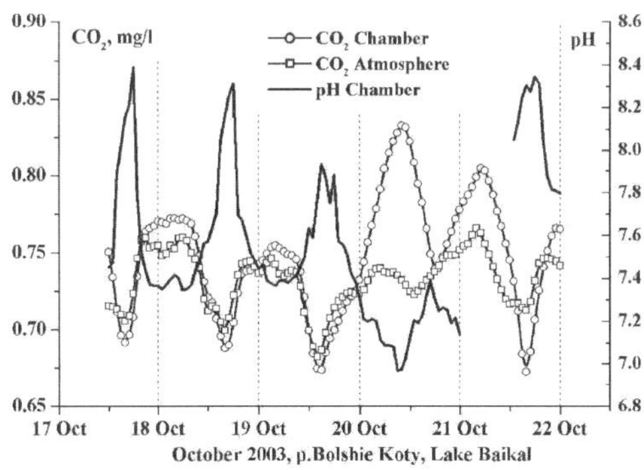

Figure 3. The results of measurements of CO_2 and pH of water in October 2003.

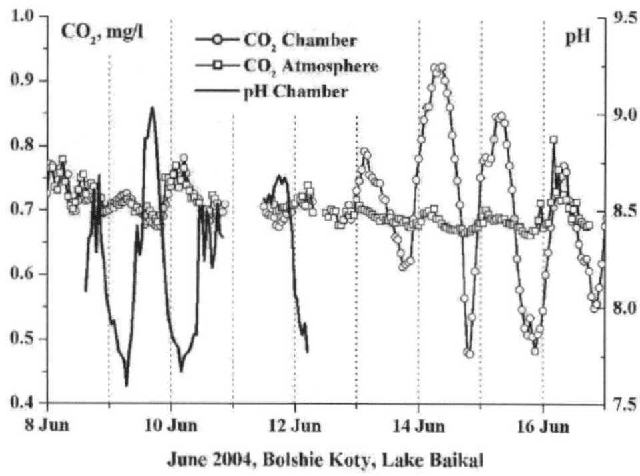

Figure 4. The results of measurements of CO_2 and pH of water in June 2004.

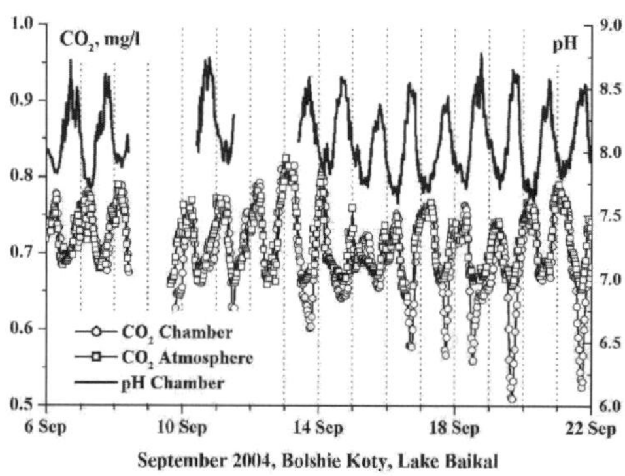

Figure 5. The results of measurements of CO_2 and pH of water in September 2004.

The most interesting result of both the October (Fig. 3) and June (Fig. 4) measurement cycles is the significant flux of carbon dioxide to the atmosphere from the water surface observed during acidic precipitation events (pH ~ 5). This is well seen in Figure 3 from October 20 to 221 and in Figure 4 from June 14 to 15. Although in this case the process can be clearly understood and explained, it should be noted that the role of acid precipitation in the atmospheric carbon dioxide budget has not yet been taken into account, at least not for the region under consideration.

Another chamber was applied in September of 2004 for a more accurate estimation of the CO_2 flux, as this chamber can be ventilated hourly. An example of the data obtained with this system is shown in Figure 6, while the mean diurnal behaviour of pH in the surface water in the chamber is shown in Figure 7. The diurnal behaviour of both the intensity and the direction of gas flux and corresponding variations of CO_2 content in water are clear from a comparison of the figures.

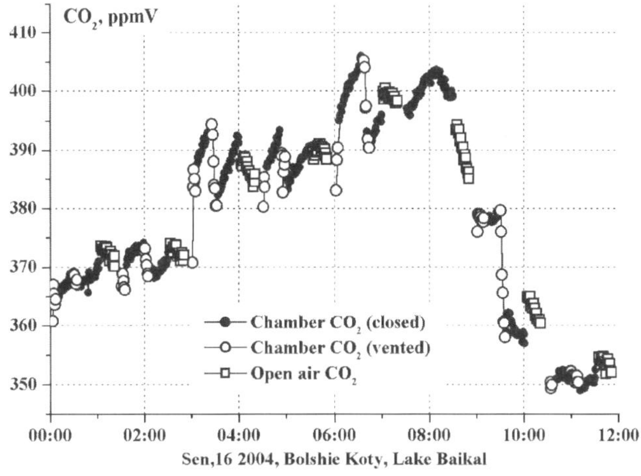

Figure 6. A portion of the signal record in the ventilated chamber (solid circles – closed chamber, open circles – open chamber, squares – the atmospheric concentration at 1 m above the water surface).

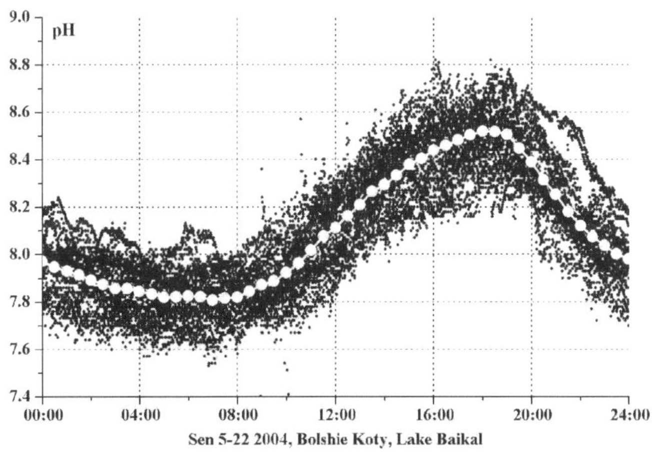

Figure 7. Mean diurnal behavior of pH in water.

3. ESTIMATION OF THE CO_2 FLUX ACROSS THE WATER SURFACE BY THE SEMI-IMMERSED VENTILATED-CHAMBER METHOD

The flux (in this case) is the change of the mass of gas (m[μg]) in an isolated air volume (v[l]) above an area (s[m^2]) of the water surface due to emission/absorption. Pure CO_2 at a pressure of one atm weights 44 g/mol. The mass of carbon dioxide in 1 litre (denoted as C_{CO2}) at pressure P and volume fraction c is:

$$C_{CO_2}[\mu g/l] = \frac{1.96[g/l] \cdot P[mmHg]}{760[mmHg]} \cdot c[ppmV] \tag{1}$$

The mass of gas M_{CO_2} in the chamber of the volume V is:

$$M_{CO_2}[\mu g] = C_{CO_2}[\mu g/l] \cdot V[l] \tag{2}$$

Thus, 238.2 mg of CO_2 are contained in a chamber having a volume V=320 l at the atmospheric content of carbon dioxide, where c = 400ppmV and atmospheric pressure P = 720 mmHg.

The use of the "ventilated" chamber provides the possibility of more accurate flux estimations, with an example shown in Figure 8.

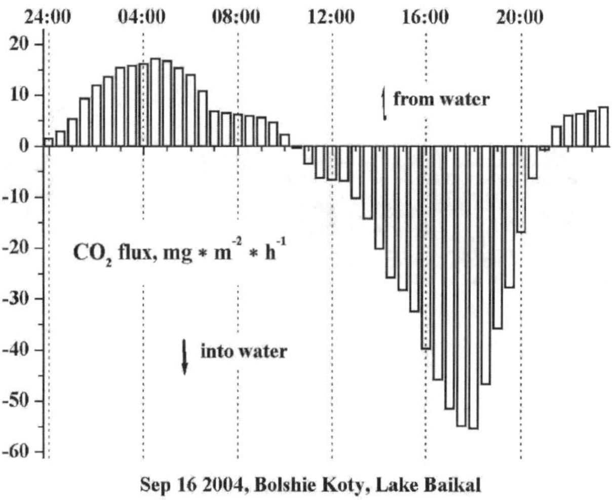

Figure 8. An example of diurnal CO2 flux across the water surface.

4. THE CHEMICAL COMPOSITION OF NEAR-SURFACE WATER IN GAS EXCHANGE EXPERIMENTS

A series of round-the-clock observations of some hydro-meteorological and hydro-chemical parameters were carried out simultaneously with CO_2 measurements. Measurements were performed every 3 hours at 2 stations situated in the littoral zone of the lake at a water depth of 2 and 5 m. Temperature, pH and the concentrations of dissolved oxygen, carbon dioxide and bicarbonate ion were measured in the near-surface water layer (5cm). Measurements of pH were carried out with an "Econics-Expert-001" ph-meter (accuracy of ±0.02 pH units), the oxygen concentration was measured with the Winkler technique with a relative error of 0.3% (the sensitivity of the method is 0.05 mg O_2/l), and the bicarbonate ion concentration was measured with the potentiometric method with an accuracy of ~ ± 2%. The carbon dioxide concentration was determined with the titrimetric method (the accuracy was ±10%) and the concentration of free CO_2 was calculated using the HCO_3^- concentration, water temperature and pH reduced to in situ temperature. The biogenic elements were determined using colorimetric methods: silicone as silicic-molybdenum heteropolyacid, phosphates with the formation of complex phosphorous-molybdenum acid, and nitrates by liquid chromatography (with an accuracy of ±5%).

Water samples were collected in September not only at the far site (marked as "far" in graphs) but also between the chambers ("close"). Water temperature differences (up to 9°C) and, possibly, the occurrence of water plants cause the observed difference in gas geochemistry (Fig. 9).

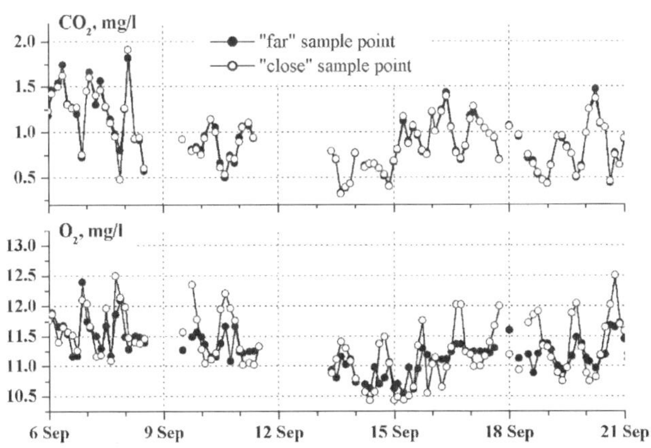

Figure 9. Chemical composition of near-surface water at "close" and "far" sites.

The main factors determining the diurnal dynamics of dissolved gases are the processes of photosynthesis and the destruction of organic substance. Emission of oxygen and absorption of not only carbon dioxide but also biogenic elements occur as a result of the synthesis of organic substances by water plants. Breakdown of the organic substances via heterotrophic organisms leads to the emission of carbon dioxide and biogenic elements as well as the consumption of oxygen. Diurnal variations of these processes depend on the duration and level of illumination, which results in the timing of diurnal maximums and minimums of the chemical concentrations (Fig. 10). Absorption and emission of gases across the water surface to maintain phase equilibrium is not rapid enough to smooth the kinetically faster O_2 and CO_2 exchanges caused by biological reactions. The diurnal dynamics of dissolved gases and biogenic elements is favored by slow vertical water exchange under the ice during the winter and in the period of thermal stratification (epilimnion) during the summer. Convective and dynamic mixing and turbulent diffusion during both spring and fall homothermy lead to a decrease in the amplitude of diurnal concentration variations of dissolved gases and biogenic elements. In addition, water temperature and lake roughness also affect the amplitude of diurnal chemical variations.

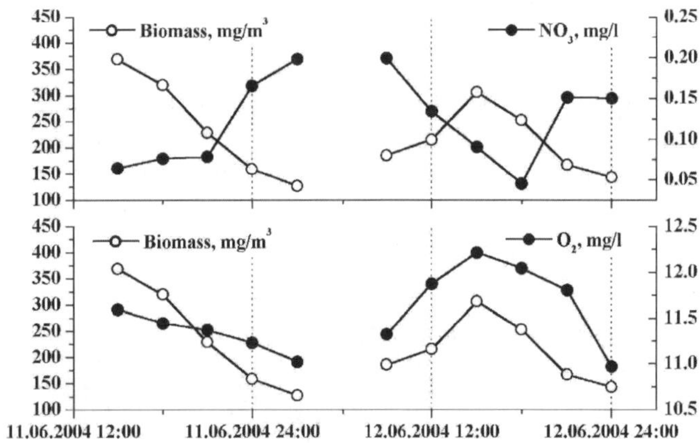

Figure 10. Diurnal dynamics of total biomass (phytoplankton and picoplankton) and the concentrations of nitrates (a) and oxygen (b) (data on biomass of phyto- and picoplankton were kindly presented by Prof. G.I. Popovskaya and Dr. O.I. Belykh).

Thus the diurnal behaviour of chemical components results from the superposition of variations in the external hydro-meteorological conditions and the biological rhythms of production and consumption of organic matter and (for dissolved gases) with the physical-chemical processes of gas absorption and emission across the water surface.

In general the diurnal O_2 variations are inverse to that of the biogenic elements and CO_2. The absolute content of O_2decreases during the night and in the morning, while the contents of free carbonic acid, nitrates and phosphates increase. Minimum concentrations of CO_2, NO_3^- and PO_4^{3-} correspond to daytime O_2 maximums. The timing of extreme O_2 and CO_2 values during the diurnal cycle depends on the relative intensities of photosynthesis and organic matter destruction. The amplitude of the diurnal concentration variations of the studied components increases with increasing photosynthesis intensity and decreases with increasing organic matter consumption. Temperature variations, maxim of water masses and gas exchange with the atmosphere do not affect the timing of diurnal concentration maximums and minimums.

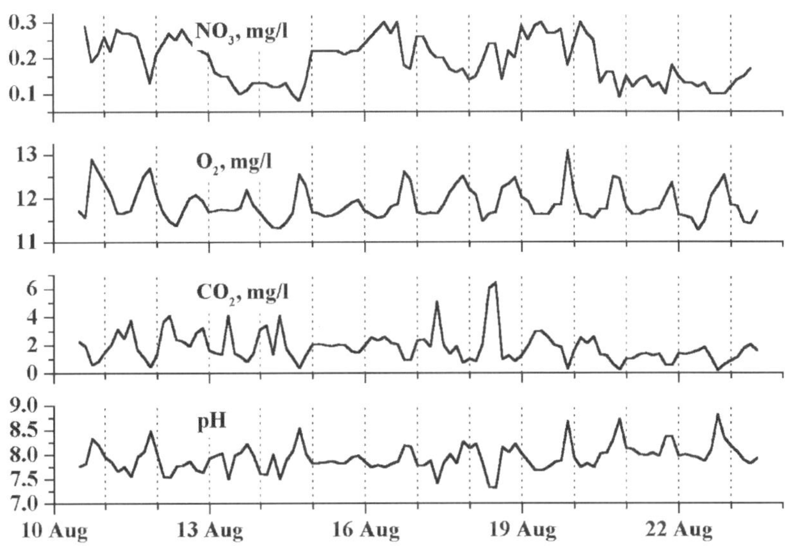

Figure 11. Diurnal aqueous chemistry behaviour in the littoral of southern Baikal in August 2003.

Data analysis has shown that the temperature maximum of the water surface is observed between 3–6 p.m. in the summer. The greatest pH value, O_2 concentration, and minimum CO_2 and biogenic element concentrations are recorded at the same time (Fig. 11). Enrichment of biogenic elements and CO_2 and the decrease in pH and O_2 concentrations are observed during the night and in the morning (6–9 a.m.). The concentration maximums and minimums in the winter are observed at 6 a.m. and 3 p.m., respectively (Fig. 12). The maximum amplitude of the diurnal variations was observed in June

2004 (ΔO_2=1,7 mg/l, ΔCO_2=2,1 mg/l), and the minimum was observed in the winter (ΔpH=0,1 unit, ΔO_2=0,47 mg/l, ΔCO_2=0,55 mg/l).

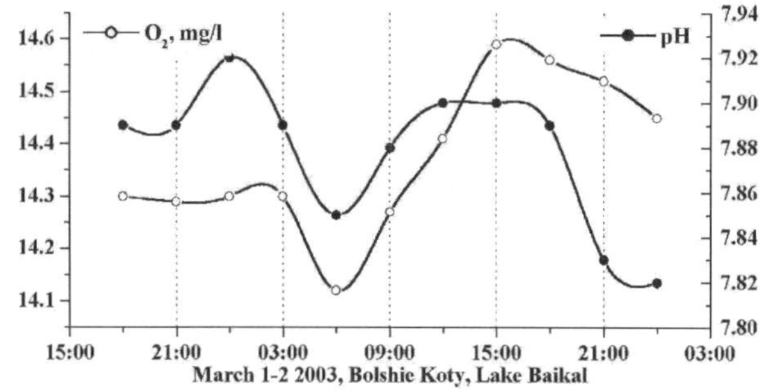

Figure 12. Diurnal behavior of oxygen and pH in March 2004.

The relative estimate of the CO_2 flux is calculated using the difference between the measured and equilibrium concentrations of CO_2 (Fig. 13).

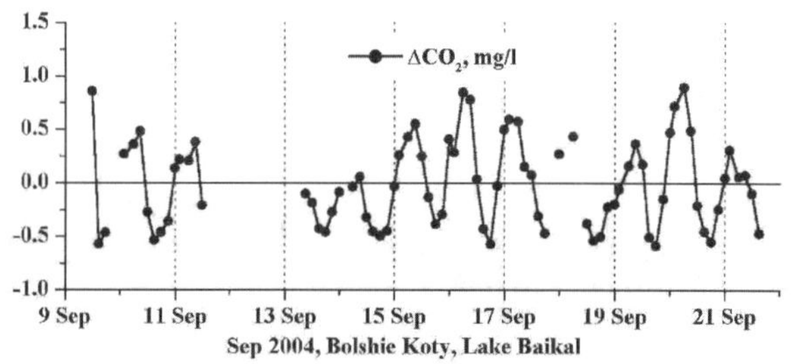

Figure 13. Difference between the measured and equilibrium concentrations of CO_2.

5. CONCLUSIONS

Thus complicated diurnal and day-to-day periodic oscillations of pH, CO_2, O_2 and biogenic elements occur in the near-surface water of Lake Baikal. They are caused by hydro-meteorological and hydro-biological factors. Deviations of the dissolved gas concentrations from their values in equilibrium with the atmosphere should cause the observed short-period variations in gas exchange intensity with the atmosphere, and the variations of the state of the carbonate system of the trophogenic zone of the lake.

ACKNOWLEDGEMENTS

The work was supported in part by program No. 3 of the Department of Earth Sciences RAS "Water resources, dynamics and preservations of underground water and glaciers", project " The study of pollution of the atmosphere and near-surface water of Lake Baikal" as well as SB RAS expedition grant "Baikal".

REMOTE SENSING AND GIS FOR SPATIAL ANALYSIS OF ANTHROPOGENIC CARBON OXIDE EMISSIONS

Yu.M. Polishchuk and O.S. Tokareva
Institute of Petroleum Chemistry, RAS, 3 Academichesky Ave., Tomsk, 634021, Russia

Abstract: Western Siberia hosts one of the main oil-bearing basins in Russia, producing more than 70 % of the country's oil. It is known that oil field flares release carbon oxides into the atmosphere which contribute to the greenhouse effect. This work details the development of a Geographical Information System (GIS) technique which uses satellite image processing to assess the spatial irregularities of anthropogenic carbon oxide emissions. This approach, which tries to make a quantitative assessment of the impact of atmospheric pollution on forest ecosystems, is based on calculating forest ecosystem areas (cells) which are inside atmospheric pollution zones. Particular issues related to the modeling of atmosphere pollution caused by oil field flares are considered. Pollution zones were revealed by standard modeling of contaminant dispersal in the atmosphere. Polluted ecosystem cells were calculated on the basis of oil production volume.

Key words: remote sensing, geographic information systems, GIS, carbon oxides, gas flares, environmental impact assessment, oil field

1. INTRODUCTION

Western Siberia hosts one of the main oil-bearing basins in Russia, producing more than 70 % of the country's oil. Oil reserves in Western Siberia have been estimated to exceed 35 – 40 billion tons, and thus production levels will remain high for at least the next two or three decades. It is known that about half of the gas associated with the produced oil is burnt in flares, with the subsequent release of carbon and nitrogen oxides, hydrocarbons and soot into the atmosphere. Once released the carbon oxides

S. Lombardi et al. (eds.),
Advances in the Geological Storage of Carbon Dioxide, 47–55

contribute to the greenhouse effect, soot blocks the stomas of conifer needles that leads to the drying of forests (especially dark-coniferous ones due to a long cycle of needle replacement), while nitrous oxides form acid precipitation and oxidants in leaves which reduce plant viability (Polishchuk et al., 1999). It is for these reasons linked to atmospheric pollution that Agan forests near the largest Samotlor oil field in Western Siberia are dying.

At present, atmospheric pollution caused by oil field flaring is the main environmental impact of oil production (Fedyunin, 1996; Polishchuk et al., 1999)). Solving the ecological problems created by oil production in western Siberia, perhaps through CO_2 geological sequestration, requires a better understanding of the spatial structure of anthropogenic carbon oxide emissions. A geographic information system (GIS) approach to analysing the spatial structure of carbon oxide emissions is discussed in this paper, extending previous work which assessed the impact of atmospheric pollution on the natural environment (Polishchuk and Tokareva, 2000, Polishchuk et al., 2002). This approach to environmental impact is based on the combination of health and geochemical methods (Polishchuk et al., 2002) for the quantitative estimation of the forest-swamp ecosystems polluted in western-Siberian oil-producing areas. This approach, however, requires ecosystem maps, which are expensive and labour-intensive to produce with traditional methods.

The use of satellite data decreases the costs of producing thematic maps, however their application to assess environmental impact caused by oil production is limited in the scientific literature. In addition the application of GIS and GIS-technologies in combination with satellite imagery to study the effects of pollution on forest-swamp ecosystems is even less common. The application of mid-resolution space images which cover an area of up to 300-350 thousand km^2 is more preferable to research ecosystem structures of large territories.

In this regard, the main purpose of this paper is to address remote sensing and GIS methods as applied to both the spatial analysis of oxide emissions and the assessment of oil production impact on the natural environment in western Siberia. The paper presents some results related to the effects of oil-field flaring on forest-swamp ecosystems in Western Siberia.

As shown in Figure 1, the spatial distribution of atmospheric pollution in Western Siberia is very irregular. One anomalous zone is located near the towns of Nizhnevartovsk and Strezhevoi, caused by gas flaring in the surrounding oil fields. The detailed analysis of this irregular spatial distribution of oxide emission and environmental impact assessment can be carried out using different cartographic and computer modeling techniques, which are described in the following section.

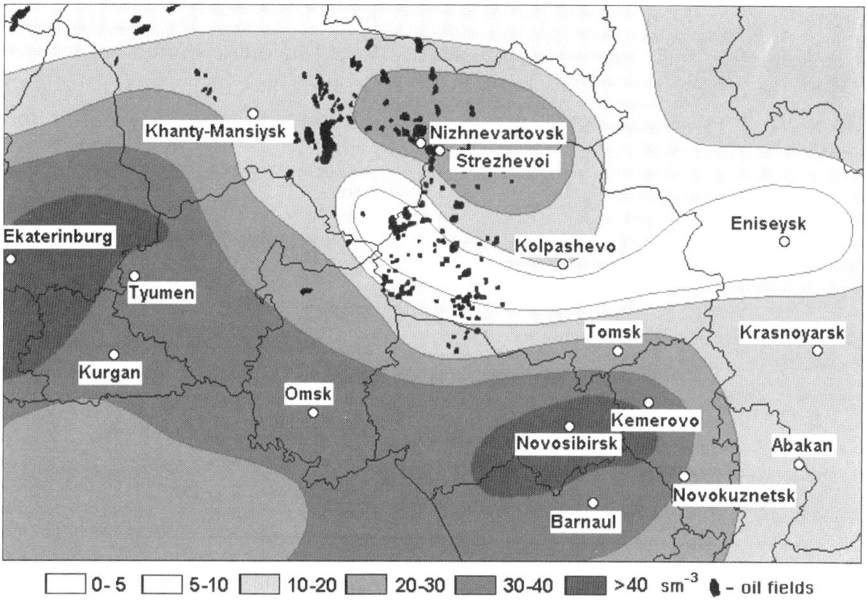

Figure 1. Map of atmospheric pollution distribution in western Siberia

2. GIS MODELLING

Typically environmental assessment is based on the application of health and safety rules, even if they are not adapted to ecosystem and ecosystem biodiversity. Instead one of the best approaches to assess the impact of chemical pollution on vegetation biosystems is to determine the areal distribution and size of polluted ecosystem complexes. That is why the complex approach (Polishchuk and Tokareva, 2000; Polishchuk et al., 2002) to environmental impact assessment used in this paper is based on the combination of health and geochemical methods. Such an approach, however, requires large volumes of environmental and remote sensing data on the anthropogenically-impacted areas, data which cannot be processed without the use of GIS and GIS-technologies. Application of GIS tools, which permit multidimensional data analysis, simplifies the procedure of ecosystem forecasting and environmental impact assessment of carbon oxides and other gas contaminants.

The general structure of the Geo-Information Modeling System (GIMS), given in Figure 2, consists of the following units. "Database" includes environmental and other quantitative information about the study area. The "Digital map subsystem" hosts the area maps, including a 1:1,000,000 scale digital map which forms the basis of the subsystem as well as other topographical maps of other scales and different thematic digital maps. The "Modeling subsystem" is the most dynamic part of the GIMS, as it is used to calculate atmospheric contaminant concentrations in order to identify the spatial distribution of atmospheric pollution.

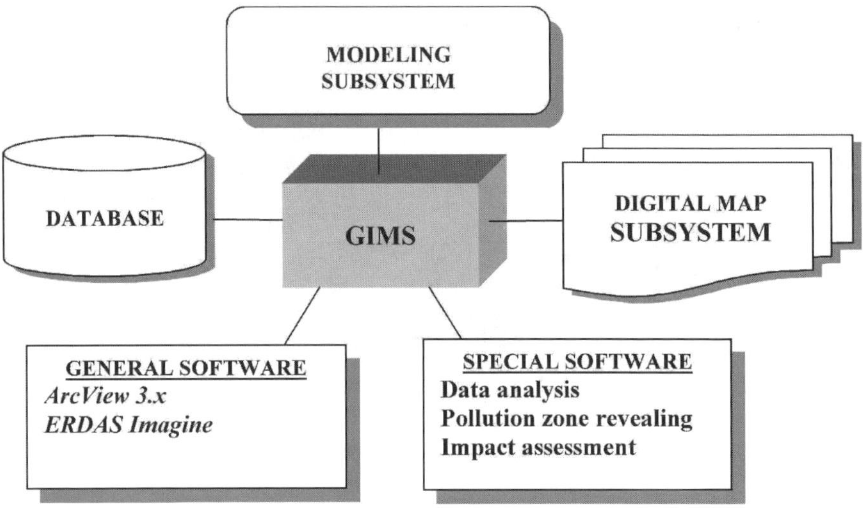

Figure 2. General structure of the geo-information modeling system.

As is shown in Figure 2 "General software" consist of the GIS software ArcView 3.x and ERDAS Imagine, while "Special software" consists of various components to assess the impact of atmospheric pollution on the ecosystem. One of the main components are program tools to identify pollution zones by analysing modeling data from the "Modeling subsystem" and to identify multiyear-averaged pollution zones. Air pollution zones (Polishchuk and Tokareva, 2000) are determined for different pollutant concentrations by modeling the pollutant dispersal in the atmosphere.

Another component of "Special software" is a computer program which assesses impact. By overlaying pollution-zone outlines on the ecosystem map it was possible to calculate the areas of the polluted complexes using GIS tools. This code, developed with the program language Avenue, calculates the relative areas of ecosystem types occurring in a pollution zone. The program consists of some standard functions and is activated directly within the ArcView interface. The user must first active the thematic layer containing the pollution zone of interest and then start the calculation of the relative areas. The layers containing ecosystem types for which the user wants to make calculations will be accommodated in the list of the program.

The program calculates the area of every ecosystem type occurring in a pollution zone, and then determines the its percentage relative to the full ecosystem type area or the area of a satellite image. Calculation results are represented graphically on the monitor screen, while the final impact assessment results regard the analysis of polluted ecosystem areas normalized to area. Clearly these relative areas will depend on the concentration of the pollutants and the volume of the oil produced. The results of this study are given below in the next sections.

3. SATELLITE IMAGE PROCESSING AND MODELING OF POLLUTION ZONES

Satellite images are widely applied to the study of environmental status. High and middle resolution images obtained from the Russian satellite Resource–O1 were used in this paper. Satellite image processing involves the following procedures:

1. Validation of remote sensing data.
2. Classification and interpretation.
3. Vectorization.
4. Thematic layer formation.

Ecosystem maps, forest-use plans and field data may all be used for validation and classification. All basic forest-swamp ecosystem types are represented within the pilot study territory (PT). The 1:1,000,000 scale digital map was used as a topographic base map for the GIS. The middle resolution satellite image classification was performed using ERDAS Imagine software and forest-use maps as a classification scheme of the pilot study territory. Classification results of the PT ecosystem types were then moved to the entire studied area in the satellite image, which below will be referred to as a "scene". A fragment of a three-zone satellite image (spatial resolution = 140 m) from the Russian satellite "Resource - O1" (scanner MSU-SK) was used as the scene. Image vectorization was performed using ERDAS Imagine and then vectorized data were exported to a GIS ArcView format. The relative areas of each ecosystem type were determined using the vectorized satellite image. The ratio of the areas of various vegetation types characterize the ecosystem structure of the territory.

As mentioned above, the pollutant impact assessment is based on the definition of the pollution zones and their subsequent plotting on the computer map or satellite image. To identify the pollution zones it is necessary to model pollutant distribution in the atmosphere. Nowadays there are many different models of air-chemical dispersion. The present work has applied the OND-86 method, which is the most commonly used approach in Russian environmental articles. It allows for the user to calculate chemical concentrations at any point in the study area using the distance from the emission source, year-averaged volume of emitted gas and average local meteorological conditions. Modeling of air pollutant dispersal consists of the following steps:

1. Determination of the locations of pollution sources using GIS and computer maps. GIS allows a user to make spatial selections of map objects and thus one can select pollution sources in the study area.
2. Input sources and chemical properties from the "Database". The user interface allows one to construct queries to the "Database" using dialog forms and properties of selected air pollution sources and chemicals.
3. Calculation of chemical concentrations using the air pollutant dispersal model.

Issues regarding the outlining of pollution zones were considered in previous papers (Polishchuk and Tokareva, 2000; Polishchuk et al., 2002). In summary this work involves the following steps:

1. Air pollution dispersion modeling.
2. Determination of pollution zone boundaries.
3. Visualisation of the pollution zones.

To outline the pollution zones it is necessary to determine chemical concentrations throughout the study area. This is accomplished by dividing the region into a spatial grid consisting of individual cells, the size of which depends on the map representation scale. Every considered air pollution source and chemical are accounted for via dispersal simulation in every cell. The next step is to determine the boundaries of the areas, while the final step involves the plotting of these boundaries on the computer map or satellite image using GIS tools, as is shown in Figure 3.

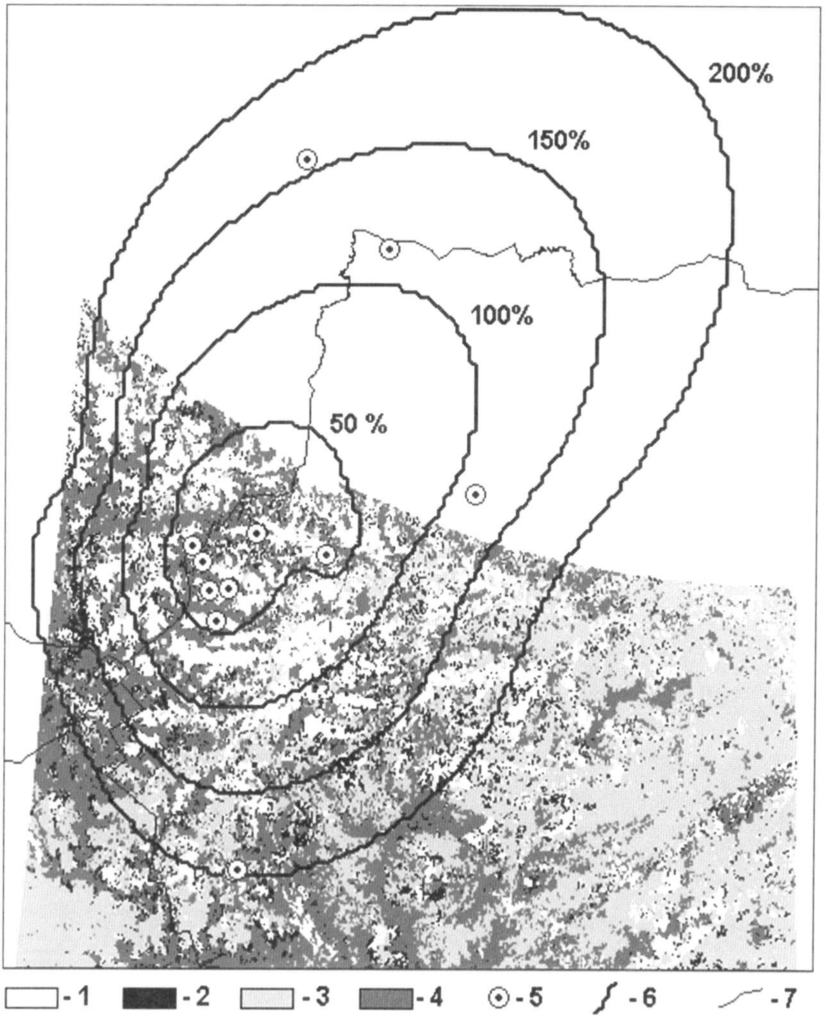

Figure 3. Ecosystem-type map after image processing. 1 – dark-coniferous forest, 2 – pine forest, 3 – small-leaved forest, 4 – swamp, 5 – gas flare, 6 – air pollution zone, 7 – oil fields.

4. IMPACT ASSESSMENT OF CARBON OXIDE EMISSIONS ON SIBERIAN FORESTS

The results of an environmental impact assessment obtained on the basis of satellite images are presented below. First a digital map of the natural ecosystem structure was constructed, and then the size of the various polluted ecosystem types (normalized to the area of corresponding ecosystem type) were calculated. These relative areas depend on both the pollution level and the volume of oil production, to which the volume of burnt accompanying gas is connected. The dependency of the relative areas of polluted ecosystems on the volume of oil production, calculated for fixed levels of air pollution by carbon oxide emissions, are represented in Figure 4. Results of the calculations for oil fields of the Vasugan group, under concentration, are $0{,}3*10^{-3}$ mg/m^3.

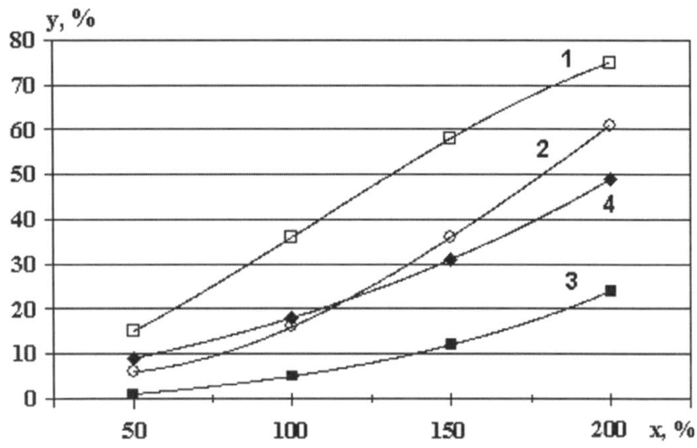

Figure 4. Relative areas of polluted vegetation types, depending on the volume of oil production at the Vasugan oil field. 1 – dark-coniferous forest, 2 – pine forest, 3 – small-leaved forest, 4 – swamp.

5. CONCLUSIONS

Study results have shown that the geo-information approach to assess ecological impact caused by chemical atmospheric pollution using remote sensing data allows the objective definition of areas having significant ecological loads on specific ecosystems. The described procedures, involving satellite image processing, makes it possible to transfer the results from the classification of one satellite image section onto other sections of a satellite image showing regions which are difficult to access. Such an approach allows the analysis of various vegetation types and ecosystems over large territories using satellite imagery. This approach may be used to carry out regular monitoring of changes in the ecological load, both depending on oil recovery levels and ecosystem types. The paper demonstrates the opportunity to study the spatial distribution of anthropogenic carbon oxide emissions and to assess its environmental impact.

ACKNOWLEDGEMENTS

This paper is the result of research performed within Integration projects № 137, 138, and 169 of the Siberian Branch of the Russian Academy of Sciences. This work was also performed with EC INCO Copernicus 2 support (Project ISIREMM, contract ICA2-CT-2000-10024), EC INTAS support (Project ATMOS: A Scientific WWW Portal for the Atmospheric Environment, contract: INTAS-00-189), and RF President's grant support for leading scientific schools (grant № SS-1008.2003.5).

REFERENCES

Fedyunin, V.A., 1996, Environmental Protection at "Tomskneft", *Oil industry*, **11**, pp. 87-89.

Polishchuk, Yu., Kozin, E., Ryukhko, V., Tokareva, O., 1999, Geoinformation Systems for Regional Environmental Studies, *Proceedings of SPIE*, 3983, pp. 572 – 577.

Polishchuk, Yu., Tokareva, O., 2000, Geoinformation analysis of atmosphere pollution impact on the ecosystem of Siberian oil-producing territories, *Proceedings of SPIE*, 4341, pp. 571 – 577.

Polishchuk, Yu., Tokareva, O., and Bulgakova, I., 2002, Space Images Processing Methodology for Assessment of Atmosphere Pollution Impact on Forest-Swamp Territories, *Proceedings of SPIE*, 5026, pp. 232 – 236.

THE SOURCES OF CARCINOGENIC PAH EMISSION IN ALUMINIUM PRODUCTION USING SODERBERG CELLS

L.I. Kurteeva[1], S.V. Morozov[2] and A.G. Anshits[1]
[1]*Institute of Chemistry and Chemical Technology, SB RAS, K.Marx str. 42, Krasnoyarsk 660049, Russia;* [2]*N.N.Vorozhtsov Novosibirsk Institute of Organic Chemistry, SB RAS, Lavrentyev av. 9, Novosibirsk 630090, Russia*

Abstract: Specific yields were determined for tars, benzo(a)pyrene, diberzo(ah)anthracene and benzo(ghi)perylene from the main PAH sources in aluminium production using Soderberg cells with anode pastes based on medium- (MP) and high- (HP) softening-point pitches. It was evaluated that the most significant source of carcinogens from the Soderberg cell is the open hole during the process stage of repositioning the current-carrying studs. This phase of the work contributes about 60-70% of the total carcinogenic PAH emissions to the atmosphere. It is shown that a replacement of MP with HP results in a decrease of the specific tar yields from all sources, but it does not reduce the specific yield of the carcinogenic PAHs and therefore the carcinogenic danger of this technology.

Key words: aluminium production; anode pastes; carbonisation; carcinogenic PAHs; coal-tar pitches; Soderberg cell.

1. INTRODUCTION

About 2M tons of aluminium is currently generated by the plants in Siberia. Unfortunately only a minor part is generated using the currently environmentally-appropriate "prebaked" technology, which ensures low specific yields of tar that contain carcinogenic polycyclic aromatic hydrocarbons (PAHs) (0.06 kg/t Al based on data from the Pechiney Company). Instead the majority of aluminium is generated in Soderberg cells with self-baking anodes, which generates the highest carcinogenic risk

S. Lombardi et al. (eds.),
Advances in the Geological Storage of Carbon Dioxide, 57–65.

in aluminium production. The specific tar yields are considerably higher for aluminium productions that use Soderberg technology, producing 0.21 kg/t Al with a 'dry' gas cleaning system (Reynolds Company) and up to 0.65 kg/t Al with a 'wet' gas cleaning system at the Krasnoyarsk aluminium plant (Kaiser Aluminium & Chemical Corporation) (Johnson and Lobachev, 1998). The reduction of carcinogenic PAH emissions is ecologically very significant for large centres of aluminium production using Soderberg technology (Krasnoyarsk, Bratsk, Irkutsk, Novokuznetsk).

When determining specific yields of tar and carcinogenic PAHs using Soderberg technology, emissions from the gas cleaning system and gas escapes from under the gas skirt are traditionally considered while the process stage of current-carrying stud repositioning is not. It is important to evaluate the contribution of two major sources of atmospheric PAH emissions: 1) gases formed during slow (0.01-0.1°/min) carbonisation of 'dry' anode paste and anode body formation; and 2) gases released from the open hole during stud repositioning, when rapid heating (70-100°/min) pyrolysis of the studhole anode paste takes place, and 'secondary' anode formation.

The use of high-softening-point pitches (HP) instead of medium-softening-point pitches (MP) to prepare the anode paste is considered one of the ways to reduce the environmental danger. Such a replacement is associated with economic and technological advantages due to a decrease in anode paste consumption and a decrease in specific tar yields, which should in turn result in a decrease of carcinogenic PAH yields. However, it has not been proven that a switch from MP to HP will decrease the carcinogenic danger. As such it is important to compare the carcinogenic PAH yields of carbonisation products of MP, HP and anode pastes based on MP (APMP) and HP (APHP).

The objectives of this work are to evaluate the contributions of the main sources of carcinogenic PAH emissions and to compare their carcinogenic danger during aluminium production using Soderberg cells.

2. EXPERIMENTAL

Samples of medium- and high-softening-point coal-tar pitches were used under laboratory conditions. The elemental composition and technical characteristics of the pitches are presented in Table 1. Samples of industrial 'dry' anode pastes were used which contained petroleum coke calcined at 1,200°C and 30.5 wt.% MP or 27 wt.% HP. Studhole anode paste was also used, which contained 37 wt.% MP or HP (APMP and APHP).

The experimental techniques for simulating the formation of slow (1°/min) and rapid (70°/min) carbonisation were described by Koptyug et al. (1997) and Anshits et al. (2001). Extraction of industrial dust and carbon residues was described by Kurteeva et al. (2002). Analysis of PAHs in pyrolysis tars and extracts was carried out using the U.S. EPA method number 610, using a micro-capillary quartz column with an SE-54 immobilized silicone phase (12 m x 0.2 mm x 0.3 μm). The overall content of anthracene and phenanthrene (A+Ph) was determined, as well as the contents of the following individual PAHs: fluoranthene (Fl), pyrene (Py), benz(a)anthracene (BaA), chrysene (Chr), benz(b)fluoranthene (BbF), benz(e)pyrene (BeP), benz(a)pyrene (BaP), dibenz(ah)anthracene (DBA), benz(ghi)perilene (BP). The sum of the PAHs (ΣPAH) was determined as a summary content of A+P and 9 individual PAHs.

Table 1. The elemental composition and technical characteristics of the studied pitches.

Parameter	High-softening-point pitch	Medium-softening-point pitch
Softening point, °C	110-113 (Metler)	75-77 (ring-rod)
Content, wt.%:		
Insoluble matter:		
In toluene	25.6-26.4	26.0-28.0
In quinoline	6.0-7.0	10.0-13.8
Volatiles	56	55-58
Ash	0.16-0.20	0.10-0.20
Content, wt.%:		
C	92.46-93.01	91.66-92.48
H	4.00-4.40	3.90-4.10
N	1.00-1.08	1.52-1.75
S	0.27-0.36	0.16-1.27
O (determined by difference)	1.44-1.78	1.46-1.76
C/H	1.75-1.94	1.80-1.88

3. SIMULATION OF SODERBERG ANODE FORMATION UNDER LABORATORY CONDITIONS

Simulation of industrial carbonisation conditions permits an estimate of the ecological danger caused by the various process stages when using different feedstock. As such a laboratory study of MP, HP and APMP, APHP was carried out under conditions which simulate both regimes of Soderberg anode formation: slow (1°C/min) carbonisation of 'dry' anode paste and formation of the anode body, and rapid heating (70°C/min) carbonisation of the studhole anode paste and formation of a 'secondary' anode. To take into account random feedstock variations a required number of experimental data was treated statistically.

The dynamics and content of the main products (gases, tar containing PAHs, carbon residue) during slow (1°/min) carbonisation of MP and HP have been studied. During pitch carbonisation (1°/min) the formation of the main (H_2, CH_4) (Fig. 1) and minor gases (C_2H_6, C_3H_8, CO) were observed. Intensive gas generation was observed in the low- (480-530°C) and high- (640-750°C) temperature intervals, with rate peaks occurring at 460-500°C and 720°C. The generation of ethane and propane was only observed in the low-temperature interval. Within the high-temperature interval the peak gas generation rates for MP were higher compared to those for HP. For both pitches a steady hydrogen generation rate was observed within the temperature range of 500-800°C (Fig. 1).

The thermochemical conversion processes were accompanied by thermal effects, predominantly with heat absorption in the low-temperature range (460-520°C) and heat release in the high-temperature range (680-720°C) (Fig. 1). For MP there was a retarding (by 50-100°C), a cessation in growth, or even a decrease in temperature in the reaction layer of a pitch sample as compared with the furnace temperature in the low-temperature range. The temperature rise rate increased in the high-temperature range. The thermal effects here are considered minor during HP carbonisation (Fig. 1). The main tar release occurs from 640 to 750°C, at the same time as the gases, while the main release of carcinogenic PAHs (BaP, DBA, BP) occurs within 700-800°C (Fig. 2).

Thus the study of the dynamics of the main products formed during MP and HP slow (1°/min) carbonisation indicates two intervals of intensive gas generation: the low- and high-temperature areas, with peaks of gas generation rates at 460-500°C and 720°C along with the simultaneous generation of tar and carcinogenic PAHs within the high-temperature interval.

During the current-carrying stud repositioning stage, anode paste is placed into the open hole. The hole bottom temperature is 700-850°C, and thus rapid pyrolysis of the anode paste takes place along a front of high temperature heating.

The laboratory study showed that the tar yields during rapid heating carbonisation of APMP and APHP are about 10 times higher (and yields of carcinogenic PAHs are about 5-30 times higher), than with slow carbonisation of both anode pastes (Fig. 3). Based on this it was shown that the most significant source of atmospheric emissions of carcinogenic PAHs is the open hole, where rapid heating carbonisation of the studhole anode paste takes place.

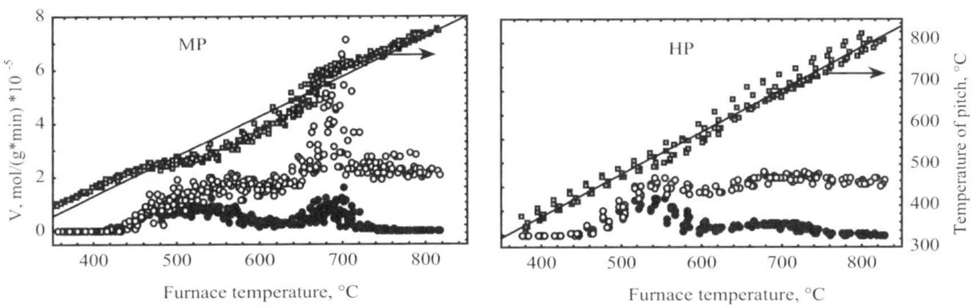

Figure 1. Variations in the formation rate of hydrogen (°), methane (•) and temperature of pitch samples (□) with furnace temperature ramping (heating rate 1°C/min) during carbonisation of MP (sampling 16) and HP (sampling 7).

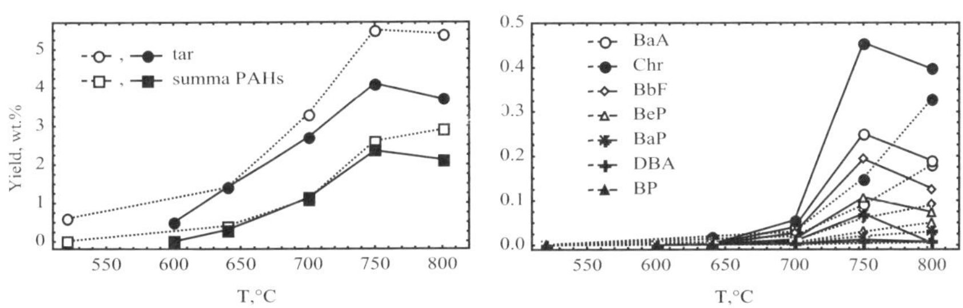

Figure 2. Dependence of tar and PAH yields on carbonisation (heating rate 1°C/min) temperature of MP (---) and HP (—).

It was established that the tar yields during slow and rapid heating carbonisation of HP and APHP are reduced approximately 1.5 times as compared with MP and APMP. These results verify data by Mirtchi et al. (1995) and Sertakov et al. (1998) regarding a decrease in tar yields when using HP instead of MP. However, the yields of BaP and other carcinogenic PAHs are not reduced (Fig. 3), and thus the replacement of MP with HP does not result in a decrease in the yield of carcinogenic PAHs and, therefore, the carcinogenic danger of Soderberg technology.

Under actual industrial conditions, tars containing carcinogenic PAHs are generated in an adsorbed state on the surface of dispersed particles such as soot, alumina (Lane, 1989). In order to understand how well the laboratory experiments represent true industrial conditions, a comparison was made of the profiles of PAHs generated from anode paste pyrolysis tars in the laboratory and the profiles of extraction tars of industrial dusts and slimes from the Krasnoyarsk aluminium plant (KrAZ), for both carbonisation regimes. Earlier these profiles were analysed by Anshits et al. (2001), and it was shown that they are practically identical. Based on these results one can conclude that the laboratory data can be used to assess the carcinogenic danger of slow and rapid heating stages of Soderberg anode formation.

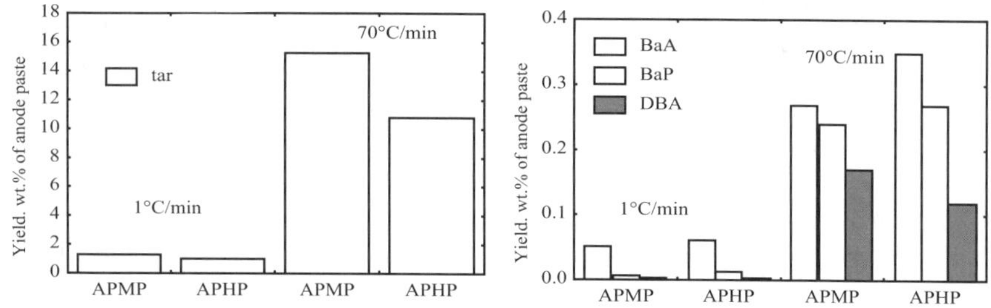

Figure 3. Yields of tars and carcinogenic PAHs of slow (1°/min) and rapid (70°/min) heating carbonisation of anode pastes based on the medium- (APMP) and high- (APHP) softening-point pitches. Sampling for the slow carbonisation : 8 for APMP, 9 for APHP; sampling for the rapid heating carbonisation : 3 for APMP, 5 for APHP.

4. EVALUATION OF INDUSTRIAL SOURCES OF PAH EMISSIONS UNDER THE CONDITIONS OF SODERBERG ANODE FORMATION

Specific yields of tar and carcinogenic PAHs generated from the main technogenic sources from KrAZ when using APMP and APHP were calculated using the Methodika VAMI (1988). This method uses anode paste consumption values, the efficiency factors for all gas cleaning system devices, the permeability of the gas skirt, and the results of laboratory research into carbonisation of pitches and anode pastes based on these pitches.

As an example, Figure 4 schematically illustrates the balance of specific yields of tar and the most dangerous carcinogenic PAHs (BaA, BaP and DBA) from the main sources of the KrAZ Soderberg electrolyser when using APMP and the 'wet' gas-cleaning system. The local sources of tars containing PAHs include:

- emission from the hole during stud repositioning;
- gas escape from under the gas skirt;
- emission of the 'wet' gas-cleaning system;
- slimes during slime storage.

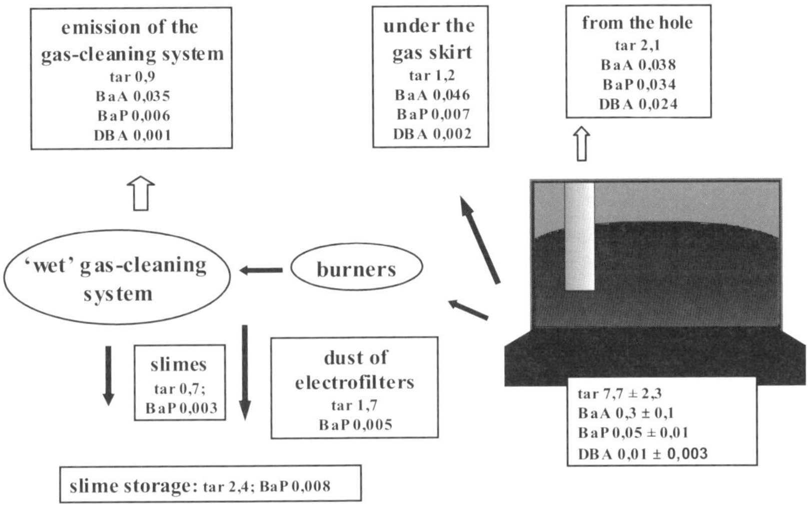

Figure 4. Specific yields of tar and carcinogenic PAHs from the main sources of the KrAZ Soderberg electrolyzer using anode paste based on medium-softening point pitch and 'wet' gas-cleaning system.

Rapid carbonisation of the studhole anode paste determines the emissions from the hole, slow carbonisation of 'dry' anode paste determines other emissions.

It is interesting to compare the specific yields of carcinogenic PAHs generated from the main local sources, i.e. from the hole during stud repositioning, gas escape from under the gas skirt and emission of the gas-cleaning system. The calculated specific yield values of tar and carcinogenic PAHs (BaA, BaP and DBA) from the main sources when using anode pastes based on MP and HP, and 'wet' and 'dry' gas-cleaning systems, are presented in Table 2.

These results indicate that the most dangerous local source of carcinogenic PAHs is emission from the open hole at the process stage of stud repositioning, which is about 60-70% of the total atmospheric emission. It stipulated low specific tar yields in aluminium production using the Soderberg technology of the Reynolds Company, Kaiser Aluminium & Chemical Corporation and other firms. In particular, when using HP-based anode paste the specific yield of BaP from the hole is 0.037 kg/t Al while the total atmospheric emission is 0.048 kg/t Al, which includes emissions from the hole, gas escape from under the gas skirt, and emissions from the 'dry' gas cleaning system. Traditionally this source has not been considered in calculations of specific yields of tar. Thus, the calculated total specific yield of tar when using APMP is 2.1 kg/t Al, excluding emissions from the hole (Table 2), which corresponds to data reported for the Krasnoyarsk aluminium plant (2.09 kg/t Al - Zhuravlev and Petushkov, 1998). When using HP, the specific tar yield is 0.9 kg/t Al excluding emissions from the hole (Table 2). There is a good correlation with data from the KrAZ-Kaiser-VAMI project (Johnson and Lobachev, 1998) – 0.39-0.65 kg/t Al. The total emission of carcinogenic PAHs in this project is reduced due to the use of the 'dry' gas cleaning system, and not due to the replacement of MP with HP.

Table 2. Specific yields (kg/t Al) of tar and carcinogenic PAHs from the main sources of technogenic waste produced by the Krasnoyarsk aluminium plant when using anode pastes based on MP (APMP) and HP (APHP).

Technogenic emission	APMP				APHP			
	tar	BaA	BaP	DBA	tar	BaA	BaP	DBA
From the hole	2.1	0.038	0.034	0.024	1.5	0.049	0.037	0.017
From under the gas skirt	1.2	0.046	0.007	0.002	0.8	0.048	0.009	0.002
From the gas cleaning system:								
'wet'	0.9	0.035	0.006	0.001	0.6	0.036	0.007	0.0016
'dry'	-	-	-	-	0.1	0.006	0.002	0.0002

A replacement of APMP with APHP in aluminium production using Soderberg cells results in a decrease in the yield of tar from all local sources. However, in this case the yield of BaP and other carcinogenic PAHs from these sources is not reduced.

5. CONCLUSIONS

The laboratory research into slow and rapid carbonisation of MP and HP, and anode pastes based on these products, has been studied. The dynamics and the content of the main products formed (gases, tar containing PAHs, carbon residue) during slow (1°/min) carbonisation of MP and HP have been studied.

It was shown that during rapid (70°/min) carbonisation of anode pastes, the yields of tar are about 10 times higher and yields of carcinogenic PAHs are about 5-30 times higher than with slow carbonisation . Based on this, the most significant source of carcinogenic danger of Soderberg cells was found to be the open hole during the process stage of repositioning the current-carrying studs. Despite the finding that this stage contributes 60-70% of the carcinogenic emissions of PAHs to the atmosphere, this source is traditionally not considered.

Specific yields were determined for tars, BaA, BaP and DBA from the main sources of PAHs in aluminium production using vertical stud Søderberg cells with anode pastes based on MP and HP. It was shown that use of HP in the anode pastes results in a considerable decrease of specific tar yields from all sources, as compared with MP, while the specific yields of BaP and other carcinogenic PAHs does not change significantly. A replacement of MP with HP in aluminium production using Søderberg technology does not reduce the carcinogenic danger of this production technique.

REFERENCES

Anshits, A.G., Kurteeva, L.I., Tsyganova, S.I., Suzdorf, A.R., Anshits, N.N., and Morozov, S.V, 2001, The Estimation of the Carcinogenic Substance Emission Using Medium- and High-Softening-Point Pitches in Aluminium Production in Soderberg electrolyzer, *Chemistry for Sustainable Development.* **9**, pp. 345-352.

Johnson J.A., and Lobachev S.V., 1998, Krasnoyarsk Aluminium Anode Paste Plant: Modernization and Phase 1 Conversion to Dry Anode Operation, *Light Metals.* pp. 721-728.

Koptyug, V.A., Anshits, A.G., Savinov, V.I., Suzdorf, A.R., Morozov, S.V., Kurteeva, L.I., Vereshchagin, S.N., Frizorger, V.K., Anshits, N.N., and Krak, M.I., 1997, Ecological risk of aluminium production in Soderberg electrolyzers. Formation of the Main Products of Thermochemical Conversion of Medium-Softening-Point Pitches during Slow Pyrolysis and Estimation of Their Carcinogenic Danger, *Chemistry for Sustainable Development.* **5**, pp. 553-567.

Kurteeva, L.I., Tsyganova, S.I., Morozov, S.V., Anshits, N.N., Suzdors, A.R., Plekhanov, V.P., and Anshits, A.G., 2002, Study of the Stage Determing the Emission of Carcinogenic Polycyclic Aromatic Hydrocarbons during the Carbonisation of the medium-softening-point pitches, *Chemistry for Sustainable Developmen.* **10**, pp. 431-441.

Lane, D.A., 1989, In: *Chemical analysis of polycyclic aromatic compounds*, T. Vo-Dinh, ed., John Wiley & Sons, New York, p.31.

Metodika VAMI, 1988, *Method of waste determination in aluminium production using vertical stud Søderberg cells*, *VAMI*, Leningrad, pp.1-30. (in Russian).

Mirtchi, A.A., Proulx, A.L., and Castonguay, L., 1995, Reduction of the PAH Emissions for Horizontal Stud Soderberg Potrooms, *Light metals.* pp. 601-605.

Sertakov, N.N., Aleksandrov, A.I., and Lazarenko, O.Yu., 1998, Influence of Different Factors on the Quality of Anode Paste, *Nonferrous Metals.* **6**, pp. 41.

Zhuravlev, Yu.A., and Petukhov, M.P., 1998, Economical Aspects of Modernization of Production in Krasnoyarsk Aluminium Plant, *Nonferrous Metals.* **5**, pp. 53-58.

PART II

PERMAFROST CO_2 STORAGE

The Potential for using Permafrost Terrains
for the Geological Storage of Carbon Dioxide

DISTRIBUTION OF PERMAFROST IN RUSSIA

Frozen ground and CO_2 sequestration

V.P. Melnikov and D.S. Drozdov
Earth Cryosphere Institute SB RAS, P/O 1230, Tyumen, Russia (E-mail: melnikov@ikz.ru, ds_drozdov@mail.ru)

Abstract: Vast areas of the Russian north is underlain by permafrost, and many of these areas also host exploitable oil and gas reserves as well as various industries which are sources of atmospheric CO_2 pollution. It has been suggested that permafrost in this region could be used to host or to isolate man-made CO_2 in deep geological reservoirs, however an in-depth understanding of permafrost distribution, morphology, structure and stability is required in order to examine its feasibility and long-term safety. The present work outlines the spatial distribution of permafrost in Russia, defines the various formation histories, mechanisms and structures, and discusses the permafrost types which may be best suited to CO_2 sequestration.

Key words: Cryolithozone / permafrost distribution or extent / lithology / cryopeg / talik / sub-permafrost CO_2 sequestration

1. INTRODUCTION

Cryolithozone occupies 2/3 of the Russian Federation territory, including northern and north–eastern regions of Russia. Within these areas cryotic rocks containing ice inclusions make up a part or all of the near-surface part of the earth's crust. To the south, outside the cryolithozone, only seasonal soil freezing occurs.

The cryolithozone areas are primarily under-developed regions with inclement climatic conditions. The economics of these regions are basically dependent upon either traditional activities which have little affect on the geo-ecology of the area, or on natural resource development. The geocryological aspect is particularly important for petroleum and natural gas industries, since the most significant fields and prospects are situated in the

S. Lombardi et al. (eds.),
Advances in the Geological Storage of Carbon Dioxide, 69–80.

permafrost zone (fig.1). Permafrost conditions affect not only industrial and economic projects, but also surrounding infrastructure related to transport, heat-and-energy, housing, municipal services and so on.

The mineral resource and infrastructure industries involve the release of considerable volumes of waste, including mineral processing tailings, liquid and gas wastes. The majority of waste-gas production (i.e. CO_2) is caused by the carbonic acid industry. The production and disposal of all types of wastes in the permafrost zone should be carried out taking into account the geocryological factor.

In most cases permafrost complicates resource recovery and waste management (loss of stability of thawing ground, propagation of disastrous exogenous processes, landscape degradation), but sometimes it can act as a favorable factor (providing safe foundations, forming aquicludes and reducing contaminant migration). In order to evaluate permafrost in regards to CO_2 sequestration one must consider the entire complex geocryological system. The basic characteristics of permafrost are its thickness, continuity, temperature, ice content, lithological and thermal ground properties, and the exogenous geological processes dependent on permafrost. Groundwater salinity frequently determines the rock's status under sub-zero temperatures and the formation of so-called cryopegs within these rocks.

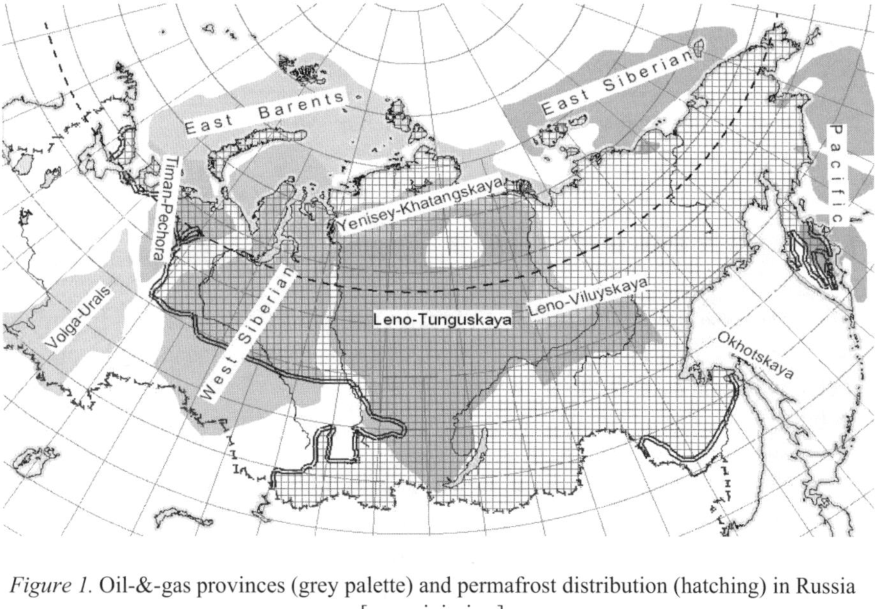

Figure 1. Oil-&-gas provinces (grey palette) and permafrost distribution (hatching) in Russia [www.igirgi.ru].

Permafrost stability is bound to complicated destructive exogenous geological processes. Generally anthropogenic impact greatly accelerates these processes, especially when melting and degradation occurs within ice-rich ground and rocks of so-called "icy complexes" or "massive ice beds".

2. PERMAFROST ORIGIN, ICE CONTENT

Permafrost is a natural geological formation. There have been several periods of cooling during the geological evolution of the Earth which caused the formation of frozen strata in the Earth's crust. Polar migration and continental drift give evidence of ancient freezing in various sites, including those that occur nowadays in equatorial latitudes. The most ancient freezing period during the Proterozoic (more than 2 billion years ago) corresponded to the Huronian glaciation. This was followed by Riphean, Vendian and Paleozoic glaciations. The permafrost underlying the Antarctic ice sheet (the age of the ice which permanently "flows" to the surrounding ocean doesn't exceed a few hundreds of thousands of years) is probably the relict of Oligocene-Miocene glaciation that began 25 million years ago. The modern global distribution of permafrost (fig.2) started to form during the same period (i.e. during the Miocene). The most ancient permafrost relics of the last glaciations correspond to the upper Pliocene – Pleistocene (1.8 - 0.7 million years).

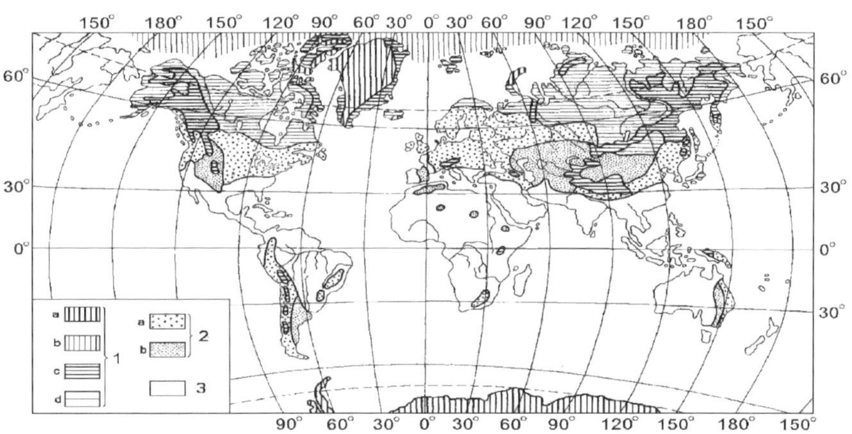

Figure 2. The world permafrost map: 1 – permafrost: a) under glaciers, b) in glaciers, c) in mountains, d) in plains; 2 – seasonally frozen ground: a) in wet climatic zone, b) in dry climatic zone; 3 – no freezing [Geocryology of the USSR: European territory of the USSR / Ed. E.D.Ershov. – Moscow: Nedra, 1988, p.15].

Frozen soils and rocks are regarded as geological formations which are characterized by negative temperatures and which contain non-freezing interfacial water and the ice that cements granular material and fills pores and cracks. Ground of any genesis, mineral composition, grain size (clastic, sandy, clay soils) and lithification (massive and fissured bedrocks) can be found in the frozen state. When ice is inter-granular it is considered a specific rock-forming mineral included in rock texture and structure descriptions, whereas ice bodies (such as various buried ices, wedge and segregation ices, massive ice beds) are regarded as mono-mineral geological bodies. Occasionally cryotic, but iceless, rocks are referred to as cryotic dry rock or dry permafrost.

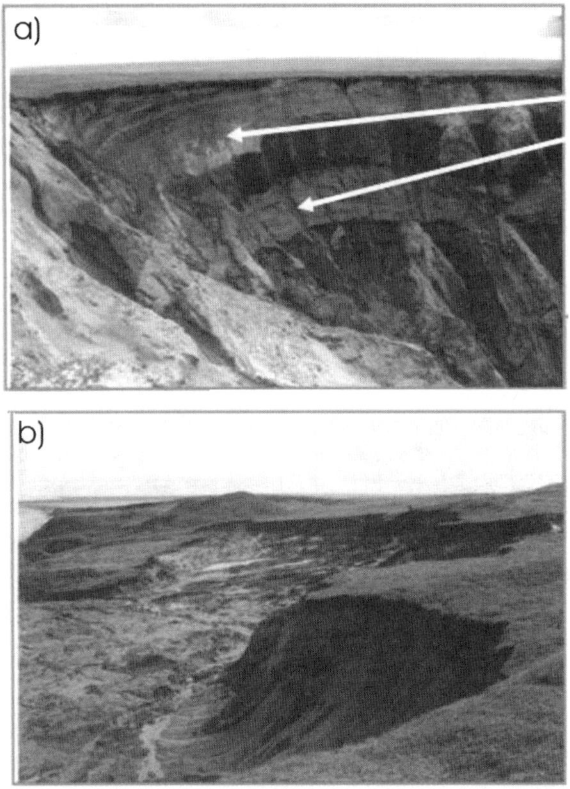

Figure 3. Photographs showing surficial landscapes in permafrost regions. In (a) arrows highlight ground ice layers at the Yugorskii peninsula, while in (b) thermokarst and thermoerosion in picryogenic ice-rich soils at the Yugorskii peninsula. Photos by A.I.Kizyakov.

The frozen strata are subdivided into syngenetic and epigenetic according to the time of their formation versus the time of accumulation of the mineral and biogenic components. For the most part the frozen strata have had an epicryogenic origin, i.e. previously accumulated or formed units had frozen during a cooling epoch. The ice content inside this ground shows an approximate correlation with the water content present before freezing. In some cases conditions are created for the redistribution of moisture throughout the section as a result of moisture addition to the freezing front throughout the aquifers. In these cases ice beds from tens of centimeters to tens of meters are formed inside the freezing rock strata (fig. 3).

Syngenetically frozen strata have frozen in parallel with mineral and/or biogenic rock accumulation. They include recent Upper-Pleistocene and Holocene clay and silt deposits of fluvial and littoral plains, shelf deposits (for example, in Yakutia), aeolian-alluvial-deluvial slopes of arctic rock masses, and depressions of large northern lakes. The moisture content in this ground, freezing while accumulated, is essentially unlimited. The ice accumulates in abundance within so-called "cryostructures" and often forms wedge ices. Their total volume can reach 80% and more (fig.4).

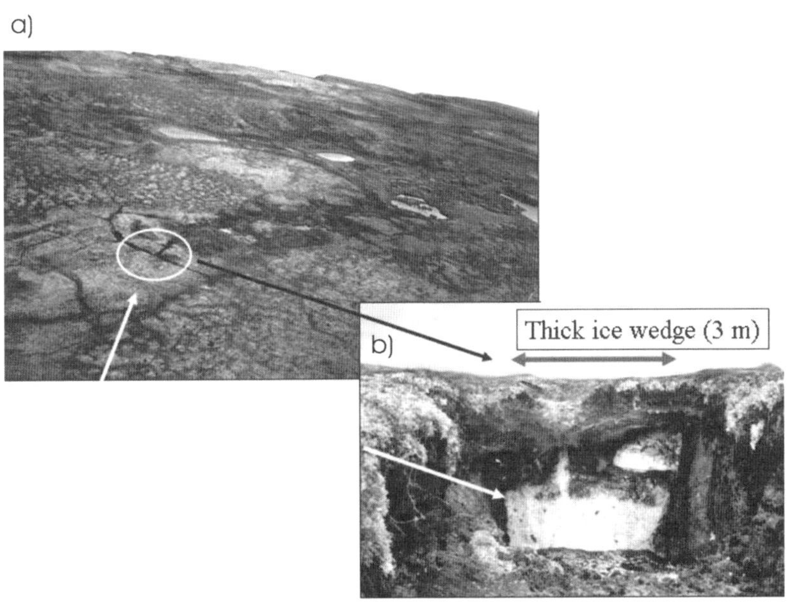

Figure 4. Photographs showing thermokarst in ice wedges at the Bolvanskii Peninsula, with (a) giving an areal view from a helicopter while (b) shows a cross-section of one of the ice wedges. Photos by M.Z.Kanevsky.

To a lesser extent the state difference (frozen/unfrozen) affects bedrock, and their cryogenesis is always epigenetic. Ice within fissures and cavities do not significantly influence rock strength, however permeability is changed appreciably, with frozen fractured saturated stratum being transformed into an aquiclude. The presence of spatially restricted taliks leads to active water exchange and enhanced leaching across them . The cryotic dry rocks only occur in the uppermost part of a geological section, (typically only 1-3 m below surface in granular rocks) and thus are of interest with respect to CO_2 sequestration only as an upper aquiclude above a CO_2 underground storage facility. In this case most attention should be paid to the chemically active residuals, because under the conditions of talik water exchange and in the presence of dissolved CO_2 fast reactions can take place which could lead to increased porosity and enhanced CO_2 escape out of the storage.

Pebbly, gravelly and sandy ground (except silty sands) in frozen or unfrozen states don't vary greatly from each other in ice content, but the difference in conductivity and strength is rather considerable. In the frozen state it is a strong foundation which can take an almost unlimited construction load and it is a stable aquiclude providing an effect barrier to gas and water migration even at moderate thicknesses of the frozen horizon.

The moisture content in frozen clastic and sandy ground is usually slightly higher than in thawed ground. However at an aquiclude's boundaries (for instance, thawed aquicludes – clays) massive ice beds up to several tens of meters thick (fig.3) can be generated while freezing sandy strata. In such places the total thickness of ice layers can reach 80% of the overall frozen strata thickness within near-surface zones which are intersected by constructions. If human engineering works affect the massive ice beds, they may initiate the potentially disastrous geocryological processes of thermokarst and thermoerosion, thereby destroying landscapes as well as nearby engineering infrastructure.

Silty and clayey ground attract the special attention of geocryologists, constructors and other experts, as they can accumulate significantly greater quantities of water during both epi- and syncryogenic freezing than in their thawed state. This is the result of their mineral and granulometric composition which ensures the abundant migration of moisture to the freezing front. Lenticular, layered, reticulate or honeycomb cryostructures are formed, and the mineral ground-mass is dissected by an ice net and layers. Their thickness and spaces between layers range from sub-millimetre to several centimetres. The ground-mass is also divided into polygons by macrocracks and filled syngenetic ice wedges. Several meters thick wedges can form a multilayer structure having a total thickness of tens of meters. Air filled pores are essentially absent and ice content can reach more than 80%. In this case thermokarst and thermoerosion lead to disastrous results.

Thus the frozen rock is of no interest as a material which can host the carbon dioxide. The ice-saturated ground is gas and water impermeable, explaining why it is impossible to create conditions for solution or chemical absorption (for example, as gas hydrate) of carbon dioxide in the frozen rocks. The cryotic dry rocks are extremely rare and are of no practical value. Therefore, it is reasonable to examine permafrost not as the enclosing medium for CO_2 sequestration but rather as a unit which can enhance the isolation of CO_2 in deep geological units.

3. PERMAFROST DISTRIBUTION, TEMPERATURE

According to its geographical position the cryolithozone is subdivided into subaeral, shelf and subglacial. The subaeral cryolithozone is the most widespread and is usually made up of surface-laying permafrost. The shelf cryolithozone is typical for arctic seas of the Siberian sector of the Russian Federation. The subglacial permafrost underlay glacial covers on the northern islands and in highlands. Strictly speaking permafrost is not restricted to the polar territories, but rather it extends both under glaciers and on iceless mountain ridges in lower latitudes (fig. 5 and also fig.2).

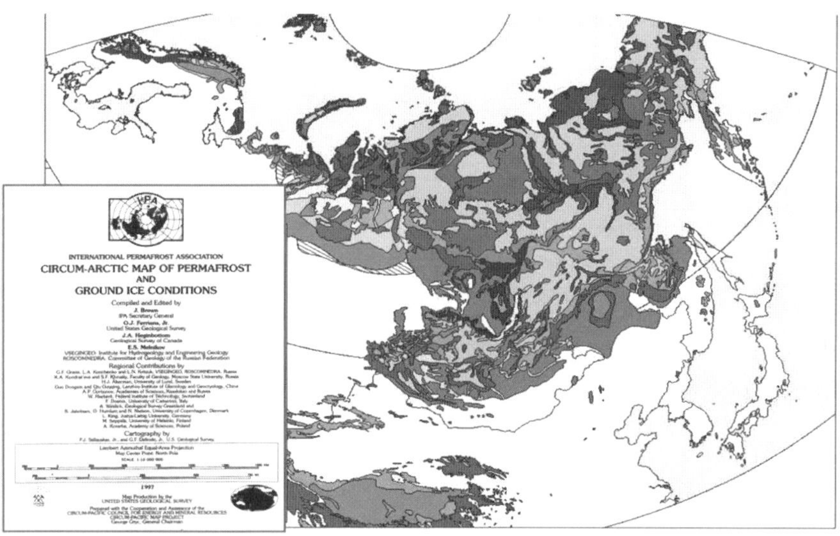

Figure 5. Circumpolar map of permafrost. Eurasian sector.

Alpine permafrost can be found in the Caucasus, in Mongolia and in Tien Shan. In Tien Shan (i.e. at the latitude south of the Caspian Sea and Japan) the alpine permafrost areas are more than the whole Northland. Nevertheless alpine permafrost in low latitudes is rather an exotic event, and thus northern and eastern regions are the dominant areas of permafrost distribution.

Permafrost can be subdivided into 4 types based on unit continuity: i) continuous (frozen rocks occupy 90-100% of the territory); ii) discontinuous (50-90% permafrost); iii) island or sporadic (10-50% permafrost); and iv) isolated patches (0-10% permafrost). Their distribution is controlled by strict latitudinal (for plains and mountains) and altitude (for mountains) zonality (fig. 6). The boundary of the continuous permafrost distribution shifts more and more southward as it moves from west to east. Permafrost temperature decreases in the same direction (from –2 to –100° C and below), while its thickness increases in the plains (from the first tens of meters to 300 - 500 m and more) and in mountain ridges (up to 1500 m).

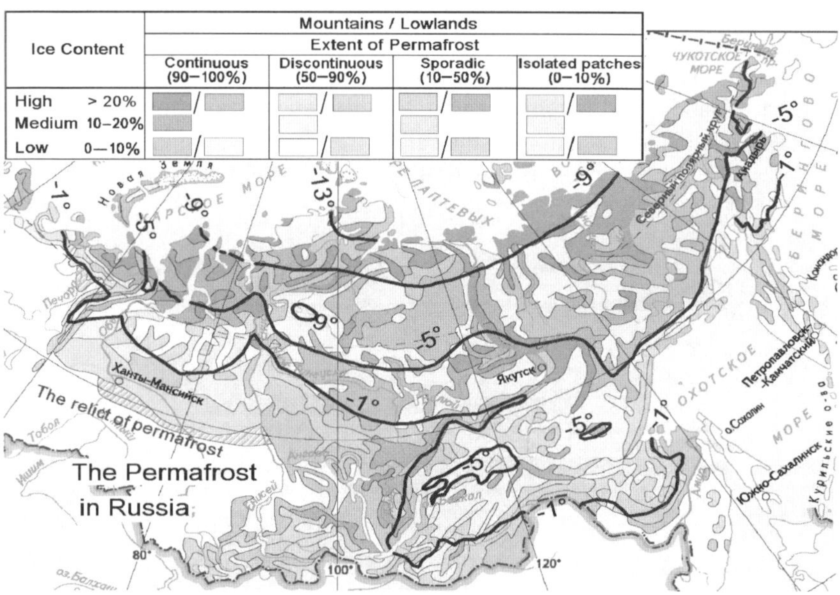

Figure 6. Permafrost in Russia: Space distribution, iciness and temperature at the depths of zero seasonal amplitudes. [by E.S. Melnikov, K.A.Kondratieva, G.F.Gravis].

Local natural and human factors are superimposed on the zonal and altitude regularities of the permafrost distribution, thus highlighting the local changeability of the geocryological conditions. For example, taliks can appear in continuous permafrost in river valleys and lake depressions due to the defrosting effect of the hydrosphere, while frozen sites can disappear completely in the sporadic permafrost zone. Various natural and man-made cooling factors (for instance, wind erosion or snow removal) can cause the formation of permafrost islands even outside the cryolithozone.

The continuous permafrost zone is the largest (fig.6). On the plains of the Russian European north it stretches out along the coast in a 100 to 200 km wide strip. Its southern limit coincides approximately with the polar circle in the ranges of the West-Siberian Lowland, and it reaches the latitude of Lake Baikal in eastern Siberia and Yakutia.

The discontinuous permafrost zone stretches out in a very narrow strip in the European north, and then in western Siberia it widens up to several hundreds of kilometres. Eastward discontinuous permafrost is basically localized to the intermountain troughs or to the coastal plains south of the 60th parallel.

Sporadic permafrost and isolated patches of permafrost extend up to the latitude of the polar circle in Europe and up to the 60th parallel in western Siberia. In the more eastern regions it extends to the southern frontier of the Russian Federation.

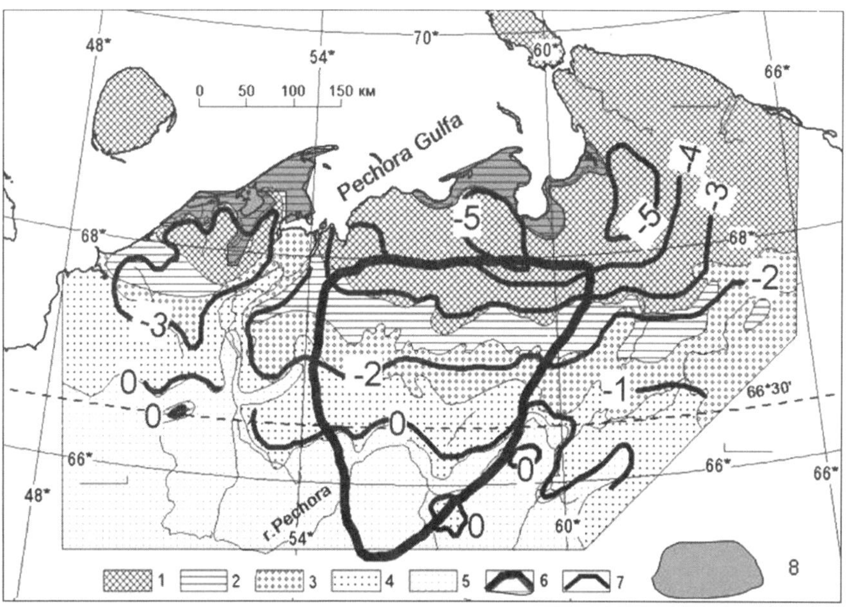

Figure 7. The extent and temperature of permafrost in the Timan-Pechora oil and gas province: 1-4 –permafrost types: 1–- continuous, 2 – discontinuous, 3 – sporadic, 4 – isolated patches; 5 – non frozen and thawed soils; 6 – permafrost relict contour; 7 – permafrost temperature isograms; 8 –cryopegs.

Relict permafrost bedding occurs at a depth of tens to hundreds of meters below the earth's surface, often separated by thawed (lateral) ground horizons from the overlying modern superficial permafrost. The relict permafrost can form a single stratum, layers or separate fragments. The vast lobe of relict permafrost in the south of western Siberia is well seen on fig.6, while the zone of relict permafrost in the European north is well seen on fig.7.

Relict permafrost has been the result of cyclic changes in climate and the alternation of warm and cool epochs, with many being due to the heritage of ancient Upper-Pleistocene cooling epochs. However it would be incorrect to assume that relict permafrost layers are a single dimensional formation. The long-term history of its evolution has resulted in the formation of numerous taliks that have broken down the single whole into many isolated patches (fig.8).

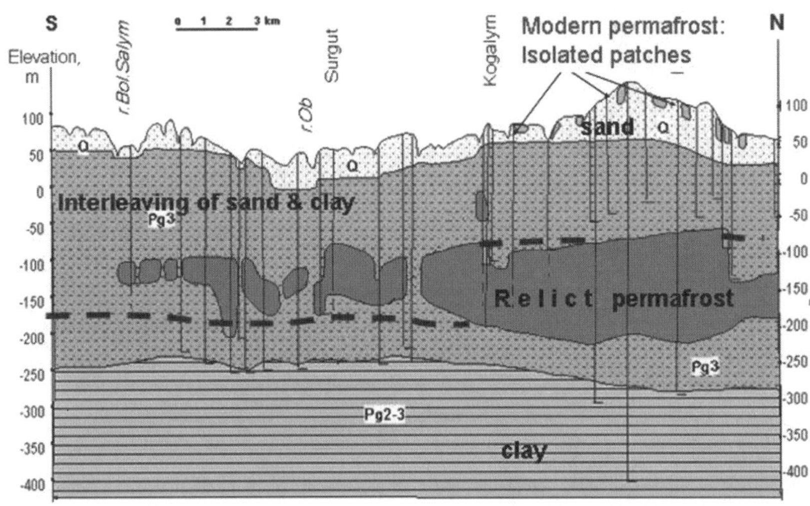

Figure 8. Relict permafrost. The thickness of relict permafrost decreases gradually from the north to the south. Local isolated subsurface patches of modern (Holocene) permafrost can be found to the north of Kogalym city. Their thickness could achieve 20-30 m. [by G.V.Malkova].

Continuous permafrost is a good aquiclude, and it is a reliable foundation for construction if the proper technologies are applied. The active exchange between supra- and sub-permafrost ground waters takes place in the zone of discontinuous permafrost, and pre-thawing is required before building at temperatures close to 0°C. Sporadic and island permafrost is also a complicating factor from a hydrogeological and engineering point of view.

The sites of unfrozen rocks in all types of permafrost are referred to as taliks, which can be subdivided into closed, open and lateral taliks. Their formation has been conditioned both by geological and paleoenvironmental evolution and by recent warming factors. So-called "cryopegs" occupy a unique place in talik classification, as they are unfrozen sites and horizons, despite sub-zero temperatures, within rocks hosting saline waters. Predominantly these are either deep horizons of ancient rocks under many-hundred-meters thick permafrost strata saturated with brines, or the shallow-laying bedrocks or sedimentary rocks with marine saline waters along the northern coasts (fig. 7). The presence of cryopegs testifies to the importance of geochemical and hydrogeochemical factors in the distribution of frozen and unfrozen rocks in the cryolithozone.

4. PERMAFROST AND CO_2 SEQUESTRATION

Since in most cases frozen ground is impermeable and can't be considered as a possible natural reservoir for man-made gases and liquids, only taliks, cryopegs and sub-permafrost aquifers are of interest for CO_2 burial.

Not considering here any possible ecological consequences, let us address the main possible variants of CO_2 burial in diverse geocryological conditions. Continuous permafrost is practically an ideal aquiclude and it theoretically permits CO_2 injection into sub-permafrost porous reservoirs. Sites which have a domal structure of the permafrost base and without taliks are most favourable. The low reservoir temperature, the high geostatic pressure, as well as the extra pressure of the injected gas could create auspicious conditions for CO_2-oriented gas hydrate formation. This may essentially increase the sub-permafrost reservoir effective capacity and ensure the high stability of the sequestered CO_2.

In the view of CO_2 burial the sites with discontinuous permafrost must be examined carefully, paying special attention to the selection of permafrost masses undisturbed by taliks.

The sites within sporadic or isolated permafrost are not very promising for CO_2 sequestration due to the possibility of active water- and gas-exchange between relatively deep horizons and the surface. The cryogenic factor may, however, play a positive role in the presence of an ideal lithologic-and-petrographic aquiclude (for example, a thick unit of dense clays and argillites), providing the minimal reservoir temperatures at the depth of CO_2 burial and the opportunity for proper gas hydrate formation.

With respect to CO_2 burial cryopegs should attract particular attention. On the one hand cryopeg distribution areas can coincide with areas of open taliks formed by both thermal and chemogenic processes. This fact reduces the insulating reliability of the waterproof stratum. On the other hand the temperature of the cryopeg-reservoir is 1 to 20° C lower than the temperature of the fresh-water reservoir, which is a significant point when one considers the possibility of underground gas hydrate formation. In addition, one can expect that some part of the injected CO_2 will react with the cryopeg brines and generate stable carbonate minerals. The last two factors are important if one wishes to examine the problem of CO_2 sequestration in the European north of Russia, where active prospecting and the development of continental, shelf and island oil-and-gas fields are being performed today.

CHARACTERISTICS OF PERMAFROST IN SIBERIA

A.D. Duchkov
Institute of Geophysics SB RAS, 3, pr. Koptyug, Novosibirsk, 630090, Russia

Abstract: Permafrost is a huge natural "thermometer" sensitive to significant changes in climate. During the Late Pleistocene permafrost reached as far south as 48-49 °N, covering Siberia. During Holocene warming the extent of the permafrost has greatly decreased. This process may now be intensified due to global warming of the climate during the 20-21 centuries. Siberian meteorological stations show that the last 30-40 years are characterized by an air temperature increase of 0.02-0.05 °C per year. If this warming trend is maintained the overall temperature increase in northern Siberia may reach 1.2-2 °C by 2050. In this case the southern permafrost border will be moved to the north by 300-400 km. Calculations were made to estimate possible changes in permafrost parameters under the influence of climatic variations. In some blocks it is possible that frozen rocks will thaw from the surface to a depth of 12-17 m by 2100.

Key words: Russia, Siberia, thermal and phase state of permafrost, temperature of rocks, meteorological stations, climate warming, temperature field modeling, degradation of permafrost

1. INTRODUCTION

Permafrost occupies about 80% of Siberia, the north-eastern part of Asia. Within its boundaries one can find large cities and industrial complexes. Permafrost stability is mainly governed by temperature. The changes in temperature (due to climatic variations or anthropogenic factors) may result in either permafrost degradation or stabilization. In this article we consider modern data collected on the temperature field of Siberian permafrost.

S. Lombardi et al. (eds.),
Advances in the Geological Storage of Carbon Dioxide, 81–91.

Temporal changes in the temperature of the upper rock layer are basically determined by air temperature variations (i.e. climate). Long lasting negative air temperatures on the Earth's surface result in ground cooling and the formation of frozen rock layers (permafrost). Terrestrial heat flow limits the depth extension of the permafrost .

Geocryological studies have shown that the permafrost in Siberia was mostly developed in Late Pleistocene, during the epoch of Sartanian glaciation (18-27 thousand years ago) (Kondratieva et al., 1993). The average air temperature at that time was about 8-10 °C less than at present. Under these favourable conditions permafrost spread over Siberia, with its southern border reaching 48-49° N. Subsequently the spatial extent of the permafrost was greatly reduced. In Western Siberia the southern permafrost border was displaced to the north up to 60°N. In Eastern Siberia the permafrost has almost the same spatial coverage but has likely been reduced in thickness. Heat transfer occurs extremely slowly in rocks, and thus a present temperature field of frozen rocks contains ("remembers") anomalies related to the most significant climate variations of the last epoch. As such permafrost is a huge natural "thermometer" sensitive to significant changes in climate.

2. STRUCTURE AND TEMPERATURE FIELD OF PERMAFROST

2.1 Structure of permafrost

Permafrost reaches the greatest thickness (over 1 km) in Yakutiya, central Siberia (Duchkov and Balobaev, 2001), where it forms the most cooled block of lithosphere in Northern Eurasia. Terrestrial heat flow (*q*) in this area does not exceed 30 mW/m^2. In other areas of Siberia, where the heat flow reaches 50-70 mW/m^2, permafrost thickness does not exceed 400-600 m.

Three types of permafrost are observed in Siberia (Fig. 1). In northern areas there is continuous permafrost which begins directly at surface and which may reach considerable depths. To the south one can observe broken or insular permafrost, where blocks of frozen rock may be separated by isolated thawed zones (taliks). For example in western Siberia one can observe isolated blocks of frozen rock (buried or relic permafrost) south of 60°N. The average air (*Ta*) and surface (*Ts*) temperatures determine the type of permafrost. The value *Ts* is usually defined in thermal physics as an annually averaged rock temperature at a depth of 15-20 m. In northern areas *Ts* is higher than *Ta* by 3-7 °C (depending on latitude) because of the shielding effect of snow cover and vegetation. Continuous permafrost is formed in those areas where negative values of *Ta* and *Ts* are observed. Permafrost decay begins from the surface when *Ts* becomes positive (see Fig. 1).

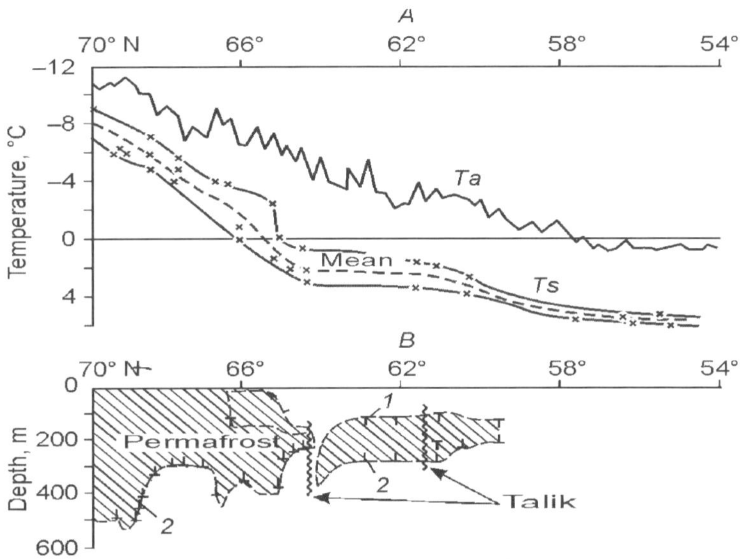

Figure 1. Modern parameters of western Siberia permafrost along a northerly profile (Duchkov and Balobaev, 2001; Duchkov et al., 1995). A – Plots of air temperature (Ta) and ground temperature (Ts); B – Distribution of the upper and lower permafrost boundaries.

2.2 Modern temperature field of permafrost

Temperature sections (temperature distribution with depth below ground surface) may be divided into two groups typical for stable and unstable permafrost (see examples in Fig. 5). The upper part of stable permafrost is cooled between -6 to -15 °C. At certain depths the temperature begins to increase due to the influence of terrestrial heat flow. It reaches zero at a depth of between 200-600 m.

In blocks of unstable (warm) permafrost the temperature in the entire frozen rock layer is not much below zero, on the order of - 0.2 to -0.5 °C. Unstable permafrost is usually formed in blocks of sedimentary rocks saturated with fresh water. In this case the frozen rocks contain ice, and the bottom permafrost boundary is a phase boundary. Such a situation is found in the Meso-Cenozoic depressions of Siberia. The speed of a moving phase boundary is about one order of magnitude slower than the speed of a moving thermal wave. Therefore the temperature field of the permafrost doesn't have enough time to follow the varying conditions at surface (in comparison with dry rocks or those saturated with mineralised waters). Therefore blocks of frozen rocks with a non-stationary temperature field and a thickness which is not appropriate to modern climatic conditions are preserved here for some time. The non-stationary temperature field is fixed by a break in the thermograms (temperature-depth profiles) at the depth of a phase boundary location.

2.3 Permafrost zoning on the basis of temperature field stability

Heat flows below and above the phase boundary (temperature close to 0 °C) are connected to each other by the Stefan relation (Duchkov et al., 2000):

$$q_p = q + Q \cdot W \cdot V$$

where q is the heat flow determined in thawed rocks underlying a permafrost layer; q_p is the heat flow in frozen rocks, Q is the melting heat for ice, W is the ice content in rocks, and V is the speed of phase boundary movement (speed of freezing/thawing).

The heat flow ratio ($n = q_p/q = 1 + Q \cdot W \cdot V/q$) is used as a criterion of temperature field stationarity (Duchkov et al., 1995; Balobaev, 1991; Duchkov and Balobaev, 2001). In the stationary case $V = 0$ and the coefficient $n = 1$. For temperature increases at surface (climate warming and permafrost decay) we get $V < 0$ and $n < 1$. For decreasing Ts (climate cooling and permafrost growth) we get $V > 0$ and $n > 1$. Thus the coefficient n may serve as a tool for estimating the stability of modern temperature fields in permafrost and predicting its evolution in the near future. It is obvious that the method is applicable only if a phase transition occurs at the bottom of the permafrost (e.g. for the Meso-Cainozoic depressions). For its realization one needs heat flow values above and below the phase boundary.

This method was used for zoning permafrost in depressions of western and eastern Siberia (Fig. 2). The most detailed picture is given for western Siberia (Fig. 2A) where the permafrost, according to parameter *n,* consists of two sharply differing parts – one northern and the other southern. The boundary between them passes approximately along the latitude of the Polar Circle. Climate warming will mostly affect a rather narrow strip of the continuous permafrost located to the north of the Polar Circle, where n is small ($n < 0.5$). To the far north the permafrost forms a monolithic body whose temperature steadily remains negative. Here n is 0.5-0.8, showing some evidence of permafrost decay which corresponds to raising of its lower boundary. The similar ratio between q_p and q is also established in the limits of the Viluyi depression (Fig. 2B), though here there is not enough experimental data. Zoning of the other eastern Siberia depressions was made using isolated thermal logs and thus is quantitative in nature. The permafrost of eastern Siberia as a whole is more cooled and resistant to climate warming. Today the entire Siberian region is characterized by a decaying type of permafrost (everywhere n < 1), and thus global climate warming may speed up the process of permafrost decay. This rate could be estimated given data on possible scenarios for climate warming in the 21st century.

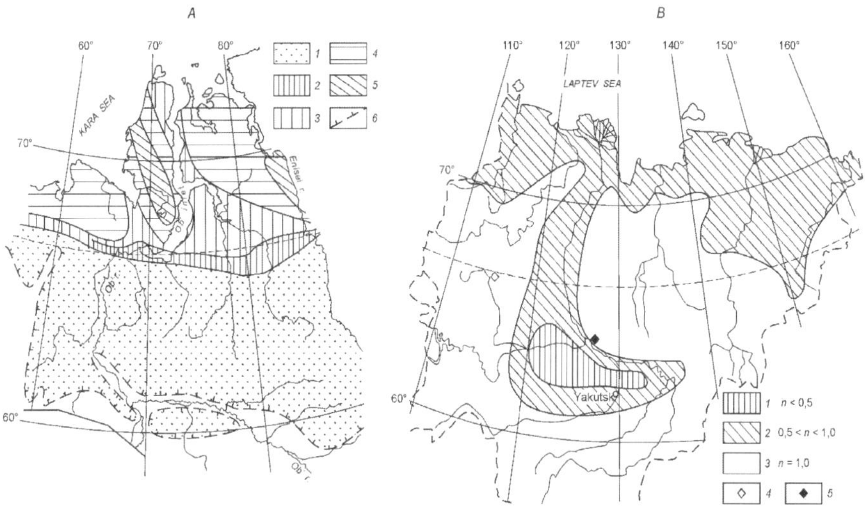

Figure 2. Schematic zoning of the permafrost temperature field for western (A) and eastern (B) Siberia according to its stability level (Duchkov and Balobaev, 2001). A: 1 - n =0-0.1; 2 – n = 0.1-0.2; 3 – n =0.2-0.5; 4 - n = 0.5-0.8; 5 - n = 0.8-1; 6 – the southern permafrost boundary . B: 4 – site with permafrost thickness of about 1.5 km; 5 – site with talik.

3. SCENARIOS OF TEMPERATURE CHANGE AT THE EARTH SURFACE

As was mentioned earlier, permafrost evolution is affected by significant (in amplitude) and long-term variations of air and surface temperatures (*Ta* and *Ts*). Many meteorological stations in Siberia have been performing measurements of these parameters for more than 100 years, and thus these data can be used for estimating modern climatic cycles.

3.1 Model of surface temperature change

In meteorological records from Siberia one can see an increase in air and ground temperatures throughout the area during the last 40-50 years (Duchkov et al., 2000; Skachkov, 2001; Pavlov and Anan'eva, 2004).

Annual average Ta values , however, have different increasing rates in different areas of Siberia (see Fig. 3), ranging from 0 up to 0.08 °C/year. The average trend for all of Siberia is 0.04 °C/year.

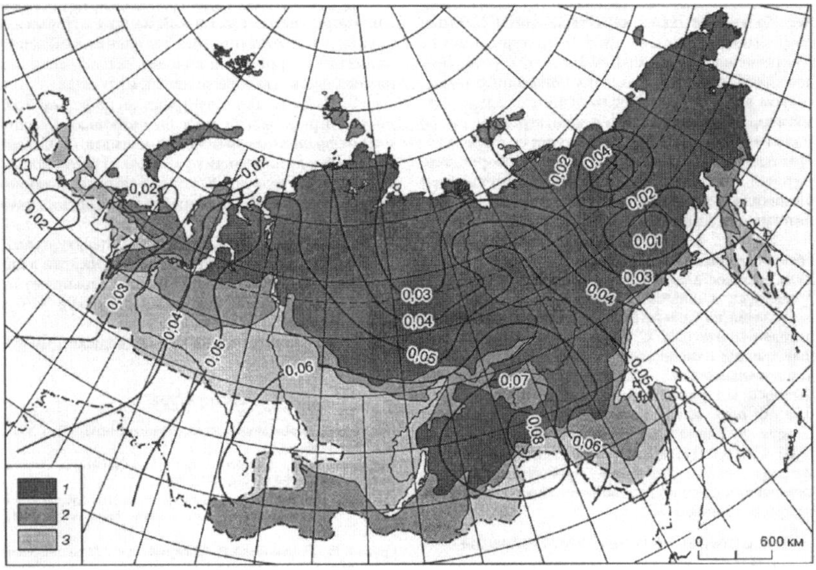

Figure 3. Map of increasing Ts trends (°C/year) derived from observations at meteorological stations in 1960-2000 (Pavlov and Anan'eva, 2004). Types of permafrost: 1 - continuous, 2 - faltering, 3 - island.

It is interesting to note that present trends for Ta show a good correspondence with those predicted earlier by radiation-circulation models of global climate warming due to a doubling of atmospheric CO_2 concentrations (Hansen et al., 1983; Manabe and Wetherald, 1975). If the present day climate-change trends are continued into the future it is possible to develop a model (scenario) of surface temperature (Ts) changes in Siberia during the 21st century (Table 1).

The model is based on the joint analysis of information available on changes in Ta and Ts values and also CO_2 contents in the atmosphere, as well as the coupling of these two parameters (Duchkov et al., 1994). This model predicts a surface temperature increase in Siberia during the 21st century of about 3-6 °C, depending on the latitude. This forecast is used to estimate possible temperature field changes in permafrost, and is based on the assumption that climate warming will continue due to CO_2 inflow into atmosphere. This in turn will result in fast permafrost decay.

Table 1. Possible increase of a surface temperature (Ts) in Siberia by global climate change during the 21st century, °C (Duchkov and Balobaev, 2001 a, b; Duchkov et al., 1994).

	2000	2020	2040	2060	2080	2100
Latitude						
50	0.0	0.4	1.0	1.5	2.1	2.8
55	0.0	0.4	1.3	1.8	1.4	3.1
60	0.0	0.6	1.3	2.1	3.0	4.0
65	0.0	0.8	1.5	2.5	3.7	4.9
70-75	0.0	1.0	1.9	2.2	4.6	6.4

3.2 Dynamics of permafrost in the 21st century

We used the scenario of climate warming (Table 1) for modeling changes in the thermal and phase condition of permafrost in the 21st century. As was mentioned earlier permafrost decay starts when *Ts* becomes positive (see Fig. 1). *Ts* values reflect reaction of the upper permafrost layer to warming. Due to warming the zero isotherm of *Ts* will move to the north, opening new areas where permafrost becomes unstable and starts to thaw. Surface temperatures at different sites and different moments in the 21st century may be estimated while adding values from the Table 1 to modern values of *Ts*.

The starting time of permafrost decay is defined as a time when the upper permafrost boundary begins to come off a layer of seasonal freezing (depth > 2-4 m). Climate warming defines only the time when the permafrost thawing starts but the complete decay of the frozen block is governed by laws of thermal transfer and may last for hundreds or thousands of years. The described method was used for drawing a map (Fig. 4) of predicted decay of the upper permafrost layer that may take place during the first half of the 21st century (Pavlov and Gravis, 2000). If the scenario is realized the surface temperature within the permafrost region will increase considerably (about 1.2-2 °C) by 2050, and the depth of the seasonal thawing layer will increase by 20-30 %. The southern permafrost boundary will recede to the north by several hundreds of kilometres.

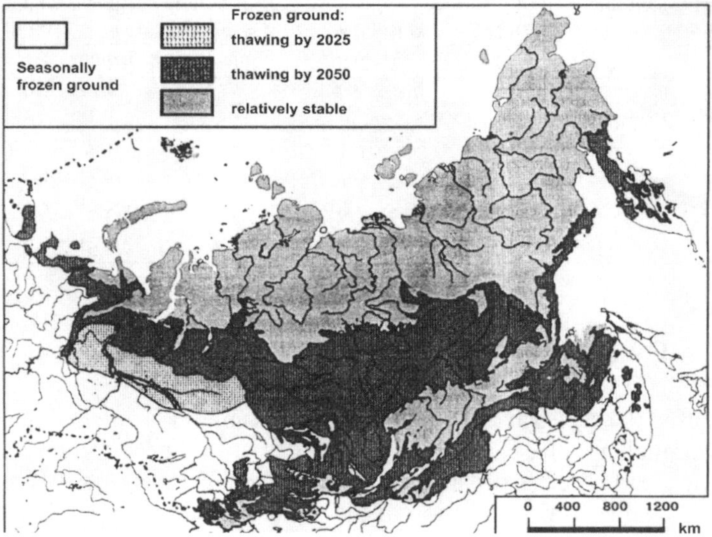

Figure 4. Prediction of permafrost evolution in Siberia due to climate warming (Pavlov and Gravis, 2000).

3.3 Modeling changes of temperature field

Transformation of the permafrost temperature field caused by Ts variations may be modeled numerically while solving a 1-D equation of heat conduction for some real sections. In the presence of a phase transition at the permafrost borders one should use two heat conduction equations for two blocks: frozen and thawed. The effective difference-iterative method for solving one- and many-dimensional Stefan type problems was used (Balobaev, 1991). Information used for modeling included present day thermal logs, data on terrestrial heat flow values and physical properties of rocks, yielding predicted Ts changes in the 21st century (according to Table 1). Temperature distributions with depth were calculated for several future time values starting from the year 2010 till 2100 (Fig. 5).

Calculations were made for 20 sites. Some numerical results corresponding to blocks of continuous permafrost in rocks containing ice (western Siberia) are shown in Figure 5. One can see that climate warming and a rise in surface temperature should result in warming up and thawing of frozen rocks in the surface layer. But the character of the temperature field transformation is different in different permafrost blocks. By the end of the 21st century there will only be a rise of permafrost temperature in the far north (Fig. 5A), and not actual thawing. However one can see that warming may create favourable conditions for an increase in the seasonally thawing layer. During the same time the blocks of non-stationary permafrost may thaw to a depth of 12-17 m (Figs. 5B and 5C).

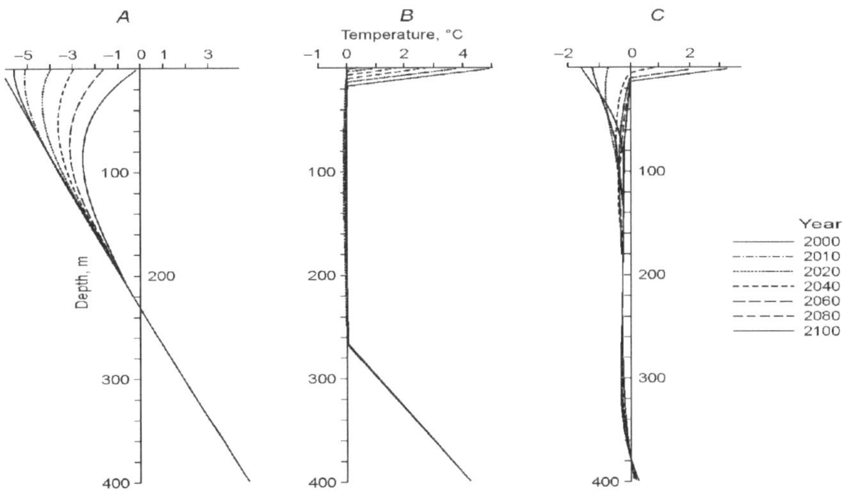

Figure 5. Predicted change of permafrost temperature field caused by climate warming in the 21st century (Duchkov and Balobaev, 2001 a, b). A - northern cooled stable permafrost (Arcticheskaya area, peninsula Yamal); B - non-stationary permafrost (Kostrovskaya area); C - non-stationary permafrost (Urengoiskaya area). Modern distribution of temperature - curves marked as "2000".

4. CONCLUSIONS

The present paper has provided some information on the long-term evolution of the temperature field in permafrost of Siberia. During the Late Pleistocene the southern permafrost border reached 48-49° N. During the Holocene the permafrost area has decreased and this process continues today. The permafrost decay may be intensified due to global climate warming in the 21st century. Regime temperature measurements at meteorological stations confirm an increase in the air temperature of Siberia during the last 40-50 years (average rate 0.04 °C/year). If modern trends are preserved then the surface temperature may increase by 1.2-2 °C by 2050. The southern border of continuous permafrost may be moved to the north by hundreds of kilometres. Numerical modeling of the permafrost temperature field has shown that an increase in *Ts* (ground surface temperature) will result in the formation of significant temperature anomalies in the upper rock layer, as can easily be seen with temperature monitoring techniques. In the far north regions values of *Ts* will remain negative up to the end of the 21st century and the permafrost will not start to decay. Here the only process will be an increase in the temperature of the frozen rocks (within the depth interval 0-260 m) due to conductive heat transfer, but no phase transition processes. In contrast in non-stationary permafrost blocks the upper permafrost layer (upper 12-17 m) will undergo thawing by 2100. Further analysis is required to better quantify estimations of the discussed thermal processes in permafrost. But one important process clearly shown in this paper is that fast thawing in the upper layer of frozen ground will occur in the case of climate warming. This process is rather serious as it may cause destructive influence on all engineering constructions: pipelines, roads, petroleum and gas stations and other buildings. Also permafrost decay will promote the release of natural or human-sequestered gases (CO_2, CH_4) preserved in frozen rocks.

ACKNOWLEDGEMENTS

The author wishes to thank colleagues from different institutes of the Siberian Branch of the Russian Academy Sciences for providing experimental data used in this work and the Directors of the NATO-ARW "Advances in CO_2 geological sequestration in eastern and western European countries" for the opportunity to make this presentation. The research was supported by Integration projects of the RAS (No. 13.16 and 13.17) and SB RAS (No. 121).

REFERENCES

Balobaev V.T., 1991, Geothermy of frozen zone of North Asia lithosphere, Novosibirsk: Nauka. p. 194 (in Russian).

Duchkov, A.D., and Balobaev, V.T., 2001a, Geothermal studies of permafrost response to global natural changes. In: *Permafrost response on economic development, environmental security and natural resources*. Paepe, R., and Melnikov, V., (Eds.), Kluwer Academic Publishers. Dordrecht. pp. 317-332.

Duchkov, A.D., and Balobaev, V.T., 2001b, Evolution of a thermal and phase condition of Siberian permafrost. In: *Global changes in the natural environment – 2001*, Dobretsov, N.L., and Kovalenko, V.I. (Eds.). Novosibirsk: Publishing house of the SB RAS: Geo p. 79-104. (in Russian).

Duchkov, A.D., Sokolova L.S., Pavlov, A.V., 2000, An estimation of modern changes of temperature of air and ground in Western Siberia, *Cryosphere of Earth*, **4** (1), pp. 52-59. (in Russian).

Duchkov, A.D., Balobaev, V.T., Devyatkin, V.N., and Sokolova, L.S., 1995,. Geothermal model of western Siberian permafrost, *Geologiya i Geofizika* (Russian Geology and Geophysics), **36** (8), pp. 70-79. (in Russian).

Duchkov, A.D., Balobaev, V.T. Volod'ko, B.V., Devyatkin, V.N., Lysak, S.V., Puzankov, Yu.M., Nozhkin, A.D., Sokolova, L.S., Berèzkin, V.I., Bogomolova, L.M., Botulu, E.A., Dorofeeva, R.P., Duk, V.L., Egorov, A.S., Zedgenizov, A.N., Zuy, A.N., Kitsul, V.I.,

Koveshnikov, A.M., Kotov, A.B., Kurchikov, A.R., Medvedev, V.I., Popov, N.V., Rusakov, V.G., Smelov, A.P., Stogniy, V.V., Turkina, O.M., and Shender, N.I., 1994,. Temperature, permafrost and radiogenic heat generation in an Earth's crust of Northern Asia. Novosibirsk: Publishing house of the SB RAS: UIGGM, 141 p. (in Russian).

Hansen, J.E., Russell, G., Rind, D., Stone, D., Lacis, A., Lebedeff, S., Ruedly, R. and Travis, L., 1983, Efficient three-dimensional global models for climatic studies: models I and II. *Monthly Weather Rev.*, **111**, pp. 609-622.

Kondratieva, K.A., Khurutzky, S.F., and Romanovsky, N.N., 1993, Changes in the extent of permafrost during the late Quaternary period in the territory of the former Soviet Union, *Permafrost and periglacial processes*, **4** (1), pp. 113-119.

Manabe, S., and Wetherald, R.T., 1975, The effect of doubling the CO_2 concentration on the climate of a general circulation model, *J. Atmos. Sci.*, **32**, pp. 3-15.

Pavlov, A.V., and Anan'eva, G.V., 2004, An estimation of modern changes of air temperature in territory Siberian permafrost, *Cryosphere of Earth*, **8** (2), pp. 3-9 (in Russian).

Pavlov, A.V., Gravis, G.F., 2000, Permafrost and a modern climate, *Priroda*, **4**, pp. 10-18 (in Russian).

Skachkov, Yu.B., 2001, Tendencies of modern changes of air temperature in Republic Saha (Yakutia), Questions of geography of Yakutia. Yakutsk: Publishing house of the YaGU, pp. 26-31 (in Russian).

POSSIBILITIES OF SO_2 STORAGE IN GEOLOGICAL STRATA OF PERMAFROST TERRAIN

A.G. Anshits[1], N.P. Kirik[1] and B.V. Shibistov[2]

[1]Institute of Chemistry and Chemical Technology, SB RAS, 42 K.Marx St., Krasnoyarsk, 660049, Russia, (E-mail: anshits@icct.ru); [2]Krasnoyarsk Regional Geological Association, 62 K.Marx St., Krasnoyarsk, 660049, Russia

Abstract: This report considers the possibility of SO_2 recovery using liquid sulphur dioxide generation and the disposal of excessive liquid SO_2 in geological structures within permafrost areas. Analysis demonstrates this option allows the use of exhaust sulphur dioxide in the required volumes in order to achieve an ecologically safe level. Geological analysis shows a section of the northwestern limb of the Vologochanskaya syncline seems to be the most favorable area for disposal of excessive liquid SO_2. It is located between the South-Pyasinskaya and Bolgokhtokhskaya brachyanticlines, 30km west of Norilsk.

Key words: geological structures; liquid sulphur dioxide; metallurgical gases; permafrost areas; sulphur dioxide disposal.

1. INTRODUCTION

The problem of sulphur dioxide disposal at the plants of the Norilsk Nickel Company represents an extremely serious environmental problem. Up to 2Mtpa of SO_2 are produced by the company's plants, of which only 10% is disposed. As a result the sulphur dioxide emissions greatly exceed the maximum permissible values.

At metallurgical plants worldwide the problem of SO_2 disposal is typically solved using a well-developed process of commercial sulphuric acid production, with up to 99.5% of the sulphur dioxide being caught. As

S. Lombardi et al. (eds.),
Advances in the Geological Storage of Carbon Dioxide, 93–102.

this option is unacceptable for Norilsk due to the geographical isolation and required sulphuric acid transportation, it was decided to apply the technology of enriched sulphurous gas disposal (20-50% SO_2), and the recovery of elemental sulphur via the technology of the Outokumpu Company, Finland (Abramov et al., 1987). Originally it planned to utilize coal dust as a reducing agent, but the process was later converted to natural gas. The technology includes the high-temperature (approx. 1250°C) natural gas reduction of sulphur dioxide to elemental sulphur with the subsequent recovery of sulphur from off-gases during the Klaus' catalytic stages, with sulphur condensation after each step. The technology was first applied for flash smelter gases, and later with a modification it was used for liquid bath smelter gases, subsequently named Vanyukov smelting (VS) (Mechev et al., 1989). Sulphur production from flash smelter gases is currently down (for over 10 years - Platonov et al., 2004), and only 14% of the SO_2 produced from VS off-gases is disposed of (Ladin et al., 2003a).

The objective of the present work was to assess the possibility of using an alternative option of effluent sulphurous gas disposal in order to attain the maximum permissible emission level. This procedure involves straight condensation with the recovery of liquid sulphur dioxide and disposal of excessive liquid sulphur dioxide in geological structures within permafrost areas.

2. ASSESSMENT OF DISPOSAL OPTIONS FOR ENRICHED AND IMPOVERSIHED GASES TO REACH MAXIMUM PERMISSIBLE EMISSION LEVELS

According to the General Production Development Strategy adopted by the Norilsk Nickel Company (Geoinfo, 2003), the proportion of enriched sulphurous gases in the total volume of generating gases is expected to increase up to 80% by 2007. It is planned to dispose of these gases, and recover sulphur, using the methane technology once it is updated. According to calculations, after system updating sulphur recovery will be around 90-92%, and the off-gases generated in sulphur production will contain 1-3 vol.% SO_2 (Ladin et al., 2003a, b).

It should be noted, however, that despite a number of improvements over the last 20 years during which methane technology has been used for sulphurous gas disposal and sulphur recovery, sulphur recovery from both flash smelter gases and liquid bath smelter gases has not exceeded 60% (Burukhin et al., 2000).

The principal factors that limit sulphur recovery efficiency are (Ladin et al., 2003b, Commission Report, 1985): i) significant fluctuations in the composition and volume of smelter off-gases, which considerably hampers process optimisation; ii) high water vapour content in the gases after reduction; iii) air inflow; iv) difficulties in providing complete condensation of sulphur generated at all process stages; and v) limited equipment availability. The presence of water vapour reduces sulphur recovery at both the high-temperature reduction stage and the catalytic stages. The presence of oxygen in the gases leads to SO_3 generation, which aggravates the problem of equipment corrosion, and to deactivation of the catalyst used. Due to low technology efficiency the tail gases (after the burning furnaces) in the currently-operating sulphur production system contain up to 3-5% SO_2.

Mass balance analysis of the methane technology demonstrated that disposal of the entire volume of enriched gases, even at the calculated maximum possible efficiency (which is not likely achievable, as practice shows), does not ensure reducing rates to the maximum permissible emission levels. It is therefore necessary to further treat either the high volume tail gases generated during sulphur production at an efficiency of more than 0.98, or impoverished gas streams generated by other production sections of the plant with an efficiency of more than 0.70. As is well known, the cost efficiency of impoverished gas disposal is nearly proportional to the gas volume. It should also be noted that the utilization of natural gas as the sulfur dioxide reducing agent and as the gas flux heating agent (up to 1250°C after water treatment) will produce significant amounts of carbon dioxide, the emissions of which must also be regulated, as Russia has signed the Kyoto Protocol. How the situation would change when using high-performance processes for sulphurous gas disposal is discussed below.

Analysis shows that the use of high-performance processes for enriched gas disposal, with sulphur dioxide recovery of 98-99% and sulphur dioxide content in the off-gases of 0.5-0.6 vol.%, will make it possible to reach the maximum permissible emissions.

A high-performance process of straight condensation with liquid sulphur dioxide recovery was selected for consideration as the most promising option for enriched gas disposal. The process allows for an easy attainment of the preset sulphur dioxide concentrations in the off-gases. It is suggested that impoverished gases (until main production and enriched/impoverished gas ratios are stable) should be concentrated and liquefied along with enriched gases. A number of methods are applied in the industry for impoverished gas concentration with the recovery of concentrated or liquid sulphur dioxide.

3. SCHEME FOR GENERATING LIQUID SO_2 USING LOW TEMPERATURES

The following scheme was considered for generating liquid sulphur dioxide from gases having an SO_2 concentration of 20-50% (Figs.1, 2). The gas is compressed to 10 atmospheres, cooled to -30°C and passed through the turbo-expander where it is cooled to -78°C. Liquefied sulphur dioxide is separated after each stage. The effluent gas contains 0.6% SO_2, and sulphur dioxide recovery at the last stage is 97.6% and 99.4% for gas streams containing 20% and 50% SO_2, respectively.

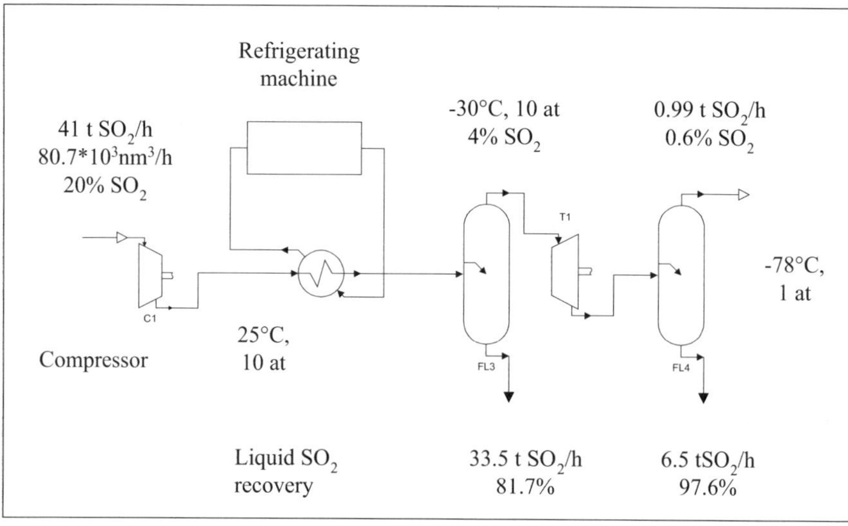

Figure 1. Schematic diagram of low temperature production of liquid sulphur dioxide, 20% SO_2 in gas.

Power inputs to generate liquid SO_2 (calculated per ton) using gas streams with 20% and 50% SO_2 at the compression and cooling stages are 205 and 78.5kW/t and 133 and 94.9kW/t, respectively. Figure 3 shows the change in the total power inputs (compression and cooling) needed to generate liquid sulphur dioxide using the proposed scheme, with liquid sulphur dioxide concentrations increasing in the considered gases. It is evident that power inputs reduce considerably as sulphur dioxide concentrations increase in the original gas stream.

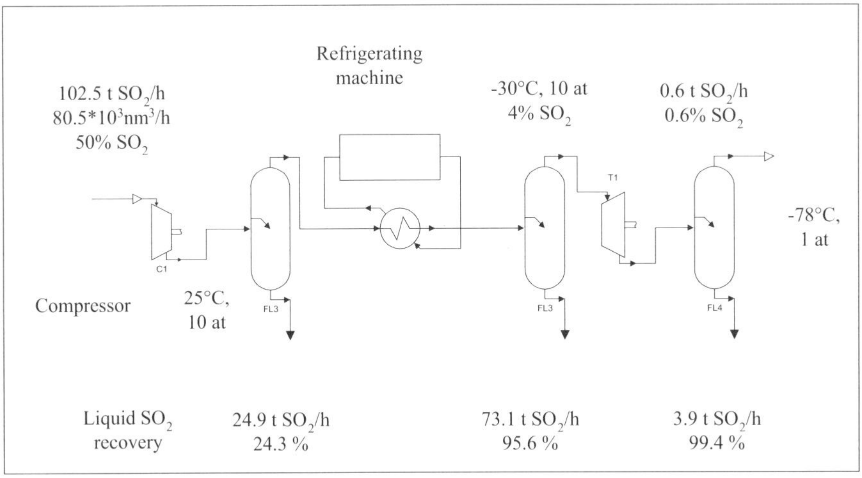

Figure 2. Schematic diagram of low temperature production of liquid sulphur dioxide, 50% SO_2 in gas.

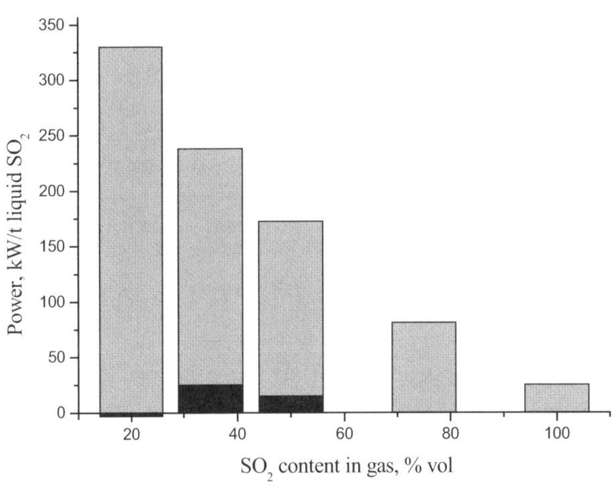

Figure 3. Power inputs for low temperature production of liquid SO_2: ■ - with heat recuperation, □ - without heat recuperation.

Power inputs can be reduced at the cooling stage by using cold recovery. The calculations based on using cold recovery for 20% SO_2 streams under the scheme shown in Figure 4 demonstrate that power inputs reduce by nearly 2.5 times at the cooling stage. Another way to conserve energy can be the use of power generated by hot gases emerging from the smelters. Power-input data based on the use of energy recovered from cooling gases when generating liquid sulphur dioxide from gas streams containing 20, 50 and 75% SO_2 is shown in Figure 3. It is evident that power inputs for compression and cooling are almost compensated with the energy generated via gas cooling.

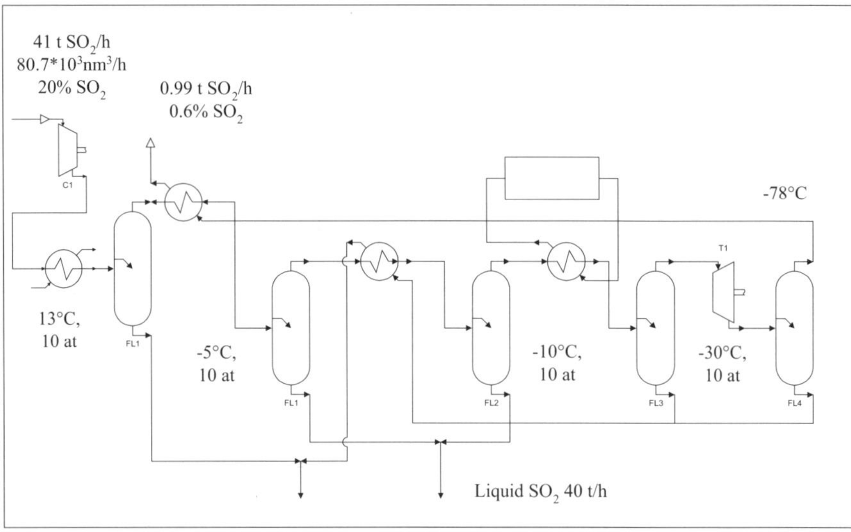

Figure 4. Schematic diagram of low temperature production of liquid SO_2 with cold recuperation; 20% SO_2 in gas.

4. USE OF LIQUID SULPHUR DIOXIDE

When creating a liquid sulphur dioxide buffer there is the desire to generate gases with a constant SO_2 concentration in order to stabilize all operating sections of the plant producing sulphur, sulphuric acid and sodium bisulfite. In the future it would be possible to utilize liquid SO_2 in metallurgical technology, e.g. oxidative leaching of pyrrhotite concentrate (Sobol and Frash, 1974). The laboratory studies of this process demonstrated the increasing recovery efficiency of both base metals and platinum metals etc., and thus liquid sulphur dioxide could become a marketable product if the situation in the market changes.

In the absence of such a market it is suggested to dispose of excessive liquid sulphur dioxide in geological structures within permafrost areas.

5. ASSESSMENT OF THE POSSIBILITY TO DISPOSE LIQUID SO_2 IN GEOLOGICAL STRUCTURES

Methods exist worldwide for the geological disposal of various waste streams, including the injection of CO_2 (Pearce and Rochelle, 1999, Bachu and Gunter, 2004), CO_2/H_2S mixtures (Bachu et al., 2003), and liquid toxic and radioactive waste (Stepanov, 2004).

The selection criteria for deep industrial liquid waste storage are relatively well developed (Stepanov, 2004). It is preferable to select water saturated units located within negative platform structures. These should have capacitance properties and be isolated from shallow groundwater and surface water with an overlying impermeable argillaceous mass. The reservoirs should be capable of holding significant volumes of waste and contain it within a limited section of the unit over a long period of time.

The selection of disposal sites for acid gases mainly depends on properties such as phase equilibrium, water content and solubility, and the generation of hydrates causing blockages in both boreholes during pumping and in compressors (Bachu et al., 2003).

Sulphur dioxide is a readily liquefied gas, forming at t = -10°C at a pressure of 760 millimetres of mercury. The critical parameters for SO_2 are T=157.5°C and P=77.8 atmospheres. Sulphur dioxide forms gas hydrates relatively easily, even at temperatures of up to +12°C. Sulphur dioxide gas-hydrate has a low generation pressure equilibrium, for example at 12.1°C the vapor pressure equilibrium is 2.3 atmospheres. Two types of solutions can be formed depending on the sulphur dioxide content in the SO_2-H_2O system: sulphur dioxide solution in water and water solution in sulphur dioxide. Depending on the sulphur dioxide concentration, SO_2 gas-hydrate is in equilibrium with one of the solutions or with ice below 12°C. There is a narrow range of water concentration in liquid sulphur dioxide when gas-hydrate is not formed in the solution between temperatures of -30 to -40°C (at 50 atm). When the water content in the mixture is higher the temperature must be maintained over 12°C to avoid gas-hydrate formation.

6. SELECTION OF LIQUID SO_2 DISPOSAL SITE IN THE NORILSK INDUSTRIAL AREA

A geological analysis of the Norilsk Industrial Area was conducted to select the most appropriate site for liquid sulphur dioxide disposal based on available data.

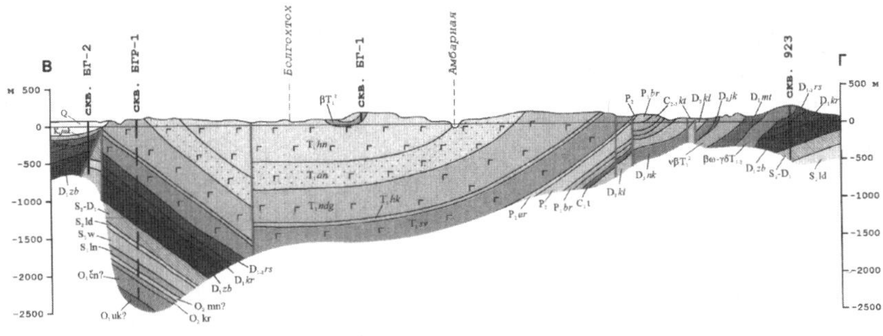

Figure 5. Geological profile of the Vologochanskaya syncline (Mezhvilk et al., 1972): T_1hl - Lower Triasian, Chachankanskaya Series, T_1sv - Lower Triasian, Syverminskaya Series, D_1zb - Lower Devonian, Zubovskaya Suite, D_1zb - Lower Devonian, Zubovskaya Suite, S_2ld_2 - Upper Silurian, Upper Ludlow Series, S_2ld_l - Upper Silurian, Lower Ludlow Series, S_1w - Lower Silurian, Wenlock Series, S_1ln - Lower Silurian, Llandovery Series, O_2mn - Middle Ordovician, Mangazeysky Series, O_2kr - Middle Ordovician, Krivolutsky Series, $O_1\check{c}n$ - Lower Ordovician, Chunsky Series.

A section of the northwestern limb of the Vologochanskaya syncline seems to be the most favorable area. It is located between the South-Pyasinskaya and Bolgokhtokhskaya brachyanticlines, 30km west of Norilsk (Fig.5). The Vologochanskaya syncline is formed by Lower and Mid-Paleozoic carbonate-terrigenous rocks and Lower Triassic volcanogenic rocks, and is stable tectonically. The size of the syncline provides enough space for a large reservoir suitable for long-term storage, while the negative structure provides a proper isolation from the biosphere. Silurian and Devonian rocks, as well as oil-saturated Lower Triassic tufogenic rocks separated by basalt covers, have good reservoir properties. The prevalence of carbonate rocks in the Paleozoic is a favorable factor for liquid SO_2 disposal. The structure of the syncline and the alternation of reservoir beds with impermeable cover beds create a multilevel target for disposal of liquid sulphur dioxide. The topography of the syncline is a vast, poorly split watershed area with an elevated width of permafrost, which serves as a fluid-resistant cover.

7. PRINCIPAL RESULTS AND CONCLUSIONS

- It was demonstrated that in order to reach the maximum permissible emission of sulphur dioxide, straight condensation of sulphur-containing gases could be used as the basic technology for enriched gas recycling, the sulphur dioxide recovery of which is 98-99% .

- The power inputs of straight condensation depends considerably on SO_2 content. Power inputs at the compression stage are 205 and 79 kW/t SO_2 liq., for input streams of 20 and 50 % SO_2, respectively. Utilization of naturally cold conditions would allow for a significant reduction in refrigerating machine costs. The use of energy generated by cooling hot gases emerging from the liquid bath smelters (e.g., use of steam turbines) would allow for compensating up to 90-100% of total power inputs.
- It is proposed to dispose excessive liquid SO_2 in the geological structures of the Norilsk Industrial Area, with isolation being accomplished via both stratigraphic traps and the overlying gas-hydrate layer at the lower edge of the permafrost. A geological analysis of the area demonstrates that there is a theoretical possibility of such disposalThe northwestern portion of the Vologochanskaya syncline, between the South-Pyasinskaya and Bolgokhtokhskaya anticlines, is proposed as a liquid sulphur dioxide disposal site.
- There are two types of geological reservoirs: aluminosilicate volcanogenic and terrigenic rocks and carbonate rocks. The first type ensures sulphur dioxide disposal without reactive interaction, while the latter ensures sulphur dioxide disposal with the formation of calcium salts.

REFERENCES

Abramov, N.P., Yeremin, O.G., Baryshev, A.A., Zaltsman, S.L., and Kuzhel, B.I., 1987, Experience of mastering of sulphur production from emission gases of flash furnaces at NGMK, *Tsvetnye metally*, 7:26-28 (in Russian).

Bachu, S., and Gunter, W.D., 2004, Overview of Acid-Gas Injection Operations in Western Canada, in: *Proc.7th Int. Conf. on Greenhouse Gas Control Technologies. Volume 1: Peer-Reviewed Papers and Plenary Presentations*, edited by E.S. Rubin, D.W. Keith and C.F. Gilboy, IEA Greenhouse Gas Programme, Cheltenham, UK,

Bachu, S., Haug, K., Michael, K., Buschkuehle, B.E., and Adams, J.J., 2003, Deep Injection of Acid Gas in Western Canada, in: *Proc. 2-th Int. Sym. on Underground Injection Science and Technology,* Berkeley, CA, Oct 22-25 (2003).

Burukhin, A.N., Kaytmazov, N.G., Dyachenko, V.T., Galantsev V.N., and Ilyukhin, I.V., 2000, Production of fertilizers in Norilsk?, *Tsvetnye metally,* **6**:63-64 (in Russian).

Commission Report, 1985, Report of Temporal Scientific and Technical Commission, Resolution of Gov. Science and Engineering Committee of USSR NN 305, 554, (unpublished).

Geoinfo, 02.04.2003; http://www.infogeo.ru/metals/news

Ladin, N.A., Danilov, M.P., Severilov, A.V., Kozlov, A.N.,. Sapegin, Yu.V, Baryshev, A.A., Derevnin, B.T., and Semenov, M.Yu., 2003a, Modernization of sulphur dioxide recovery at "Nadeshda" smelter for further introduction at smelting divisions of "MMC Norilsk Nikel"s, polar branch, in: *Book of abstracts "Metallurgy of Non-Ferrous & Rare Metals"*, Krasnoyarsk, Russia, 9-12 September, 2003 , pp. 152-153.

Ladin, N.A., Danilov, M.P., Severilov, A.V., Kozlov, A.N., Sapegin, Yu.V., Baryshev A.A., and Derevnin, B.T., 2003b, Modernization of sulpfur dioxide recovery at "Nadeshda" of "MMC Norilsk Nikel"s, *Khimicheskaya tekhnologiya,* **12**:4-6 (in Russian).

Mechev, V.V., Abramov, N.P., Sokolov, O.K., Yeremin, O.G., Baryshev, A.A., and Orlov, V.N., 1989, Development and mastering of sulphur production from metallurgical emission gases, *Tsvetnye metally,* **7**:69-72 (in Russian).

Mezhvilk, A.A., Karzeva, G.N., and Mikhalyuk, Yu.R., 1972, *Geological map of USSR 1: 200000, list R-45-XXI,XXII* (M.: Izd-vo Nedra,) (in Russian).

Pearce, J., and Rochelle, C., 1999, CO_2 storage: mineral reactions and their influences on reservoir permeability–a comparison of laboratory and field studies, in: *Proc. Fourth Int. Conf. on Greenhouse Gas Control Technologies,* Interlaken, Switzerland, 30 August - 2nd September (Pergamon Press, 1999), p.1087.

Platonov, O.I., Vasil'ev, Yu.V., Ryabko, A.G., Tsemekhman, L.Sh., and Yatsenko, S.P., 2004, Relative to the choice of reduction scheme of emission gas from Vanyukov' furnace *Tsvetnye metally,* **2**:68-72 (in Russian).

Sobol, S.I., and Frash, T.M., 1974, Sulphur anhydride in autoclave treatment of sulfide-containing raw materials, *Tsvetnye metally,* **2**:14-21 (in Russian).

Stepanov, A.N., 2004, Deep disposal of liquid toxic wastes is effective protection of nature, *Prirodnye resursy Rossii: upravlenie, ekonomika, finansy,* **1**:95-100 (in Russian).

CRYOGELS – A PROMISING MATERIAL FOR UNDERGROUND WORKS IN PERMAFROST

L.K. Altunina[1], V.A. Kuvshinov[1] and S.N. Dolgikh[2]
[1]*Institute of Petroleum Chemistry SB RAS, 3 Academichesky Avenue, 634021, Tomsk, Russia;*
[2]*Institute "Yakutniproalmaz", Stock Company "ALROSA", Mirny, Yakutia*

Abstract: Cryotropic gel-forming systems have been developed at the Institute of Petroleum Chemistry SB RAS in Tomsk, Russia. They are capable of generating gels at 0–20 °C, subsequently converting to cryogels during freeze-thaw cycles which exhibit high elasticity and adhesion to rock. Both laboratory and field experiments are reported here, including 2003–2004 pilot tests conducted to seal a leaking interval at the base of a dam at the Irelyakh hydrosystem, Stock Company "ALROSA", Yakutia.

Key words: cryogel, permafrost, viscosity, elasticity, sealing, cutoff wall, dam

1. INTRODUCTION

The use of permafrost is considered as one option for CO_2 geological storage. From this view point two issues arise:

– depending on injection depth and site conditions the CO_2 may occur as individual and mixed gas hydrates, mineral and organo-mineral complexes (clathrates and carbonates) or in other forms;

– the long-term existence of underground CO_2 storage implies permafrost conservation and consolidation.

The present paper discusses cryogels – a promising material for underground works in permafrost regions. These are polymer gels generated due to freezing and subsequent thawing of aqueous polymer solutions. The more "freezing – thawing" cycles the cryogel is subjected to the better its mechanical properties: strength and elasticity increase and formation bond intensifies (Lozinsky, 1998, 2002). Cryogels, i.e. cryotropic gels, are

S. Lombardi et al. (eds.),
Advances in the Geological Storage of Carbon Dioxide, 103–110.

peculiar polymer solutions, with a higher critical solution temperature (Rebinder, 1979; Papkov, 1974). The spatial pattern of cryogels is formed during phase transformation and is stabilized with intermolecular bonds of different nature: covalent, coordination and ionic-electrostatic. This variable bond type generates the cryogel manifold and allows their properties to be varied significantly (Lozinsky, 1998, 2002; Sergeev et al., 1978).

Cryogels based on polyvinyl alcohol (PVA) are of great interest both scientifically and from an applied viewpoint. They are widely used in biotechnology, medicine, food industry and others (Sergeev et al., 1978; Lozinsky, 1998). In many respects this interest results from the excellent mechanical, diffusional and thermophysical properties of PVA cryogels, polymer availability, its non-toxicity and biocompatibility, as well as its relatively simple gelation technique involving freeze-thaw cycling. It is possible to regulate physico-chemical properties of cryogels within wide limits (Sergeev et al., 1978; Lozinsky, 1998) by varying polymer characteristics (molecular mass, chain-length distribution, content of residual o-acyl groups, tacticity and PVA concentration in the system), solvent composition, additive nature and cryogenic process conditions (temperature and freezing duration, thawing rate, number of over freezing cycles, etc.). In practice water is used as the basic solvent for PVA. PVA is resistant to oils and fats, aliphatic and aromatic hydrocarbons, weak acids and alkalis. In addition PVA exhibits universal adhesive and bonding properties and is also atmosphere and oxidation-resistant.

2. RESULTS

2.1 Laboratory Experiments

Gelation kinetics and the rheological properties of PVA solutions of different concentrations (from 0.5 to 10 wt%) have been studied in the temperature range from –30 to +10 °C. We have also studied PVA solutions containing additives, which affect gelation kinetics and rheological properties of gels and cryogels, as well as their adhesion to rocks, in order to create flow barriers within ground constructions. Viscosity and Young's modulus were chosen as rheological characteristics.

Viscosity was measured using a "Reokinetika" vibration viscosimeter equipped with a tuning-fork gauge. Distilled water was used as a calibration liquid.

Modulus of elasticity for gels was determined based on "tension – deformation" diagrams, obtained during quasi-elastic cylindrical-sample compression using in-house developed equipment based on a micrometer and electronic scales. The modulus of elasticity was calculated as the angle of slope of the initial linear segment, where Hooke's law is observed, relative to the compression – deformation curve.

The kinetics of cryotropic gelation of 5 and 10 % PVA solutions was studied at - 32 and - 18 °C by using the modulus of elasticity as the principle monitoring parameter. Solutions were maintained at sub-zero temperatures for a pre-determined time and then they were subjected to thawing at 2 and 18 °C. The mode of thawing is a factor which significantly affects visco-elastic properties of cryogels. Slow thawing promotes generation of cryogels exhibiting high viscosity and modulus of elasticity, and thus the lower the freezing temperature the stronger the cryogels (Figs. 1 and 2). At higher thawing rates the effect of freezing temperature on the visco-elastic properties of cryogels decreases.

The effect of the number of "freeze-thaw" cycles on the properties of cryogels, prepared from aqueous PVA solutions, has been studied. The cryogels were frozen at - 18 and - 32 °C and then thawed at 18 and 6 °C. The results obtained in the study indicate that the modulus of elasticity increases with the number of "freeze-thaw" cycles and it does not depend on the thawing rate (Fig. 3). The greatest increase in elasticity was observed after 2 – 4 freeze-thaw cycles.

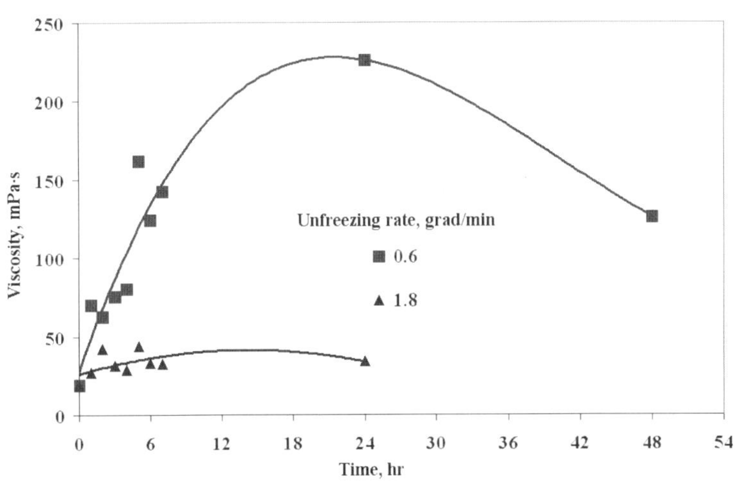

Figure 1. Viscosity change of 5 % PVA solution after thermostatting at –32 °C and thawing at different rates.

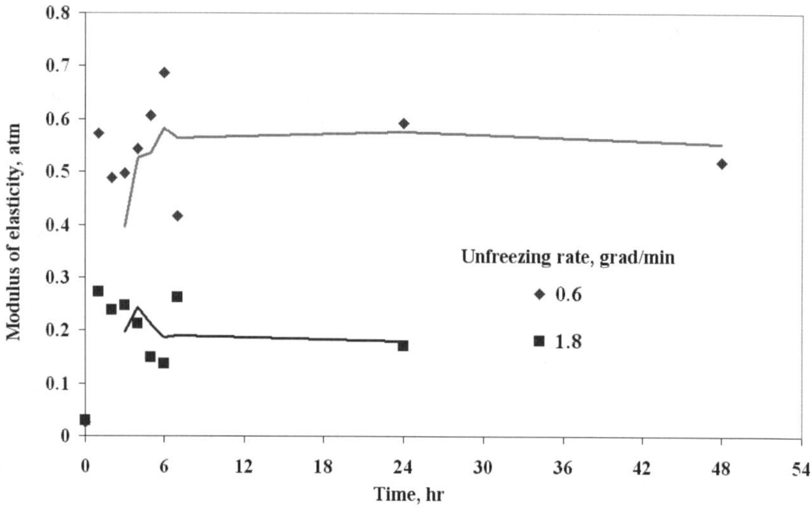

Figure 2. Elasticity changes of 5 % PVA solution after thermostatting at –32°C and thawing at different rates.

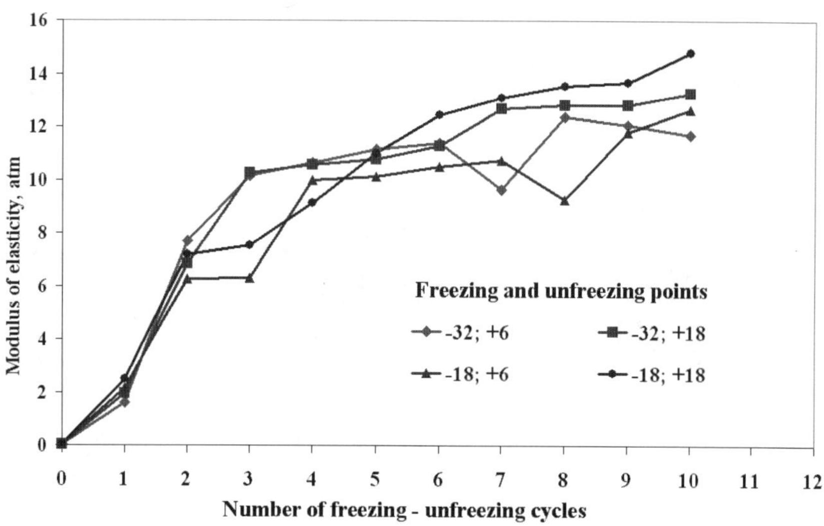

Figure 3. Dependence of the modulus of elasticity (for cryogels obtained from 10% PVA solution and which were frozen at –32 and –18 °C), on the number of freeze-thaw cycles at 18 and 6 °C.

We studied gelation kinetics of a 5 % PVA solution at +2 °C and compared the visco-elastic properties of the prepared gel with those of cryogels obtained from the same solution during freezing at –18 and –32 °C, followed by thawing at +18 °C. The study showed a significant increase in gel viscosity and modulus of elasticity at the lower freezing temperature (Fig. 4).

The viscoelastic properties of gels prepared from 5 and 7 % PVA solutions at –5, 0, +2 and +22 °C have also been studied. The gels contained electrolytes as additives, which affected gelation kinetics and the rheological properties of gels and cryogels, as well as their adhesion to rocks. Gelation kinetic studies of PVA solutions containing electrolytes as additives already showed that the gels were formed in several hours at 0 – 18 °C.

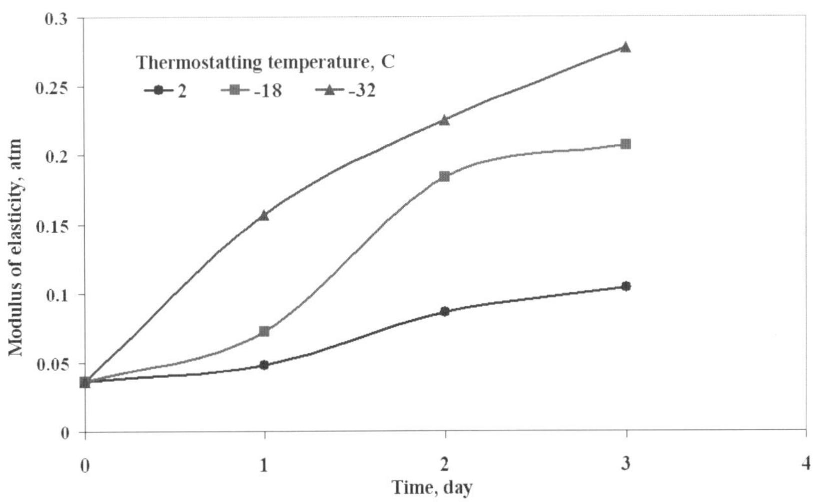

Figure 4. Dynamics of modulus of elasticity for 5 % PVA solution thermostated at different. temperatures.

Moreover subsequent thermostatting or freezing – thawing of the gels improved their strength. It has been determined that viscosity and elasticity of the solutions increased 3–5 times when a cross-linking agent was added to the PVA aqueous solutions, depending on agent concentration and thermostatting time (1 to 8 days) at +2 °C

Strength changes occurring in cryogels with different fine-filling agents were also studied, such as crushed carbonate rock (fraction 0.16 – 0.5 mm), sand and crushed core material having sandstone / argillite ratios of 2:1; 1:1 and 1:2. The highest modulus of elasticity was observed with the filling agents in a 1:1 ratio and cryogel prepared from a PVA solution with a cross-linking agent and filled with crushed carbonate rock and core material. Moreover the modulus of elasticity of the cryogel also increased with an increase in PVA and cross-linking agent concentrations. The addition of electrolyte and a cross-linking agent to the PVA solution significantly increased the modulus of elasticity in all PVA solutions – crushed carbonate rock ratios (Fig. 5).

The filtration characteristics of gel-forming systems were studied under thermobaric conditions simulating water filtration through anti-filtration barriers in constructions in the northern climatic zone. Experiments consisted of water filtration at a constant discharge rate through a laboratory model of a filtering medium.

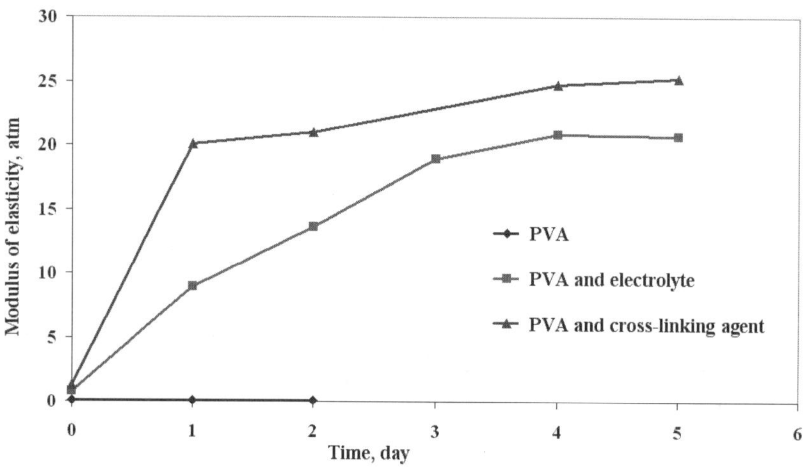

Figure 5. Increase in modulus of elasticity for PVA solutions with added electrolyte and cross-linking agents after thermostatting with crushed carbonate rock at a ratio of 1:2 at +2 °C.

The model was packed with crushed carbonate rock (fraction 0.16-0.5 mm), with permeability varying from 9.78 to 87 μm^2. A gel-forming solution containing 3 % PVA and a cross-linking agent (Fig. 6) was injected into a model consisting of crushed carbonate rock with a permeability of 87 μm^2 at 6 °C at lower pressure values (from 0.3 to 2.25 atm/m). The model was subsequently held at 6°C for 5 hours to form the gel, and then once again water infiltration experiments were continued. At pressure increases up to 50 atm/m no water infiltration was observed (Fig. 6).

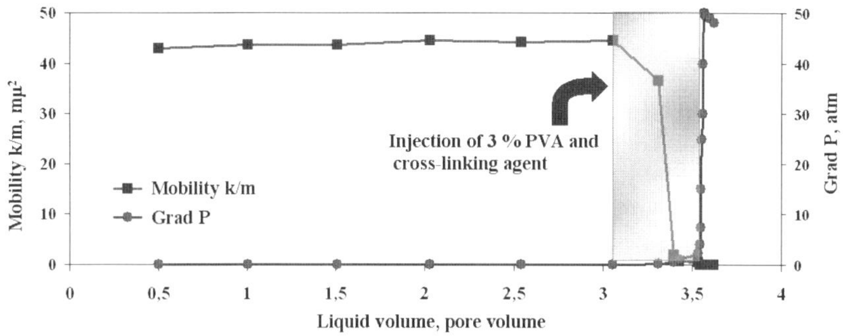

Figure 6. Creation of a cutoff wall via injection of a gel-forming solution (3 % PVA and a cross-linking agent) into a crushed carbonate rock model with permeability of 87 μm² at 6 °C.

Figure 7 presents the results of the creation of a cutoff barrier in a model consisting of crushed carbonate rock having a permeability of 10.254 μm² at 20 °C. Injection of a gel-forming solution containing 10 % PVA and electrolyte was performed at a pressure gradient of 3.5–4 atm/m. The model was subsequently held for 12 hours at 20 °C to form the gel, after which water infiltration was once again initiated.. No water infiltration was observed at pressure increases up to 140-160 atm/m (Fig. 7).

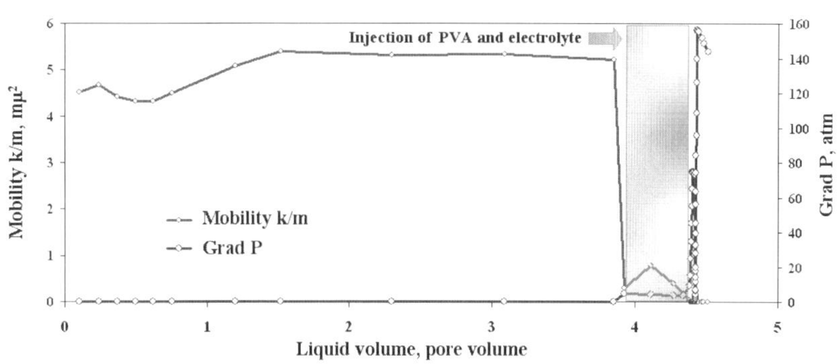

Figure 7. Creation of cutoff screen via injection of gel-forming system: 10 % PVA and electrolyte into a model of carbonate rock with permeability of 10.254 μm² at 20 °C.

Thus, PVA-based gel-forming solutions with a cross-linking agent and electrolyte as additives are capable of creating reliable cutoff barriers in carbonate rocks. The study of infiltration characteristics of soils with PVA-based gel-forming solutions under thermobaric conditions demonstrated that after gelation one could observe a sharp decrease in water mobility, even up to pressure differences higher than 100 atm/m.

2.2 Field Experiments

In order to seal a leaking interval at the base of a dam at the Irelyakh hydrosystem in Mirny City, Sakha Republic (Yakutia) we have developed the CRYOGEL-1 system based on PVA and a cross-linking agent. The system is capable of forming gels at 0–20 °C during several hours and then cryogels during the freeze-thaw process. In August-September of 2003, 51 tons of cryogel-forming solution were injected into 5 holes at the base of the dam. This injection formed a cryogel barrier that is 15m in length, 3 m thick and covers an area of about 430 m^2 at a depth ranging from 16 to 45 m. The barrier created a cutoff wall which decreased water infiltration by 150 times. The results of geophysical studies and hydrodynamic investigations in a test hole, drilled between the third and the fourth holes into which the cryogel was injected, show that the average specific water absorbing was 0.15 L/(min·m·m water column) up to 40 m deep, highlighting the positive results of the research.

In 2004 the work at the Irelyakh hydrosystem dam was continued. Based on the regulations of IPC SB RAS Stock Company, "Korona-TEK" designed and made equipment for the preparation and injection of cryogel-forming solutions. In September-November of 2004 starboard plugging of the dam, where the filtration was the strongest, was successfully carried out. A total of 70 tons of cryogel-forming solution was injected into 13 holes to create a cryogel barrier at a depth ranging from 13 to 43 meters.

In 2005 cryogel-forming solutions will be injected into more than 100 holes to create a cutoff wall along the full length of the Irelyakh hydrosystem dam.

REFERENCES

Lozinsky, V.I., 1998, Cryotropic gelation of poly(vinyl alcohol) solutions. *Uspekhi khimii*, **67**(6):, p. 641.

Lozinsky, V.I., 2002, Cryogels on the basis of natural and synthetic polymers: preparation, properties and application, *Uspekhi khimii*, **71**(6), p. 259.

Rebinder P.A. Selected works. Surface phenomena in dispersed systems. Physico-chemical mechanics. «Nauka», M., 1979, 381 p.

Papkov, S.P., 1974, Gelatinous state of polymers, *Khimia*, M., 256 p.

Sergeev, G.B., and Batyuk, V.A., 1978, Cryochemistry, *Khimia*, M., 295 p.

SUBSURFACE CARBON DIOXIDE STORAGE THROUGH CLATHRATE HYDRATE FORMATION

P. Jadhawar, A.H. Mohammadi, J. Yang, B. Tohidi
Centre for Gas Hydrate Research, Institute of Petroleum Engineering, Heriot-Watt University, Edinburgh, United Kingdom.

Abstract: Rising atmospheric emissions as a result of fossil fuel consumption is a major concern for the developed and developing countries, considering the role it plays in the greenhouse effect and hence global climate change. Various schemes for underground CO_2 storage (viz. geologic disposal into coal seams, depleted oil/gas reservoirs, salt caverns, and deep oceans) have already been reported in the literature. Subsurface CO_2 storage through clathrate hydrate formation is a novel option for the reduction of atmospheric carbon content and permanent underground CO_2 disposal over geological periods. Depths of CO_2 injection, respective pressure-temperature conditions, water salinity etc. are all important factors for successful CO_2 sequestration. Furthermore if CO_2 is injected/stored in methane hydrate reservoirs it could be possible to produce low-carbon methane energy, thereby offsetting the cost of CO_2 transportation and disposal. In this communication, we present the results of experiments carried out to understand the mechanisms of CH_4 displacement in hydrate structure by injected CO_2 and the formation of simple CO_2 or mixed CH_4-CO_2 hydrates, thereby simulating the conditions of CO_2 injection into CH_4 hydrate reservoirs. We used two sets of experimental rigs specifically designed for studying gas hydrates in porous media. They are the Medium Pressure Glass Micromodel (80 bar) for visual observation of gas hydrate formation / dissociation and distribution in porous media, and the Ultrasonic Rig (400 bar) for studying CO_2 sequestration in CH_4 hydrates in synthetic porous media.

Key words: subsurface CO_2 sequestration, permafrost oil and gas reservoirs, CH_4 hydrate reservoirs, simple CH_4 / CO_2 hydrates or mixed CH_4 -CO_2 hydrates.

S. Lombardi et al. (eds.),
Advances in the Geological Storage of Carbon Dioxide, 111–126.

1. INTRODUCTION

Gas hydrates were first identified in 1810 by Sir Humphrey Davy (Davy, 1811) and their compositions were later established by Faraday (Faraday 1823). Gas hydrates are naturally occurring crystalline inclusion compounds (clathrates) characterized by strictly determined structures for different gases (Makogon, 1997). Clathrates form when water establishes, due to hydrogen bonding, a cage-like structure around small guest molecules (e.g., light alkanes, carbon dioxide, hydrogen sulphide, nitrogen and oxygen) at low temperature and high-pressure conditions. Their formation, stable existence and decomposition depend upon the pressure, temperature and composition of the system. Gas hydrates can exist in one of the three most-common structures, I, II (Claussen, 1951; von Stackelberg and Müller, 1954; Jeffrey and McMullan, 1967) and H (Ripmeester et al., 1987) (Figure 1). Both CH_4 and CO_2 form structure I hydrate (Davidson, 1973), as do the mixtures of these gases (Adisasmito et al., 1991).

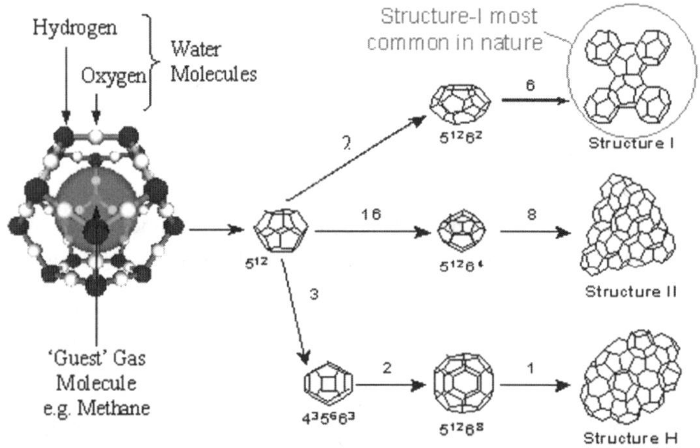

Figure 1. Different forms of clathrate hydrate structures. The numbers over the arrows indicate the number of cage types. (Modified from Sloan, 2003).

Very large sedimentary deposits of natural gas hydrates (14×10^{12} t.o.e.), whose existence was proven 30 years ago (Makogon, 1997), are inferred to occur in two types of geologic environments: permafrost regions (where cold temperatures dominate) and beneath the sea in sediments off the outer continental margins (where high subsurface pressures dominate). Both deposit types are regarded as potential energy storehouses. While methane, propane, and other gases are included in the hydrate structure, methane hydrates are the most common, and thus so are the structure-I hydrates. The methane source for clathrate hydrates formation may be either biogenic or thermogenic. Biogenic CH_4, resulting from the microbial breakdown of organic matter, form hydrates in shallow sediments. About 98% of all gas hydrate deposits come from biogenic methane sources, accumulated offshore in upper sedimentary layers. Thermogenic methane, coming from the thermal alteration of organic matter in sediment at depth, may form hydrates where deep gas migration pathways exist (Kvenvolden and Lorenson, 2001).

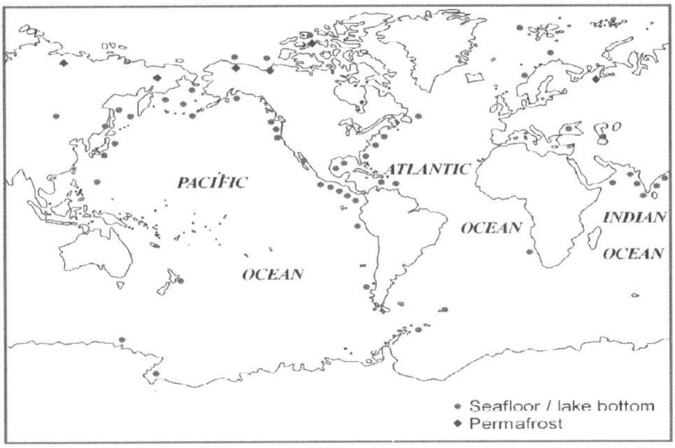

Figure 2. Worldwide occurrences of natural gas hydrate deposits (Modified from Kvenvolden and Lorenson, 2001).

The formation and distribution of gas hydrates in subsurface sediments is controlled by two opposing factors: temperature increases associated with increasing depth (the geothermal gradient) and pressure gradients. Since hydrates are generally stable under relatively low temperatures, the geothermal gradient (2-3°C/100m) limits hydrate formation and existence to relatively shallow, cooler regions of the Earth's crust. Considering seabed temperature and water salinity, hydrates are generally stable at water depths greater than 300 m in continental areas and 500 m in offshore areas. These two factors restrict the existence of clathrate hydrates in subsurface sediments to a region called the Hydrate Stability Zone (Figure 3).

1.1 Gas hydrate production and related issues

Methane gas, a low carbon energy source, can provide an energy supply for many years if produced from naturally occurring CH_4 hydrate reservoirs. The various CH_4 recovery methods (e.g. depressurisation, thermal stimulation, inhibitor injection or a combination of these methods: Sloan, 1998) have already been investigated and are reported in the literature.

These production methods may induce instability in the hydrate rich sediments. The likely mechanism is that hydrate decomposition at the base of the hydrate stability zone convert consolidated sediments into loose gas-charged sediments. This causes a decrease in the shear strength and facilitates the occurrence of landslides, tsunamis or other natural disasters, such as the 1986 Lake Nyos disaster (Rogers, 1996). After massive landslides on the continental slopes, mud volcano eruptions could cause massive hydrate dissociation and the release of CH_4 into the atmosphere, which would contribute to climate change. Natural gas pipelines laid from production platforms to shore may warm sea floor sediments and decompose surrounding hydrates. A concern of the oil and gas industry is also the fact that drilling through hydrate zones might destabilize supporting foundations for platforms and production wells. Disruptions to the ocean floor from hydrate decomposition could also result in surface slumping or faulting, endangering work crews and the environment. Loss of seawater buoyancy because of hydrate-released gas could endanger floating structures (Hovland and Gudmestad, 2001).

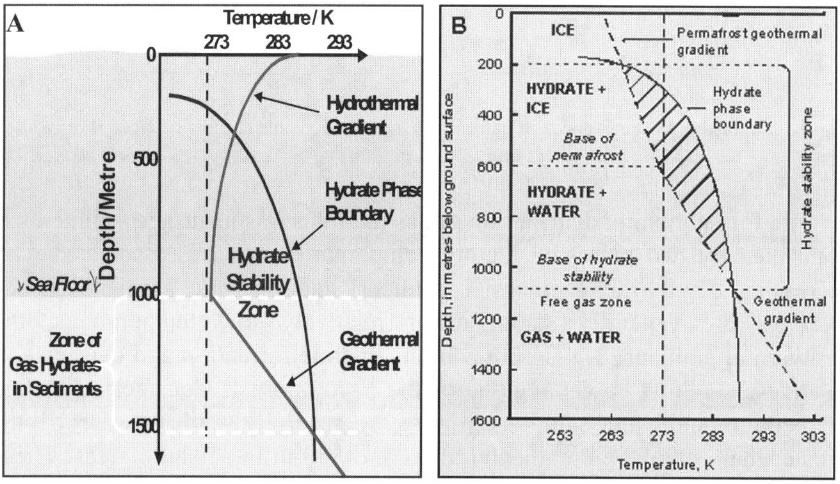

Figure 3. A. Seafloor CH_4 hydrate stability zone. B. Permafrost CH_4 hydrate stability zone.

1.2 Greenhouse gases and their mitigation

Rising levels of the greenhouse gases (mainly CO_2) in the atmosphere have been a great concern for most of the developed and developing nations because of the major climate changes that have been observed in recent decades. One of the options for mitigating the rising greenhouse gas concentrations is their sequestration and storage in subsurface formations.

The petroleum industry has been practicing the methods of sequestering CO_2 in oil and gas reservoirs for Enhanced Oil and Gas Recovery over the last few decades, with an economic incentive to partially offset the sequestration cost. Based on the same ideology, options for CO_2 disposal are being extended for Enhanced CBM recovery, Enhanced gas recovery from CH_4 hydrate reservoirs etc. We propose the option of CO_2 injection into suitably chosen geologic formations and its permanent storage through clathrate hydrate formation.

1.3 CO_2 storage through clathrate hydrate formation

Subsurface CO_2 sequestration and its permanent storage through clathrate hydrates formation is a novel method for reducing the concentration of carbon dioxide in the atmosphere. CO_2 forms stable hydrate compounds at lower temperature and pressure conditions compared to other hydrate formers, especially CH_4 hydrates (Figure 4). Since methane hydrates have been inferred to be present in permafrost and outer continental rises and slopes, and are known to have existed over many centuries, it should be possible to store atmospheric carbon dioxide as solid hydrate compounds in subsurface rock pores over many geological periods. This method is volumetrically efficient as each volume of hydrate can accommodate about 175 volumes of gas.

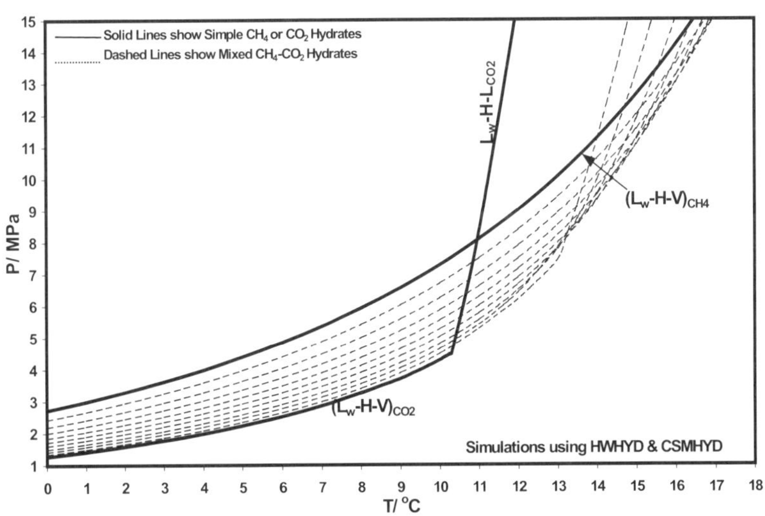

Figure 4. Phase diagram for simple CH_4, CO_2 and mixed CH_4-CO_2 hydrates.

Low temperatures, high pressures and the availability of adequate amounts of hydrate-forming gas are the necessary conditions for hydrate formation that are known to be prevalent in natural permafrost sediments in oil and gas reservoir provinces. Drilling, exploration and production activities for the exploitation of these oil and gas reservoirs have been taking place over many decades. Carbon dioxide sequestration in the sedimentary sections, which have been depleted of oil and gas volumes, would not only re-pressurize these reservoirs and consequently produce oil and gas, but would also safely store the injected carbon dioxide as clathrate hydrates in the later stages. In-place infrastructure can be used to inject the carbon dioxide into subsurface permafrost sedimentary sections (including formations other than oil and gas provinces), which would diminish the cost of CO_2 disposal. Injected CO_2 would be converted into hydrates and occupy the pore spaces of the host rock. The possibility of their atmospheric seepage would be very little as the overlying permafrost layers would act as an impermeable barrier in the form of a cap rock, thus preventing diffusion or upward migration of CO_2 through pores (Duxbury and Romanovsky, 2004; Clarke, 2001). Even in offshore sediments where oil and gas reservoirs exist, there is an increased chance of hydrate accumulations associated with them (e.g. North Sea, Indian Ocean, Taiwan).

Our experimental investigations were aimed at better understanding the processes affecting CO_2 injection into CH_4 hydrate reservoirs, thereby allowing permanent storage of CO_2 in the form of clathrate hydrates.

2. EXPERIMENTAL INVESTIGATIONS OF CO_2 INJECTION INTO CH_4 HYDRATE RESERVOIRS

Injecting CO_2 into subsurface CH_4 hydrate reservoirs would displace some of the CH_4 in the hydrate crystal lattice, converting simple CH_4 hydrates into either simple CO_2 hydrates or double CH_4-CO_2 hydrates. This process could be of particular interest for many reasons. It enables low carbon energy recovery in the form of CH_4 gas, while offsetting the cost of CO_2 transportation and storage. In addition sediment pore spaces potentially re-occupied by the CO_2 hydrate would maintain the mechanical stability of the gas-producing formations, thus preventing the possible hazard of slope failures. The most important part of this production process is that the injected greenhouse gas, carbon dioxide, is permanently and safely stored as clathrate hydrates in subsurface geologic formations over geologic times.

Oghaki et al. (1996) investigated the possibility of methane production from hydrate reservoirs in conjunction with CO_2 injection. Later, Hirohama et al. (1996), Komai et al.(1999, 2003) and Ota et al. (2004) investigated these processes. All of these studies were conducted under bulk conditions (i.e. absence of porous media). The kinetics of replacement was investigated on the micro-scale by Uchida et al. (2001) and Lee et al. (2003). Sivaraman (2003) studied the effect of CO_2 on methane recovery in the presence of porous media, with the test pressure below the CO_2 saturation pressure. McGrail et al. (2004) used micro-emulsions along with CO_2 to investigate similar processes.

2.1 CO_2 sequestration experiments on ultrasonic Rig-1

In the present work experiments into the replacement of CH_4 in hydrates by injected CO_2 were carried out under isothermal conditions (2 ^{0}C) within operating pressures just above the methane hydrate dissociation pressures, in the range of 50 psia. This test mimics reservoir conditions where replacement is observed under excess gas conditions (i.e. very little or no free water). The compositions of coexisting phases have been measured via the gas chromatography (GC) method.

2.1.1 Materials

CH_4 and CO_2 gases used in the evaluation of CH_4 replacement were purchased from Air Products PLC, with a certified purity of 99.995 vol. %. Glass beads were purchased from BioSpec Products Inc. Distilled water was used in all experiments.

2.1.2 Apparatus

The experimental set up shown in Figure 5 was used to investigate the replacement of CH_4 in the hydrate structure by the injected CO_2. It consists of a high-pressure stainless steel cell, a feed system for CH_4, CO_2 and water, and instrumentation for measuring temperature and pressure. The high-pressure cell is constructed of stainless steel with an internal pressure rating of 6,000 psia and an internal volume of 630 cc. The cell is surrounded by a coolant jacket with circulating fluids controlled by a programmable cryostat (-20 °C to 80 °C) that can be kept stable to within ± 0.05 °C. Sediments can be compacted to any given overburden pressure by means of a piston system. In this work half of the experimental cell, as shown in Figure 5, was filled with glass beads. Pore fluid pressure was controlled independently. Temperatures and pressures were monitored by means of a PRT and *Quartzdyne* pressure transducers (accuracy of ± 1.16 psia for 0-20000 psia), respectively. In addition, the pressure could be held constant for long periods of time with a high-pressure Quizix pump.

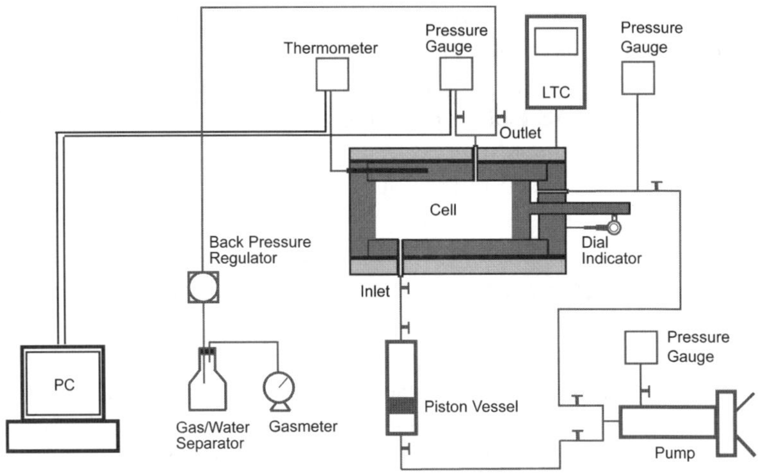

Figure 5. Experimental set-up for the CO_2 sequestration tests.

The analytical work was carried out using a Varian 3600 Gas Chromatograph configured with two thermal conductivity detectors (TCD): one is dedicated for 'parts-per-million' moisture analysis of hydrocarbon gases, while the other conducts conventional gas analysis. Composition measurement of the C_1-CO_2 binary gas samples fall into the conventional gas analysis category, and the separations were performed on a 3m x 1/8" x 2mm i.d. column, using an oven temperature program of 20 ºC to 250 ºC, ramped at 15 ºC per minute, with no isothermal periods. Detector temperature was 300 ºC, and the 6-port 'Valco' injection valve was maintained at laboratory temperature, with no heating applied. For calibration purposes several CH_4-CO_2 gravimetric standards were prepared in-house to cover the range of concentrations encountered; this was done using a small, stainless steel pressure vessel of around 15cm^3 and weighing with a 'Mettler' "Mass Comparator" balance (accurate to ± 0.001g). From these the appropriate relative molar response factors for the CO_2 were calculated across the required range of concentrations, with the CH_4 and CO_2 mole fractions then calculated using the derived response factors.

2.1.3 Experimental procedure

These preliminary tests were carried out under excess gas conditions for their simplicity, though most methane hydrate reservoirs are believed to be under excess water conditions. The test consisted of the following three steps: i) methane hydrate formation; ii) injection of CO_2 in the methane hydrate-vapour system; and finally, iii) measurement of methane recovery through replacement in hydrates by CO_2 over gas chromatograph. Most of the water and methane was utilized for the methane hydrate formation. The test was conducted in a synthetic porous media (glass beads) under thermodynamic conditions where both CO_2 and CH_4 hydrate are stable. Also the system pressure was kept below the saturation pressure of CO_2.

In this experiment, glass beads of 0.5 mm diameter were used to act as the reservoir sediments. Glass beads were placed into the cell to occupy 50% of the total cell volume, and then saturated with distilled water. CH_4 was introduced in the cell at room temperature so as to have the system pressure at a value higher than the three-phase equilibrium pressure of methane hydrate at that temperature. After gas injection, the test system was left at about 20 °C and 2000 psia overnight to stabilize and reach phase equilibrium. The following day, normally, the test system was directly cooled down to the target temperature by a cooling bath to form hydrates. During the initial stages of cooling the system pressure continues to decrease for a number of hours. However, as the methane hydrate begins to form a sharp decrease in cell pressure is observed, indicating that CH_4 is being absorbed from the headspace into the hydrate structure. Growth of methane hydrates was continued even before the injection of CO_2 into the cell. The results of CH_4 hydrate formation tests are presented in Table 1.

Table 1. Test conditions and details of methane hydrate formation.

Test	**T /^{0}C**	**P / psia**	**n_1**	**n_h**	**n_2**	**n_w**
1	2	520 to 485	1.815	0.524	1.291	0.395

In Table 1 n_1 is the moles of methane injected into the vapour phase before hydrate formation while n_2 represents the moles of methane remaining after hydrate formation, n_h shows the amount of the methane (moles) enclathrated in the hydrates, and n_w are the moles of free water remaining after CH_4 hydrate formation and before CO_2 injection in the cell.

The cell temperature was then increased to a value within the CH_4 hydrate stability region so that the cell system could be monitored for changes in vapour phase composition. Methane in the vapour phase was subsequently partially replaced by carbon dioxide in two to three injection cycles while keeping the internal pressure higher than the three-phase (liquid-vapour-hydrate) equilibrium pressure for methane hydrates. On completion of the CO_2 injection step, the pressure of the system was adjusted to the target conditions for the investigation of methane replacement by CO_2 in hydrates. CH_4 hydrates, along with 2% free water, were allowed to soak with the CO_2-CH_4 gas mixture for a certain period. For each equilibrium condition, samples were withdrawn once every 24 to 72 hours, and analyzed with the GC for the possible rise in methane content.

2.1.4 Results and discussions

Initial composition of the gas phase was 30.33% CH_4 and 69.67% CO_2 at the beginning of replacement step. Pressure and temperature conditions were such that CO_2 would remain in gaseous state. During the initial 20 hours of the process the methane recovery rate was high, which may be attributed to the simultaneous enclathratisation of both CH_4 and CO_2. Carbon dioxide has a higher affinity towards occupying cavities in the hydrate structure than does CH_4, and thus most of the free water could be utilised for CO_2 hydrate formation. It is more likely that the rate of CO_2 going into the hydrate state would be higher than CH_4. A small decrease in system pressure was observed during this period, and no rise in temperature was observed (which could be due to large heat capacity of the system). Headspace CO_2 started replacing CH_4 in the hydrate lattice from this point onwards. It is expected that the methane hydrates in the immediate vicinity of the CO_2-CH_4 headspace would be initially converted into CO_2 hydrates. Slow diffusive mass transfer of CO_2 through the porous sediments drove further methane replacement and continued to rise over the next 80 hours. Most likely mixed CH_4-CO_2 hydrates were formed at this stage, which slowed down the exchange during the latter stage. For replacing methane in hydrates, CO_2 took 206 hours to recover 8.3 mole% of methane. This replacement rate in porous media was high compared to the results published by Hirohama et al. (1996). These authors recovered 6.9 mole% of methane in the gas phase over 800 hours when CO_2 was used to replace methane in hydrates in the presence of a methane hydrate - water system (i.e. bulk conditions). The possible reason for the higher replacement rate could be due to the high gas-solid interface available in the porous media.

The replacement reaction is started when the cell pressure is controlled under isothermal conditions. CO_2 probably starts to replace methane in the hydrates within the first layer near the methane hydrate – gas mixture interface. Later replacement of CH_4 in the host hydrate lattice by CO_2 remains a diffusion-limited process. Replacement processes were continued until termination of the tests.

2.2 CO_2 storage investigation with a visual glass micromodel

A Visual Micromodel was used to study the potential for the underground sequestration of carbon dioxide in methane hydrate reservoirs. Phase behaviour of reservoir fluids in porous media have been extensively studied earlier using micromodels (Sohrabi et al., 2000), while its potential application in gas hydrate studies has been demonstrated by Tohidi et al.

(2001). Pore scale studies were aimed towards providing insight into gas hydrate growth from dissolved gas (CO_2-water) and gas hydrate distribution / cementing characteristics of grains in THF-, CO_2- and CH_4-water systems. Later Anderson et al. (2001) produced visual information on phase distribution in porous media, and hydrate-grain cementation for CH_4-water and CH_4-CO_2-water systems. Here we present the results of experiments conducted using the Medium Pressure Micromodel for studying the effect of CO_2 injection on already-existing methane hydrates in porous media.

2.2.1 Experimental investigation

Two micromodel rigs are currently in operation: a medium pressure set-up (1204 psia) and a high-pressure set-up (5947 psia). The central glass micromodels consist of an etched glass base-plate topped with a sealed glass cover plate (Figure 6). Either a geometrically designed network of pores, tubes or reproductions of actual thin sections of real sediments can be used to construct the micromodels by etching with hydrofluoric acid. The cover plate has an inlet and an outlet, which allows fluids to be pumped through the enclosed pore network using small-volume piston vessels or a precision *Quizix* pump (Figure 7).

In both set-ups the glass micromodels are mounted in a vessel that exerts an overburden pressure and are surrounded by coolant jackets controlled by temperature-controlled baths. Temperature is measured by a probe mounted in the overburden cell, and transducers measure pressure on the model inlet and outlet lines. Temperature can be kept stable to within ± 0.05 °C. Temperatures and pressures are monitored by means of a PRT and *Quartzdyne* pressure transducers (accuracy of ± 1.16 psia for 0-20000 psia), respectively.

Magnifying cameras are mounted above the models, with illumination being provided by cold light sources. Because the micromodel pore structure is only one pore thickness deep it is possible to clearly observe phase changes and fluid flow behaviour. Digital video footage is recorded during the experiment, and the pictures presented here are ones recorded either from video footage or camera clippings.

Existing equipment has been employed in the study of a wide variety of hydrate systems pertaining to various scenarios, from hydrates in sub-sea sediments to flow assurance. The technique provides novel visual information on the mechanisms of clathrate growth, micro-scale dissociation and phase distribution, with respect to pressure, temperature, wettability and fluid composition (presence of inhibitors, liquid hydrocarbons, free and dissolved gas).

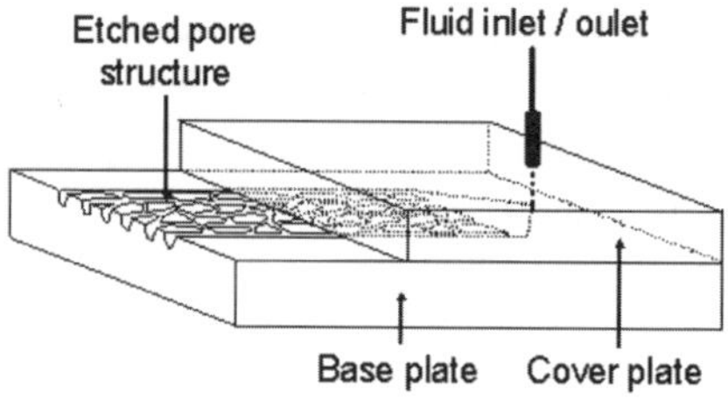

Figure 6. Glass micromodels: pore structure and operational scheme.

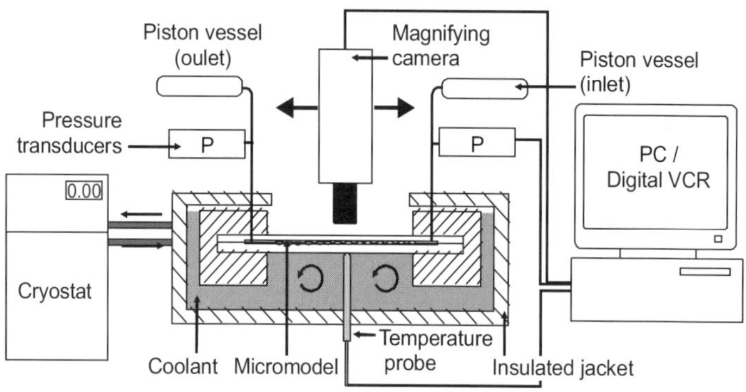

Figure 7. Glass micromodel: experimental set-up.

In all experiments the liquid (water) phase was dyed with methyl blue. Hydrates and gas exclude this dye, thus increasing the contrast between the phases, while it is not known to have any measurable effect on clathrate stability.

2.2.2 Results and discussion

Glass micromodels are used for the visual observation of the CO_2 trapping mechanism. The tests were conducted in the simple methane hydrate - water system, thereby mainly simulating reservoir conditions where excess free water exists in naturally-occurring hydrate-rich sediments.

Formation of small methane hydrate crystals occurs from free CH_4 gas, which is then followed by the injection of CO_2 into the methane hydrate-water system. Figures 8-A through D are representative micromodel images of the phase distribution prior to and after CO_2 injection. This experiment was conducted under temperature and pressure conditions where only simple CH_4 hydrates are stable, but not simple CO_2 hydrates, and CO_2 is in the liquid state. Temperature was kept constant throughout the experiment.

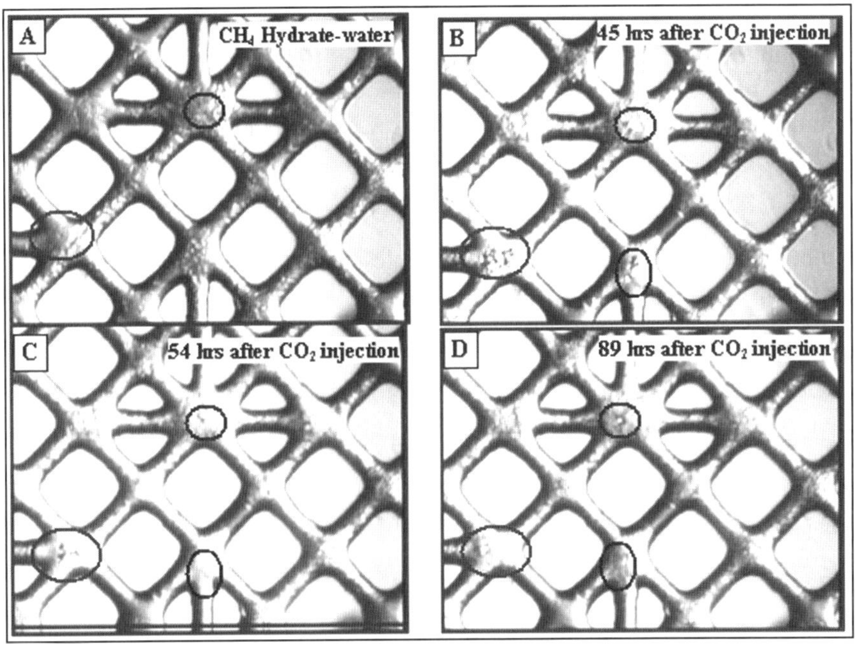

Figure 8. Micromodel images: CO_2 injection and the consequent changes in the already existing methane hydrates morphology.

Another important aspect of the subsurface storage of carbon dioxide is that favourable temperatures and pressures are not the only adequate conditions for the formation of gas hydrates in sub-sea sediments. In case of CO_2 subsurface injection, be it permafrost / seafloor storage or sequestration coupled with additional energy recovery (such as CO_2 injection into methane hydrate reservoirs), hydrate-forming gas concentration (carbon dioxide) should exceed its solubility in water in equilibrium with gas hydrates. This maximum concentration is designated by the solubility of CO_2 in the presence of gas hydrates at the given system pressure and temperature. If the CO_2 concentration in the water is higher than the solubility limit, excessive CO_2 will precipitate from solution producing gas hydrates, thereby forming new gas hydrate cells on the gas hydrate surface.

Methane hydrates were formed from dissolved gas in the region of the hydrate stability zone at 10.8 °C. Figure 8-A shows tiny methane hydrate crystals, whose morphology changes over the next 45 hours in the encircled areas. During this time, CO_2 surrounding the CH_4 hydrate crystals seems to have displaced methane in the hydrate lattice, however its release in the system is hindered by available dissolved/excess CO_2 and by the existing excess water. Once methane is released from the structure it immediately mixes with the dissolved CO_2. Since the thermodynamic conditions are favourable for clathrate hydrate formation the mixed CH_4-CO_2 gas hydrates forms. The time lapsed in this process is very small, about 8 to 9 hours, when we carefully analyze the morphology change in the encircled sections of Figure 8-C over Figure 8-B. Mixed hydrate formation occurred around the locations where methane hydrates were already present before CO_2 injection. CO_2 hydrates were formed from the gas dissolved in the water phase. The solubility of CO_2 played an important role in the formation of CO_2 hydrate from the water phase, which resulted in a concentration gradient and the subsequent diffusion across CO_2 hydrate-water interface. The re-formation of the mixed-gas hydrates was nearly complete after 89 hours (Figure D). However under real reservoir conditions the heat released from CO_2 and mixed CH_4-CO_2 hydrate formation could result in local temperature increases and the dissociation of methane hydrates. This possible phenomenon was not investigated in this phase of the work. Nevertheless, the glass micromodel proved to be an invaluable facility for generating micro-scale data to better understand gas hydrate formation and dissociation mechanisms.

3. CONCLUSIONS

The experimental results of CO_2 sequestration in CH_4 hydrate reservoirs and subsequent CH_4 recovery (under excess gas conditions) were obtained by using two types of experimental set ups. The experimental investigation using the Ultrasonic Rig imitated CO_2 injection in a methane hydrate reservoir under excess gas conditions. An increase in CH_4 composition in the headspace gas samples indicates that CH_4 enclathrated in the host hydrate lattice was partially replaced by the injected CO_2. Injection of CO_2 below its saturation pressure in this test produced 7.97-mole% methane in 206 hours. The experimental results show that the CH_4 recovery rate in porous media was higher than previous results obtained by Hirohama et al. (1996) under bulk conditions.

The tests in the Visual Glass Micromodel showed that CO_2 injection into sediments containing methane hydrates under excess water conditions was converted into either CO_2 hydrates or mixed CO_2-CH_4 hydrates. However, it should be noted that these tests were conducted under constant temperature conditions, which may not be the case in real reservoirs due to the release of heat as a result of CO_2 and/or mixed CH_4- CO_2 hydrate formation.

ACKNOWLEDGEMENTS

The Scottish Higher Education Funding Council provided the financial support for this work, which is gratefully acknowledged. The authors wish to thank Mr Jim Pantling for construction and maintenance of the experimental equipment. One of authors (Mr Jadhawar) wishes to thank the Institute of Petroleum Engineering and the Centre for Gas Hydrate Research for financial support. The authors would also like to acknowledge Ross Anderson, Rod Burgass and Alastair Reid for their help with the experiments.

REFERENCES

Adisasmito, S, Frank, RJ, and Sloan, ED Jr, 1991, Hydrates of Carbon-Dioxide and Methane Mixtures. *Journal of Chemical Engineering Data* **36**(1): 68-71.

Anderson, R, Biderkab, AB, Tohidi, B., and Clennell, MB., 2001, Visual Observation of Gas Hydrate Formation in Glass Micromodels. *EAGE 63rd Conference & Technical Exhibition* — Amsterdam, The Netherlands, 11 - 15 June, 2001.

Clarke, J, 2001, Gas Hydrate Deposits in Oil-Gas Provinces of Russia. *Internet Geology News Letter* No. **80**.

Claussen, WF, 1951, A second water structure for inert gas hydrates. *Journal of Chemical Physics* **19**:1425-1432.

Davidson, DW, 1973, Clathrate Hydrates, in: *Water: A Comprehensive Treatise,* F. Franks, ed., Plenum Press, New York, pp.115-163.

Davy, H, 1811, Phil. Trans. Roy. Soc. London **101**, 1 (Quoted in Sloan, 1998).

Duxbury, N, and Romanovsky, V, 2004, Permanent Sequestration of Emitted Gases in the form of Clathrate Hydrates. *NASA TechBriefs*.: NPO-30256, 55.

Faraday, M, 1823, Phil. Trans. Roy. Soc. London **113**, 160 (Quoted in Sloan, 1998).

Hirohama, S, Shimoyama, Y, Wakabayashi, A, Tatsuta, S, and Nishida, N, 1996, Conversion of CH_4-Hydrate to CO_2-Hydrate in liquid CO_2. *Journal of Chemical Engineering of Japan*; **29**(6):1014-1020.

Hovland, M, and Gudmestad, OT, 2001, Potential Influence of Gas Hydrates on Seabed Installations, in: *Natural Gas Hydrates- Occurrence, Distribution and Detection,* CK Paul and WP Dillon, eds., Geophysics Monograph 124, American Geophysical Union, Washington, pp. 307-315.

Jeffrey, GA, and McMullan, RK, 1967, The Clathrate Hydrates, in: *Progress in Inorganic Chemistry*. Cotton, F.A, ed., Interscience, New York, pp. 43-108.

Komai, T, Kawabe, Y, Kawamura, T, and Yoon, J, 2003, Extraction of Gas Hydrates using CO_2 sequestration. In: *International Offshore and Polar Engineering Conference*, Hawaii USA, 2003.

Komai, T, Yamamoto, Y, and Ogha, K, 1999, Dynamics of Reformation and Replacement of CO_2 and CH_4 gas hydrates. In: *Third International conference on Gas Hydrates*, Salt lake City, Utah, USA.

Kvenvolden, KA, and Lorenson, TD, 2001, The Global Occurrence of Natural Gas Hydrate, in: *Natural Gas Hydrates- Occurrence, Distribution and Detection,* Paul, CK and Dillon,

WP, eds., Geophysics Monograph 124, American Geophysical Union, Washington: pp. 3-18.

Lee, H, Seo, Y, Seo, YT, Moudrakovski, IL, and Ripmeester, JA, 2003, Recovering methane from solid methane hydrate with carbon dioxide. *Angewandte Chemie-International Edition* **42**(41):5048-5051.

Makogon, YF, 1997, *Hydrates of Hydrocarbons*. PennWell Books, Oklahoma.

McGrail, BP, Zhu, T, Hunter, RB, White, MD, Patil, SL, and Kulkarni, AS, 2004, A New Method for Enhanced Production of Gas hydrates with CO_2. In: *Gas Hydrates: Energy Resource Potential and Associated Geologic Hazards*, AAPG Hedberg conference, Vancouver, Canada, 2004.

Oghaki, K, Takano, K, Sangawa, H, Matsubara, T, and Nakano, S, 1996, Methane Exploitation by Carbon Dioxide from Gas Hydrates- Phase Equilibria for CO_2 - CH_4 mixed hydrate system. *Journal of Chemical Engineering of Japan* **29**(3): 478-483.

Ota, M, Abe, Y, Watanabe, M, Smith, RL, and Inomata, H, 2005, Methane Recovery from Methane Hydrate using pressurized CO_2. *Fluid Phase Equilibria*, Vol 228-229: 553-559, PPEPPD proceedings, Available online Nov 2004. .

Ripmeester, J.A., Tse, J.S., Ratcliffe, C.I., and Powell, B.M., 1987, A new clathrate hydrate structure. *Nature*: 325, 135.

Rogers, RE, Yevi, 1996, Gas hydrate theory explains lake Nyos disaster. In: *Proceedings of The 2nd International Conference on Natural Gas Hydrates*, Toulouse, France, 1996.

Sivaraman, R, 2003, The potential role of hydrate technology in sequestering carbon dioxide. *Gas TIPS*. Fall 2003 . http://www.gastechnology.org/

Sloan, ED Jr, 1998, *Clathrate Hydrates of Natural Gases*. Marcel Dekker Inc., New York.

Sloan, ED Jr, 2003, Fundamental Principles and Applications of Natural Gas Hydrates. *Nature* **426**:353-359.

Sohrabi, M, Henderson, GD, Tehrani, DH, and Danesh, A, 2000, Visualization of Oil recovery by Water Alternating Gas (WAG) injection using High Pressure Injection Using High Pressure Micromodels-Water-Wet System. *Proceedings of the 2000 SPE Annual Technical Conference and Exhibition,* Dallas, Texas, 1-4 October, SPE 63000.

Tohidi, B, Anderson, R, Clennell, B, Burgass, RW, and Biderkab, AB, 2001, Visual Observation of Gas Hydrate Formation and Dissociation in Synthetic Porous Media by Means of Glass Micromodels. *Geology* **29**(9): 867–870.

Uchida, T, Takeya, S, Ebinuma, T, and Narita, H, 2001, Replacing Methane with CO_2 in clathrate hydrate: Observations using Raman Spectroscopy. In: *Proceedings of Greenhouse Gas Control Technologies,* East Melbourne..523-527.

von Stackelberg, M, and Müller, HR, 1954, Feste Gas Hydrate II., *Zeitschrift fuer Elektrochem* **58**:25-31 (Quoted in Sloan, 1998).

PART III

NATURAL ANALOGUES OF CO_2 STORAGE

The Use of Natural Accumulations of Carbon Dioxide or other Greenhouse Gases to Predict the Eventual Evolution of Man-Made Storage Reservoirs

WHAT CAN WE LEARN FROM NATURAL ANALOGUES?

An overview of how analogues can benefit the geological storage of CO_2.

J. M. Pearce
British Geological Survey, Keyworth, Nottingham NG12 5GG, UK

Abstract: The timescales needed for the geological storage of carbon dioxide (CO_2) are potentially thousands of years. Therefore, before large-scale underground CO_2 storage can take place, it will be necessary to demonstrate that the processes are well understood, risks to the environment and human populations are low, and environmental disturbances can be minimised. One way of demonstrating that CO_2 can remain trapped underground for geologically significant times is to provide evidence from existing naturally occurring accumulations. These accumulations occur in a variety of geological environments and many can be demonstrated to have retained CO_2 for periods longer than those being considered for CO_2 storage. This fact will build confidence in the concept with non-specialist policy-makers, environmental NGOs and the public. Studies of natural analogues can be used to further validate predictive geochemical and geomechanical models, increasing confidence in these models to predict how CO_2 will behave during and after storage, helping to determine how much of the CO_2 will be permanently trapped through mineral reactions. The results have identified that kinetic reaction data need to be improved. It is unlikely that in reservoirs similar to those investigated here, significant mineral trapping can be expected, except over long geological timescales. Natural accumulations can be used to test methodologies for monitoring CO_2 leakage that may be appropriate for use above repositories, both onshore and offshore, to establish baseline conditions and to monitor sites at the surface during and after storage. Soil gas surveys and analyses of gas leakage rates can define how CO_2 migrates through the near surface environment. Techniques for determining the sealing capacity of caprocks have been tested on natural seals known to retain CO_2 and caprocks from future potential storage sites can be compared with these datasets.

Key words: geological storage, natural analogues, greenhouse gas emissions, risk assessment

S. Lombardi et al. (eds.),
Advances in the Geological Storage of Carbon Dioxide, 129–139.

1. INTRODUCTION

In order to address the global warming threat posed by anthropogenic greenhouse gases, the European Union member states have committed themselves, through the Kyoto Protocol, to an 8% reduction in their greenhouse gas emissions from 1990 levels during the period 2008-2012.

In the medium to long term, however, it is becoming increasingly recognised that reductions of up to 60% will be needed in order to stabilise greenhouse gas levels in the atmosphere. Such a strategy requires several parallel approaches, including more efficient energy use, increased use of renewable energy sources, reduction of reliance on fossil fuels, removal of carbon dioxide (CO_2) from the atmosphere and geological storage of CO_2.

The capture and underground storage of industrial quantities of carbon dioxide is currently being demonstrated at the Sleipner West gas field in the Norwegian sector of the North Sea (see Chadwick et al., this volume). It has been suggested that such geological storage could offer potential long-term storage of significant quantities of CO_2 that would otherwise be emitted to the atmosphere.

Although the storage duration needed may be a few hundred years, so as to reduce atmospheric emissions during the fossil fuel era, it is likely that CO_2 will remain trapped for considerably longer, geological timescales. In order not to burden future societies with unacceptable environmental consequences, it is therefore necessary to have some understanding of site performance over the long term (10^4 to 10^5 years may be an appropriate timeframe).

This requires careful integration of several different approaches. Initial site characterisation will be necessary to provide a model of the geological system, as well as identifying past system behaviour. This site characterisation will include detailed baseline surveys that can be compared with future monitoring of site performance.

The prediction of site performance over long timescales also requires an understanding of, inter alia, CO_2 behaviour within the reservoir, the possibilities and mechanisms of migration out of the reservoir, and the potential impacts of a leak on the near surface environment. The assessments of such risks will rely on a combination of predictive models of CO_2 behaviour, that should in themselves include both a consideration of fluid migration and long-term CO_2-porewater-mineralogical interactions (see for example Stenhouse et al., this volume). Such predictive models must therefore be verified through comparison with laboratory-based experiments that provide carefully constrained thermodynamic equilibrium data on specific geochemical processes (see for example Czernichowski et al., this volume). However, such experiments are necessarily limited to relatively short timescales and so other lines of evidence are required from natural systems where processes have operated for geological timescales.

This paper discusses some of the processes that have been studied at various analogue sites in western Europe during the EC-funded Nascent project (Pearce et al., 2004a) and serves as an introduction to other chapters in this book that provide much greater detail of specific studies (Lombardi et al., Chadwick et al., and Czernichowski et al., this volume; see also Schroot et al., 2005).

2. WHAT CAN ANALOGUE SITES TELL US?

The industrial demonstration projects that are now underway allow injection, reservoir evaluation and short term monitoring technologies to be refined and demonstrated. However, by their nature, such sites are very unlikely to leak. In contrast, some natural sites provide opportunities to study processes controlling how leaks may occur, their potential impacts on near-surface ecosystems and groundwaters and to develop and demonstrate methodologies for monitoring CO_2 leakage.

A key benefit of studying natural geological analogues is also their temporal and spatial scales. Many processes that must be assessed when predicting the performance of a storage site occur over long, often geological timescales and can only partially be simulated in the laboratory or observed in relatively short-term demonstrations. Studies of areas where these processes have been occurring naturally can therefore complement laboratory and demonstration studies (Pearce et al., 1996).

There have been several studies in recent years that have used natural CO_2 systems to understand long-term processes. As briefly reviewed here, and described elsewhere in this volume, Pearce et al. (2004a and references therein) have examined diagenetic processes, migration along fractures, caprock integrity and near-surface processes at several sites across Europe. Shipton et al. (2004) have studied CO_2 migration mechanisms along the Little Grand fault in the northern Paradox Basin. Stevens and Tye (2004) have studied CO_2 sources and trapping mechanisms in three CO_2 fields in the US: St John's Dome in Arizona and Texas, Jackson Dome in Mississippi and McElmo Dome in Colorado. Baines and Worden (2001) compared diagenesis in otherwise geologically similar CO_2-rich and CO_2-poor North Sea oilfields.

Techniques for monitoring CO_2 in the shallow subsurface can also be tested at naturally leaking sites. These methodologies are appropriate for use above repositories, both onshore and offshore, to establish baseline conditions and to monitor sites at the surface during and after storage. Soil gas surveys and analyses of gas leakage rates can define how CO_2 migrates through the near surface environment as well as identify migration pathways (using tracer gases like Rn and He) that may be potential routes for CO_2 escape in the near-surface (Lombardi et al., this volume). In addition, studies of the long-term geochemical interactions within the CO_2 reservoirs have been used to validate predictive geochemical models.

However, studies of analogue systems are not without their limitations. Establishing the timing of reservoir filling can often only be achieved indirectly and so precise 'dating', necessary to ascertain geochemical reaction kinetics, can be difficult. Natural systems are complex and particular care must be taken to identify those processes that are solely attributable to the presence of CO_2.

Although natural CO_2 accumulations provide greater confidence in the potential to store CO_2 for long periods, not all natural accumulations should be considered directly analogous to a storage scenario. For example, many near-surface CO_2-rich springs and seeps occur in volcanic terrains, which may not be tectonically comparable to storage in a sedimentary basin. These sites do provide many opportunities to study near-surface processes and to test monitoring methodologies. When studying natural systems to assess potential impacts on ecosystems, it is important to be aware that ecosystems may have evolved to cope, or exploit, the high CO_2 environment, which would be in contrast to an ecosystem being initially exposed to elevated CO_2 concentrations for the first time.

3. WESTERN EUROPEAN 'ANALOGUES'

Natural CO_2 occurrences are common across Europe (Pearce et al., 2004b and Figure 1) and, although they occur in a wide variety of geological settings, their distribution is principally controlled by the Cenozoic rift system and associated Tertiary volcanism. These occurrences can be classified into:

- *CO_2-rich waters both at depth and in springs.* These are often exploited for mineral waters and occur in a variety of geological settings, with the sources of CO_2 including mantle degassing, volcanic activity and thermal metamorphism of limestones.
- *Dry CO_2 gas vents ('moffettes').* These are associated with Cenozoic rifts such as the Eger and the Tyrrhenian rift systems. They are also associated with hydrothermal fields and Quaternary to Recent volcanic activity, such as in the Eiffel volcanic complex within the Rhenish Massif; the Larderello geothermal field, Italy and, of course, currently active volcanism.

- *CO_2 gas accumulations.* These occur in Cenozoic extensional basins such as within the sub-basins of the back-arc Pannonian Basin and the Florina-Ptolemais-Aminteo graben system. In addition, Mesozoic and Palaeozoic basins subject to Cenozoic tectonism such as the Triassic to Jurassic Southeast Basin of France (see below) and sub-basins within the Southern Permian Basin in Saxo-Thuringia, Germany (see below) can also host CO_2 accumulations.

The natural CO_2 accumulations in the Pannonian basin and the small gas pools in the Southeast Basin may be considered the closest analogues to a storage site in the Western Europe, since CO_2 has been trapped here for geological timescales. Many other known CO_2 accumulations are associated with volcanic regions and as such are not directly analogous to a storage site. However, they do provide opportunities to study near-surface leakage processes and the potential impacts on ecosystems on a scale not easily replicated experimentally.

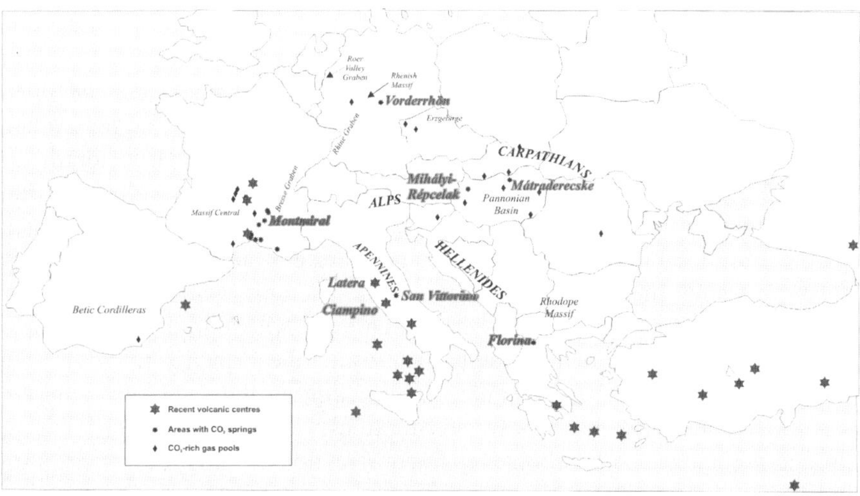

Figure 1. The locations of some natural CO_2 occurrences in Europe and sites studied during the Nascent project.

4. INTERACTIONS IN RESERVOIRS

Understanding the long-term effects of CO_2 on a reservoir is important for several reasons. In certain circumstances, CO_2 may dissolve in the reservoir pore water and react with minerals within the reservoir, which could ultimately lead to long-term trapping through precipitation of carbonate minerals. Our ability to model the geochemical and geomechanical processes that occur in the reservoir, that could ultimately influence its long-term storage performance, can be tested by modelling natural analogues of geological storage. Czernichowski et al. have discussed this in more detail (this volume). In addition, how CO_2 might migrate from the initial storage reservoir through fractures, and information about processes that could occur in fractures in limestones, has been obtained in this study.

Montmiral is one of several small CO_2 gas accumulations in the Southeast Basin of France but it is the only one currently being exploited. CO_2 is currently produced via a single production well as an industrial gas. In order to determine the CO_2-water-rock interactions within the reservoir, it was necessary to reconstitute the original brine composition, which has been evolving to increasing salinity during the lifetime of the CO_2 production. This temporal variability is due to changes in the respective volumes of discharged brine and CO_2-H_2O gas mixture (Pauwels et al., 2004). The brine composition indicates that the CO_2-water-rock system is not at equilibrium.

Diagenetic studies suggest that the introduction of CO_2 into this particular reservoir caused dissolution of feldspar, and a slight increase in reservoir porosity. The observed increase in porosity indicated an open system, i.e. the reservoir has been flushed with fresh CO_2-charged waters (Gaus et al., 2004). Even after contact times of at least hundreds of thousands of years, feldspars are still present. This illustrates that reaction kinetics can be much slower than expected, based on short term kinetic data derived from the literature. Reservoir temperature is an important parameter when assessing the storage capacity of a reservoir, with reaction rates potentially increasing by orders of magnitude where high temperatures prevail. However, to accurately constrain the kinetic rates of the geochemical reactions more detailed information on the reservoir evolution is required. No evidence was found of extensive mineral trapping of CO_2, through precipitation of carbonates. Reconstruction of the brine composition under reservoir conditions through geochemical modelling suggests that one kg of brine contained 0.86 mol CO_2 and 0.087 mol HCO_3. For comparison, the amount of dissolved CO_2 at Sleipner is 1.14 mol/kg H_2O (dependent on the fugacity, and thus temperature and pressure, plus salinity of the water). Therefore, a concentration of 0.86 mol/kg H_2O means, for an average porosity of 8%, that 3.44 g CO_2 will be dissolved in the pore waters of a volume per dm^3 rock. The total mass stored depends entirely on the volume of the reservoir. There is insufficient information on the Montmiral reservoir to estimate the volume saturated with CO_2.

5. LEAKAGE THROUGH CAPROCKS – PROCESSES AND IMPLICATIONS

The caprock or caprocks form the barriers that will prevent CO_2 from migrating out of the reservoir rock. Therefore understanding the sealing capacity of these low permeability rocks for CO_2 specifically will be necessary for site characterisation and estimating storage capacity. Capillary breakthrough measurements with CO_2 have been performed on initially water-saturated caprock samples from natural CO_2 sites (Hildenbrand, et al., 2004). The capillary displacement pressure for CO_2 is lower than that for the other gases on caprocks of equal effective permeability, i.e. for a given caprock permeability, the pressure at which CO_2 enters the caprock is lower than for nitrogen or methane. Considering the strong variability of permeability and capillary breakthrough values in natural caprocks, this effect, although discernible, is not expected to result in a substantially increased risk of capillary leakage when storing CO_2 in depleted methane- or nitrogen-dominated natural gas reservoirs.

Although diffusive loss of CO_2 through caprocks is considered negligible, the rate of potential geochemical "corrosion" of caprocks is determined by diffusion. Laboratory experiments have provided some basic information on the diffusion coefficients of CO_2 in seal rocks and have also shown evidence of chemical interactions of the CO_2 with the minerals (Hildenbrand et al., 2004). The interrelation of diffusive transport and chemical reaction of CO_2 in naturally occurring shale, marlstone and carbonate rocks requires further investigation.

Studies of fracture calcite mineralisation at Montmiral revealed that the latest millimetre-thick calcite generation formed in a CO_2-rich fluid, providing evidence of CO_2 migration above the reservoir. It is not clear how far through the overlying rocks the CO_2 has migrated. The CO_2-rich fluid also contained hydrocarbons that could have been mobilised in a similar manner to that employed during enhanced oil recovery. Geochemical modelling indicates that pressure decreases alone are insufficient to cause calcite precipitation and that a decrease in temperature is also required. The amounts of calcite precipitated however, are very small.

Continuous monitoring over deep-seated fault systems, basalt intrusions, gaps in the overlying cap rock (salt beds) and near former production wells of a natural CO_2 accumulation, in the Vorderrhön area, Germany, showed no evidence for a leak (Teschner et al., 2004). This may be because leaks, if they exist at all, release only small quantities of CO_2, which have to penetrate several aquifers where CO_2 will be naturally attenuated. The quantities of biogenic CO_2 recorded, which were produced in the shallow subsurface during the summer growing period, are much larger.

6. MIGRATION IN THE SHALLOW SUBSURFACE

The shallow subsurface may be the last barrier before CO_2 escape to the atmosphere. A detailed understanding of gas migration in this environment is therefore important to assess risk to human health and the environment.

Detailed soil gas and gas flux surveys, conducted in and around gas vents in several locations in central Italy, demonstrated how gas leaks occur over very small areas, on the order of a couple of metres, but that elevated CO_2 concentrations occur as a large halo around the actual vent, due to lateral migration in the unsaturated zone. This work is discussed in more detail by Lombardi et al. (this volume).

Accumulations of shallow gas (mainly composed of methane) are known to occur in various parts of the North Sea and their study can provide a better understanding of gas migration in the shallow subsurface (Schroot et al., 2005).

7. ASSESSING THE POTENTIAL IMPACTS OF A LEAK

To be able to evaluate site performance, some performance criteria will be needed, which will require a clear understanding of the potential impacts of a leak, in order to best inform policy makers and the public at large.

The San Vittorino Plain to the north of Rome, Italy, is a thinly populated area which has seen two major effects caused by the high-volume leakage of CO_2: modified groundwater chemistry and sinkhole formation. A clear correlation exists between high concentrations of CO_2, anomalous groundwater chemistry and the location of sinkholes. Modified groundwater chemistry resulted in a 5 to 10 times increase in most major and trace elements in surface springs, wells and water-filled sinkholes in the vicinity of a number of known fault structures. It is believed that acidic gases have risen along high-permeability pathways in the faults, causing the dissolution of carbonate and possibly silicate minerals. In spite of the increased ionic content of these altered ground waters a comparison with drinking water standards indicate that they are still considered safe for human consumption.

Closely linked to the formation of the anomalous waters at San Vittorino is the creation of various sub-circular collapse structures, many of which are now filled with water. These features, which pose a hazard for local infrastructure, were likely formed by acidic dissolution and the removal of fine-grained material by flowing ground water, followed by collapse due to low to moderate strength earthquakes. As such, the formation of such features will be highly dependent on the chemical composition of the local geology and the groundwater flow rates. A geochemical monitoring station was developed to monitor possible future collapse events. This monitoring station continually measures the concentration of CO_2 and H_2S dissolved in ground water, processes the data and then sends it via modem to a remote laboratory. The development and application of this technology has shown that such stations would be effective in monitoring dissolved gas concentrations and relatively inexpensive to construct and operate. The same system could be easily modified to monitor soil gas concentrations and installed above a CO_2 storage site (e.g. near an abandoned well) as an early warning system.

Ciampino is a rapidly growing city, close to Rome, which is constructed on the flanks of an extinct volcanic complex. Due to anomalously high heat flow and the occurrence of faults, significant quantities of CO_2 are released to the atmosphere at numerous points throughout the community. Concern has been expressed regarding the safety of the local population, as a sudden CO_2 release once killed 30 cows pastured in the city limits. Soil gas surveys and a limited number of gas flux measurements were conducted throughout the area in an effort to delineate areas of high risk. The soil gas surveys indicated areas with CO_2 concentrations in excess of 70%, along with associated high values of Rn. Despite the fact that a number of new housing developments had been built above these anomalous areas, a small pilot-scale study of indoor gas concentrations yielded CO_2 values that were typically less than 1%. These relatively low values are likely due to the Italian habit of leaving their windows open to allow for an exchange of air. Although risks exist in the Ciampino area, much is being done to minimise any danger to the local inhabitants, including the use of soil gas surveys to develop zoning bylaws and to identify pre-existing residential areas which may be at risk. Education programmes are also underway to explain to the local inhabitants what simple things they can do to greatly lower any risk.

The level of risk to human health related to CO_2 leaking from natural analogues is dependent on many factors that control the generation, migration and accumulation at dangerous levels of this gas. Work conducted partially within the Nascent project on the Latera site has attempted to examine some of the geological phenomena which influence gas emanation hazards, such as the occurrence of faults and the depth to source, using geostatistical analysis and GIS techniques. In addition, soil gas data should allow the generated risk model to be calibrated against the measured gas distribution. It is expected that the method developed during this study could be applied for site assessment of locations being considered for CO_2 storage.

A detailed hydrogeochemical survey of groundwaters was performed across the whole Florina Basin to determine if any changes in water chemistry could be observed in areas containing high CO_2 concentrations. The waters close to the CO_2 field have increased Ca, Mg, CO_3 contents and high total hardness. The remaining water samples show a good quality with some increased content of some elements in a few of those samples.

REFERENCES

Baines, S.J., and Worden, R.H., 2001, Geological CO_2 disposal: understanding the long term fate of CO_2 in naturally occurring accumulations, Proceedings of: *The Fifth Greenhouse Gas Control Technologies Conference (GHGT5)*, 311-315, CSIRO, Collingwood.

Chadwick, A., Arts, R., Eiken, O., Williamson, P., and Williams, G., 2005, Geophysical monitoring of the CO_2 plume at Sleipner, North Sea: An outline review, this volume.

Czernichowski-Lauriol, I., Rochelle, C., Gaus, I., Azaroual1, M., Pearce, P., and Durst, P., 2005, Geochemical interactions between CO2, pore-waters and reservoir rocks: Lessons learned from laboratory experiments, field studies and computer simulations, this volume.

Gaus I., Le Guern C., Serra H., 2004. Natural analogues for the storage of CO_2 in the geological environment (NASCENT). WP4: Modelling of CO_2/fluid/rock interactions. *BRGM/RP-52934-FR*.

Hildenbrand, A., Schlomer, S., Krooss, B.M., Littke, R., 2004, Gas breakthrough experiments on pelitic rocks: comparative study with N_2, CO_2 and CH_4, *Geofluids*, **4**(1):61-80.

Lombardi, S., Annunziatellis, A., Beaubien, S.E., Ciotoli, G., 2005, Near-surface gas geochemistry techniques to assess and monitor CO_2 geological sequestration sites: The use of natural analogue sites in Italy as field laboratories, this volume.

Pauwels, H., Le Nindre, Y.M., Petelet-Giraud, E., Girard, J.P., Czernichowski-Lauriol, I., Gaus, I., Pearce, J.M., Shepherd, T.J., Kemp, S.J., and Bouch, J.E., 2004, Montmiral and Carbogaseous Province of France. In: *Natural analogues for the storage of CO_2 in the geological environment: report on field characterisation including soil gas surveys, characterisation of offshore shallow gas seeps, hydrogeochemistry and diagenetic studies*. J.M. Pearce, ed., British Geological Survey Technical Report, pp. 338.

Pearce, J.M., Holloway, S., Rochelle, C. A., Bateman, K., Wacker, H., Nelis, M., Studlick, J., and Shew, R., 1996, Natural occurrences as analogues for the geological disposal of carbon dioxide. *Energy Conversion and Management*, **37(6-8)**:1123-1128.

Pearce, J.M. (ed.), 2004a, Natural analogues for the geological storage of CO_2. Final report of the Nascent project. British Geological Survey Technical Report. 122 pages.

Pearce, J.M., Czernichowski-Lauriol, I., Lombardi, S., Brune, S., Nador, A., Baker, J., Pauwels, H., Hatziyannis, G., Beaubien, S.E., and Faber, E. 2004b, A review of natural CO_2 accumulations in Europe as analogues for geological sequestration. In: *Geological Storage of Carbon Dioxide*, S. Baines, and R.H. Worden (eds). Geological Society, London, Special Publication, 233, 29-42.

Schroot, B.M., Klaver, G.T., and Schüttenhelm R.T.E., 2005, Surface and subsurface expressions of gas seepage to the seabed—examples from the Southern North Sea, Marine and Petroleum Geology, **22**(4):499-515.

Shipton, Z. K., Evans, J.P., Dockrill, B., Heath, J.M., Williams, A., Kirchner, D., and Kolesar, P.T., 2005, Natural leaking CO_2-charged systems as analogs for failed geologic storage reservoirs. 695-708. In: Thomas, D.C. and Benson, S.M., *Carbon Dioxide Capture for Storage in Deep Geologic Formations*, Volume 2, 699-712.

Stevens, S.H. and Tye, S., 2004, Isotopic analysis of natural CO_2 fields: How long has nature stored CO_2? *GHGT-7: Seventh International Conference on Greenhouse Gas Control Technologies*, Vancouver, B.C., Canada, 5-9 September 2004.

Stenhouse, M.J. Zhou, W., and Arthur R., 2005, Assessment of the long-term fate of CO_2 injected into the Weyburn Field: System-level modeling of CO_2 migration and potential impacts, this volume.

Teschner, M., Brune, S., Faber., E. and Poggenburg, J., 2004, Automatic gas-geochemical monitoring stations in the Vorderrhon region, Thuringia, Germany. In: *CO_2 Leakage mechanisms and migration in the ear-surface.* S.J. Kemp, ed., Final report, WP3, Nascent project, British Geological Survey Commissioned report CR/03/196, 65 pages.

NEAR-SURFACE GAS GEOCHEMISTRY TECHNIQUES TO ASSESS AND MONITOR CO_2 GEOLOGICAL SEQUESTRATION SITES

The use of natural analogue sites in Italy as field laboratories

S. Lombardi, A. Annunziatellis, S.E. Beaubien and G. Ciotoli

Dipartimento di Scienze della Terra, Università di Roma "La Sapienza", Piazzale Aldo Moro, 5, 00185 Roma, Italy

Abstract: As is well known the long-term effects and stability of a man-made CO_2 geological storage facility is very difficult to predict with laboratory or modeling experiments due to the size and long time scales involved. Instead attractive additional sources of information are natural sites where CO_2 produced at great depths is either trapped in porous reservoirs or leaks to the surface. These sites can be considered as "natural analogues" of what may occur over geological time spans within an engineered CO_2 geological storage site. The study of these sites can be subdivided into three broad fields: i) understanding why some reservoirs leak while others don't; ii) understanding the possible effects of CO_2 should it leak into the near-surface environment; and iii) using the leaking sites to develop, test and optimise various monitoring technologies. The present article summaries many of the near-surface gas geochemistry results obtained in central Italy during the EC-funded NASCENT project (Natural Analogues for the Storage of CO_2 in the Geological Environment). These include a comparison of leaking (Latera) and a non-leaking (Sesta) CO_2 reservoirs, detailed soil gas surveys to outline migration pathways, the development of a geochemical continuous-monitoring station to study temporal changes in CO_2 concentrations, and field experiments involving the injection of a gas mixture in the shallow subsurface to outline migration pathways and to understand the behaviour of various gas species based on their different chemical-physical-biological characteristics. Put together this data provides useful information for site selection, risk assessment and monitoring techniques, which is needed if CO_2 geological storage is to become an accepted and widely-applied technology.

Key words: soil gas surveys, geochemical monitoring stations, injection tests, leaking – non-leaking natural CO_2 reservoirs

S. Lombardi et al. (eds.),
Advances in the Geological Storage of Carbon Dioxide, 141–156.

1. INTRODUCTION

The geological storage of anthropogenic CO_2 has long been considered one of the safest and most technologically feasible methods for the short-term reduction of atmospheric loading of this greenhouse gas. In fact, geological sequestration has the potential to provide mankind with a finite period of time in which to change its energy strategy (through improved efficiency of energy production, transfer and use, energy conservation measures and the eventual move from fossil fuels to renewable sources of energy) while at the same time attaining immediate CO_2 reductions

However in order to convince legislators, environmental groups, industry and the public at large that this is the proper path to follow, the method must be shown to be safe and effective over geologically significant periods of time. Traditionally laboratory and modeling experiments have been performed to try and predict the possible fate of various engineered systems, however both approaches suffer from limitations related to scaling up their results to large natural systems over thousands to millions of years. Another approach which can address these difficulties are natural analogues, in other words natural sites where the studied phenomenon has already been occurring over extremely long periods of time. In the case of CO_2 geological storage sites there are numerous locations world-wide where CO_2 is (or has been) produced at depth, with the resulting gas either being trapped in deep reservoirs (analogous to hydrocarbon reservoirs) or is leaking to surface along preferential migration pathways like faults or fractures.

Work performed during the EC-funded NASCENT project (Natural Analogues for the Storage of CO_2 in the Geological Environment) addressed this issue through the study of a number of natural CO_2 sites in Europe (Pearce, 2004). These sites were used as natural laboratories, with the analysis of water, rock and gas samples with various geochemical, geophysical and modeling approaches. The resulting database and associated studies are a unique tool for the better understanding of the various physical and chemical processes involved, as well as for the development of new tools for site assessment and monitoring of any eventual CO_2 geological storage site.

The present article summarises some of results obtained during NASCENT on three natural analogue sites in central Italy, focusing on near-surface gas geochemistry as a tool to understand the micro and macro scale migration of gas, the possible effects of CO_2 leakage on the shallow environment, and the development of tools and instruments for site selection and/or subsequent monitoring. The three site include Sesta, a non-leaking site, and Latera and San Vittorino, two leaking sites. Work involved soil gas and gas flux surveys, as well as an injection test and the development of a geochemical continuous-monitoring station.

2. METHODS

2.1 Soil gas

The probe used to collect soil gas samples consists of a modified ¼ inch, thick-walled, stainless-steel tube. The bottom end of the tube is closed with a sharp drive point, while small lateral holes are drilled along the first 5cm of this end to act as sampling ports. The other end is left open so that a septum can be attached, while two steel cylinders are welded on the outside wall to act as pounding surfaces when installing and removing the probe with a co-axial hammer. The probes were generally placed at a depth of 50 to 90 cm in order to avoid any atmospheric input. Once the probe was in place a septum holder was attached to the open end of the tube, and a needle and 60ml plastic syringe were used to draw off gas. The first two 60ml volumes were discarded to clean the system, and then a 60ml sample was drawn for field analysis of radon while a second was placed in a previously evacuated, 25ml volume, stainless-steel canister for subsequent laboratory analyses.

All soil gas samples were analysed in the field for Rn^{222} using a Scintrex EDA200 alpha scintillometer. All soil gas samples were also analysed for hydrocarbon species (C1-C4), sulphur species (COS and SO_2) and permanent gases (N_2, O_2 and CO_2) using two Fissons 8000-series bench gas chromatographs. All samples were also analysed for helium using a Varian Portatest leak detector (model 938-41); helium concentrations are expressed as ΔHe (i.e. $\Delta He = He_{sample} - He_{atmosphere}$).

2.2 Gas Flux

Gas flux measurements were conducted using the chamber technique, whereby the open end of a container was placed on the soil surface and measurements of CO_2 concentrations were made every second via a connected infra-red analyser. The non-destructive analyser was connected to the chamber via an inlet and an outlet tube such that air was recycled through the chamber via a pump in the instrument, thereby guaranteeing a constant internal pressure during the measurement. In the present study a 13 x 13 x 9 cm Plexiglas chamber was used along with a Draeger Multiwarn infrared analyser equipped with CO_2 (0-25%), H_2 (0-1000ppm) and H_2S (0-

100ppm) sensors. Measurements were conducted for no more than 75 seconds in order to ensure that the natural flux rate from the soil was not affected. Data was then transferred to a PC and the flux rate was calculated using the following formula from Hutchinson and Livingston (1993)

$$f = (C_{final} - C_{initial})\, V / \Delta t\, A$$

Where f is the flux in $m^3\, m^{-2}\, s^{-1}$, V is the volume of the chamber (m^3), A is the surface area measured (m^2) and Δt is the time difference (sec) between the first and last measurements of the CO_2 concentrations (ml/ml). Values were then converted into kg $m^{-2}\, d^{-1}$.

3. LEAKING VS NON-LEAKING CO_2 RESERVOIRS

Due to the fact that the Italian peninsula is still tectonically and volcanically active there are numerous sites where CO_2 produced naturally at depth either becomes trapped in permeable but isolated strata or migrates to surface via faults and is released to the atmosphere. Recent work on leaking (Latera) and sealed (Sesta) sites in central Italy with near-surface geochemical techniques have attempted to explain such different behaviours, despite the fact that both occur in faulted, geothermal areas. Clearly understanding these two types of systems will be highly useful for predicting the safety and long-term isolation of man-made CO_2 geological sequestration sites.

3.1 Latera

The Latera geothermal field is located in the west-central part of the Italian peninsula within the Latera caldera. Although studied since the early 1970s for geothermal energy, the area was well known long before that for carbonate rich springs and CO_2-rich gas vents which inferred the occurrence of CO_2 production / reservoirs at depth. The great age of these gas reservoirs (more than 0.1 Ma old) and the fact that local inhabitants have lived for thousands of years above these reservoirs and associated gas vents means that the Latera site is an excellent natural analogue of what might happen at surface in the unlikely event of leakage from an engineered geological storage site for anthropogenic CO_2.

Locally the geology consists of a metamorphic basement, "Tuscan" limestones, "Ligurian" flysch, volcanic units and post-orogenic sediments. Formation of the local geothermal reservoir was likely due to the combination of compressional tectonics during the Eocene to Miocene (which formed the subsurface structural high within the host Tuscan limestones - Bertrami et al., 1984; Barberi et al., 1984) and subsequent volcanic activity and tensional tectonics (which supplied the necessary fluids and high heat flow for the thermo-metamorphic formation of CO_2 - Duchi et al., 1992). Finally the collapse structures associated with the various stages of local volcanism have either remained open for fluid flow (resulting in springs or gas vents on surface) or have become sealed by fault gouge or secondary mineral precipitation (thereby isolating entire blocks and creating heat/water convection cells).

Extensive work has been performed on the Latera test site by the authors during the NASCENT project, including regional, detailed and highly detailed soil gas surveys, detailed gas flux measurements, electrical tomography surveys, gas vent characterization and aqueous geochemistry. This research has been conducted to outline major gas migration pathways and to quantify the amount of gas reaching the atmosphere. Only some of the gas geochemistry data will be discussed here; the interested reader is referred to the final reports of the project itself (Pearce, 2004) for a more extensive treatment of the data.

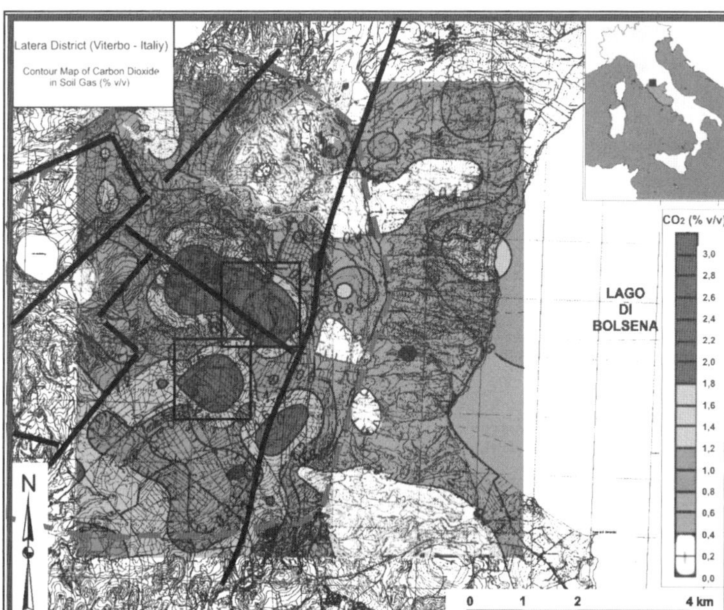

Figure 1. Soil gas CO_2 concentration distribution obtained during a regional geochemical survey in the Latera area. The two small squares in the central portion of the map indicate the location of the two detailed soil gas surveys.

A preliminary, regional soil-gas survey was conducted in the western sector of the Latera caldera with the goal of locating the fault systems which dissect the collapsed area and form the main route of gas migration towards the surface. Sampling was performed in an area of about 65 km^2 using a uniform distribution of about 2 samples/km^2 for a total of 110 samples. The CO_2 contour map shows the spatial distribution of this gas species (Fig. 1). The major anomalies occur in the central and south-western parts of the Latera caldera, in good correspondence with mapped structural elements shown in the figure. Previous geophysical investigations in these areas highlighted a series of NW-SE and NE-SW oriented discontinuities which offset the carbonate substratum and formed the structural high hosting the geothermal reservoir. Elevated values are likely due to gas migration along these geochemically active faults, which result in the clearly anisotropic CO_2 anomalies. This migration of deep (magmatic) gases is also indicated by the presence of He and Rn anomalies in the same area (not shown).

Two subsequent detailed surveys (locations shown in Figure 1) were conducted in areas which exhibited elevated soil gas CO_2 concentrations during the regional survey. As samples were concentrated around gas vents the observed concentrations are significantly higher than those measured in the regional survey. For example during the first detailed survey, CO_2 concentrations of up to 97% and CH_4 values of over 1000 ppm were measured. Radon values (up to 480 Bq/l) also exceeded the amount expected if this radioactive gas were only produced *in situ* via the decay of ^{226}Ra in the local shallow volcanic rocks. Contour maps of these gas species (not shown) highlight two clear circular anomalous zones, one next to a straight creek segment with gas bubbling at its base (fault?) and the other near a geothermal exploration well. Both of these circular anomalies correspond to vent structures that leak significant quantities of CO_2 to the atmosphere. In contrast the second detailed survey, conducted about 1km north of the first in an area defined by electrical tomography anomalies, had slightly lower mean CO_2, CH_4 and Rn values but higher He values (Table 1). Contour maps of CO_2 and He (not shown) give a similar distribution of spotty anomalies aligned parallel to the known NE-SW tectonic elements, with the major anomalies occurring in the area of significant gas vents.

Table 1. Minimum, maximum and mean values for four gas species analysed during the two detailed soil gas surveys (DS1 and DS2).

	CO_2			CH_4			Rn			He		
	Min	Max	Mean	Min	Max	Mean	Min	Max	Mean	Min	Max	Mean
DS1	0.05	97.3	9.1	0.26	1076	57.6	12.7	480	127	-540	310	-88
DS2	0.04	97.2	3.1	n.d.	1120	12.0	14.6	313	109	-913	2943	-105

Finally detailed soil gas profiles were conducted across a known gas vent occurring in the area of the second detailed profile to outline relative gas distributions and the size and form of the vent. In this work samples were collected about every 1 m and analysed for a series of different gas species, both in the field and in the laboratory.

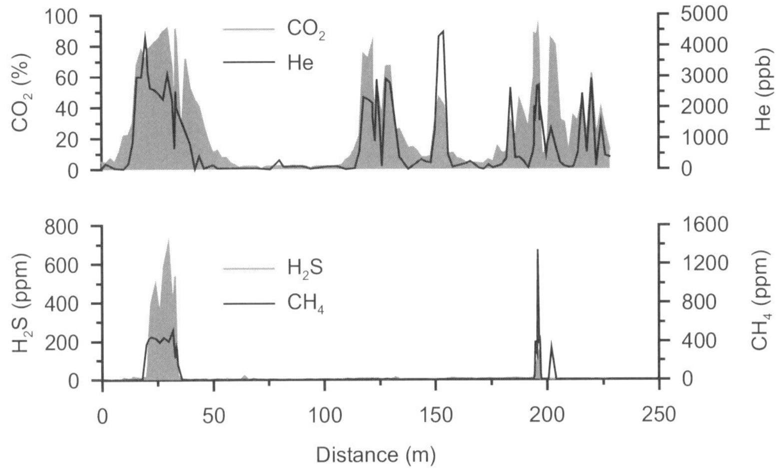

Figure 2. Soil gas concentrations of various species, measured along a profile crossing numerous gas vents in the Latera area.

Figure 2 shows the results for four of these gas types, CO_2, He, H_2S and CH_4. As can clearly be seen in this figure the survey crossed a number of gas vents, however the distribution of the various species is not consistent from one vent to another. For example CO_2 and He are the two gases with the most significant number of anomalies, with both showing a similar spatial and relative concentration distribution. In contrast H_2S and CH_4 are much more spatially restricted, with H_2S showing elevated values in the vent located around 25m while CH_4 is more concentrated in the one at 200m. The different relative distribution of CO_2 / He versus H_2S / CH_4 can be explained in terms of gas abundance and reactivity. For example He is a highly mobile and non-reactive molecule, and thus it is logical that it shows a large number of wide peaks along the entire profile. CO_2, on the other hand, which shows essentially the same distribution, is far more reactive due to its high solubility in water and its involvement in acid-base reactions. The similarity of the reactive CO_2 and the conservative He implies that the high CO_2 flux / concentration (>90%) supplies an excess mass of CO_2 at a rate which is greater than that of dissolution and eventual reaction. In contrast the relatively low concentrations and reactive nature (be it chemical or microbiological) of both H_2S and CH_4 has resulted in the consumption of these gases in the unsaturated zone. Only in the centre of vents, which have sufficiently high flux rates, is it possible for these species to migrate closer to the soil-atmosphere interface, as is observed in the two vents at 25 and 200m.

3.2 Sesta

The Sesta study site is located approximately 260 km to the NW of Rome and about 130 north of the Latera site (described above). In contrast to the Latera caldera, where there is extensive evidence of CO_2 leakage at surface, there are no such manifestations at the Sesta site. Instead the existence of a CO_2 rich gas reservoir at depth was only discovered during deep drilling undertaken for geothermal resource exploration. In terms of CO_2 geological storage it would thus be important to understand the difference between the two sites; in other words why does the Latera reservoir leak whereas the one at Sesta appear to have been sealed for hundreds of thousands of years?

The general area surrounding Sesta is of great interest as a geothermal resource, as it is characterized by one of the highest heat flux rates in Italy. The actual study site occurs within a NNW-SSE trending graben containing a Pliocene clay cover which overlies part of the geothermal field, just to the east of the Larderello geothermal field. Both the eastern and western sectors of the graben are characterized by metamorphic lithologies which are bounded by NNW-SSE and SSW-NNE normal faults. Although delimited by these boundary faults, no active faults and/or fracture systems are observed within the valley. The CO_2 gas cap is hosted within tectonic wedges and confined by a flysch nappe.

Work at Sesta included the same types of gas geochemical surveys which were applied at Latera in an effort to understand if the Sesta site is truly sealed or if it exhibits microseepage from the gas reservoir. Extensive outcrop on the boundaries of the area limited soil gas sampling to the centre of the valley where there was sufficient sedimentary cover; measurements of gas flux were conducted in areas where soil gas samples could not be taken.

A regional soil-gas survey (about 15 samples/km^2) was conducted in the study area, with the main statistical results given in Table 2. The studied gases generally show a uniform distribution and low concentrations which are comparable with other similar Italian basins, although the high CH_4 standard deviation and the strong difference between mean and median for CO_2 and CH_4 highlight the "spot" distribution of some anomalies in the data set. Most of these isolated anomalies, however, can be explained by near surface processes. For example the highest CO_2 and methane values (11% and 890 ppm, respectively) were measured close to a garbage dump and thus are most likely linked to shallow biological degradation of organic matter. Furthermore, the spatial soil gas distribution of the various species does not exhibit any clear structural alignment. This supports the idea that there are no preferential migration pathways, such as faults or fracture networks, in the centre of the valley. Radon, a commonly used fault tracer (e.g. Ciotoli et al., 1998; Ciotoli et al., 1999), has a background value of about 5 Bq/L, which is less than the calculated mean value for Italy (about 20 Bq/L) based on an unpublished database of 15,000 samples constructed by the authors. The most elevated values were observed on the eastern side of valley and are probably linked to the normal bordering graben faults. Finally helium values are almost always less than 200 ppb, with only two significant anomalies (up to 7000 ppb) along the extreme sectors of the valley, again close to the bordering faults.

Table 2. Statistics of some gas species analysed during soil gas surveys at the Sesta study site.

	CH_4	CO_2	O_2	N_2	Rn	He
samples	250	250	250	250	250	250
mean	7.71	0.68	21.18	78.12	5.01	-548.22
median	2.39	0.35	21.39	78.13	2.77	-455.7
min	n.d.	0.04	9.95	57.46	0.37	-2088
max	888.3	11.01	23.36	83.68	58.09	6823.74
quart. inf.	2.07	0.2	20.87	77.06	1.85	-1059.7
quart. sup.	2.82	0.79	21.76	79.21	6.1	-64.44
variance	3427.4	1.35	1.4	4.46	58.85	584893
std. dev.	58.54	1.16	1.18	2.11	7.04	764.78

Gas flux measurements were performed in order to complete the investigation in the areas where outcropping lithological units did not allow for the collection of soil gas samples, as well as to better evaluate the sequestration capacity of the flysch nappe which overlays the CO_2 gas cap. Table 3 compares the main statistical parameters of flux data from various Italian sites (including both Sesta and Latera) located in different geological settings. Although the limited number of measurements at each site does not allow for a true statistical comparison, as expected the volcanic areas (like Latera) have far higher flux rates compared to the sedimentary basins (like Sesta), both because of deep gas production via thermo-metamorphism and the occurrence of permeable fault structures. A good example of this can be seen by comparing the Sesta and Siena data, as the basin-wide averages are very similar, whereas the detailed flux survey values above the Siena Rapolano fault are two orders of magnitude higher (similar to the Latera and Ciampino values). These results highlight the low permeability of the flysch and sediments overlying the gas cap at Sesta and indicate that the site has efficiently sequestered the CO_2 at depth despite the occurrence of normal faults bordering the graben.

Table 3. Statistical comparison of gas flux rates from different sites located in different geological settings throughout Italy.

Geological setting	Site	count	minimum	maximum	median
Sed. basins	Sesta	25	6.0×10^{-9}	7.9×10^{-8}	2.7×10^{-8}
	Siena basin	12	n.d.	5.4×10^{-8}	1.1×10^{-8}
	Siena Rapolano fault	11	1.0×10^{-8}	1.8×10^{-5}	2.6×10^{-6}
Volcanic areas	Latera	25	7.0×10^{-8}	2.0×10^{-5}	9.0×10^{-6}
	Ciampino	5	1.8×10^{-7}	1.3×10^{-5}	3.4×10^{-6}
Tectonic basins	S. Vittorino	8	5.3×10^{-7}	5.7×10^{-5}	2.1×10^{-5}
	Fucino	15	3.0×10^{-9}	9.7×10^{-8}	2.4×10^{-8}

4. INJECTION TEST

An injection test was performed within the NASCENT project which consisted of the injection of a gas mixture into a fault structure and the subsequent monitoring of soil gas and groundwater chemistry in order to outline permeable pathways and flow rates in the subsurface. Such types of gas injection tests have been used by various researchers to better understand the dynamics of gas migration in the shallow environment (Ciotoli et al., in submission; Gascoyne and Wuschke, 1997; Lineham et al., 1996).

The gas injection test was conducted within the area of the Latera study site, at a location which has numerous gas vents and which, based on geophysical surveys and the alignment of the vents, has a fault crossing the site at depth. During this experiment a total of 8000 L of gas (40% Ar, 40% He and 20% CO_2) was injected into a faulted interval between 9 and 12 m below ground surface. Soil gas samples were collected at regular intervals around the injection borehole from a grid of fixed sample points (64) and analysed both in the field and the laboratory for the injected gas species. Ground water samples were also collected from 6 peizometers drilled to 5m depth into the shallow aquifer in order to monitor for breakthrough of the highly soluble CO_2 in the dissolved phase.

Results from the gas injection test highlight the importance of understanding the chemical and physical characteristics of the gases being monitored, as aqueous solubility, gas density and both water- and gas-phase diffusivities play a critical role in the travel times and mass attenuation features of migrating gases. The gas mixture injected at the Latera test site consisted of three rather different species, ranging from the highly insoluble, highly mobile low density helium to the very soluble, reactive and dense gas carbon dioxide, while argon has characteristics which lie between these two extremes. These differences are mirrored in the results obtained during the test, with helium arriving very early at surface within a very small area around the injection well and at high concentrations. In contrast CO_2 arrived much later, had a more diffuse distribution and was observed at much lower relative concentrations than those seen for He. In addition injected helium returned to background values within a very brief period of time, on the order of weeks, whereas the injected CO_2 appears to continue to slowly seep from the system even a month after the experiment had ceased.

A mechanism to explain these results is proposed whereby at early times the more soluble gases are stripped from the gas bubbles, enriching them in helium, and that this gas phase, upon reaching the water table, migrates rapidly through the unsaturated zone to surface due to its low density and high diffusivity. In contrast at later times the groundwater in the bubble flow path becomes progressively more saturated in CO_2, resulting in less transfer to the dissolved phase and more of this gas reaching the water table. Once in the unsaturated zone the more dense CO_2 moves laterally on top of the water table, resulting in the observed lateral dispersion, temporal attenuation and lower concentrations (due to both dilution by soil gas and removal into the aqueous phase) observed in the soil gas surveys. Lateral migration of CO_2 saturated groundwater, and eventual release of this gas to the unsaturated zone via vent activity is another possible mechanism, however this can only explain anomalies observed down gradient from the injection point. In any case, dissolved gas results confirm the importance of CO_2 mass transfer into the aqueous phase, as measured concentrations of dissolved CO_2 in observation piezometers increased by 2 to 3 times background levels in wells located down-gradient from the injection well

5. MONITORING STATION

One way to allay public concern regarding CO_2 storage safety is to develop early warning systems which can monitor gas compositions and concentrations in the near-surface environment. Increased concentrations of CO_2 or other associated gases in these phases, above natural biogenic variations, would then trigger a warning. One such station has recently been developed and deployed in the San Vittorino area of central Italy. Although the monitoring station (jointly funded by the EC NASCENT project and the Civil Protection Agency - Region of Lazio) was originally deployed to monitor for dissolved gas precursors related to CO_2-induced sinkhole formation, it could easily be deployed to monitor soil gas above a CO_2 sequestration site (for example above abandoned wells).

The San Vittorino plain is a triangular-shaped intramontane basin filled with fluvial-lacustrine sediments and local travertine deposits, and is surrounded by mountains formed by carbonate and siliciclastic successions thrusted onto various bedrock lithologies (Centamore and Nisio, 2003). The plain is bordered by normal and/or strike-slip faults and is also cross-cut by other regionally important features. The fault segments that cross the plain are not easy to identify due to the plastic nature of the filling sediments, however evidence of their existence is given by the occurrence of flowing springs, mineralised waters, gas bubbling vents and generally elevated soil gas and gas flux values throughout the valley. It has been the interaction of these deep acidic gases, like CO_2 +/-H_2S, with the spring and ground waters which have resulted in the formation of more than 30 sinkholes in the central and eastern sectors of the plain

The monitoring station uses gas permeable tubing to sample gases dissolved in groundwater via diffusion. As the internal volume of the tube represents a "head space", equilibrium concentrations will be governed by Henry's Constant. The use of gas permeable tubing has been shown to be effective for other gas species (e.g. Jacinthe and Groffman, 2001; Jacinthe and Dick, 1996). The gas permeable tubing is placed in a piezometer (installed above a microgravity minimum in the town of Vasche) and is linked at surface with the monitoring station, which houses CO_2, H_2S and H_2 sensors, a continual power supply and an external data relay system. The in-house developed software system is programmable to sample at different intervals, however a sample time of once every 4 hours was chosen to allow the sampling tube to re-equilibrate with the surrounding water. Recorded data can be down-loaded from the laboratory via modem.

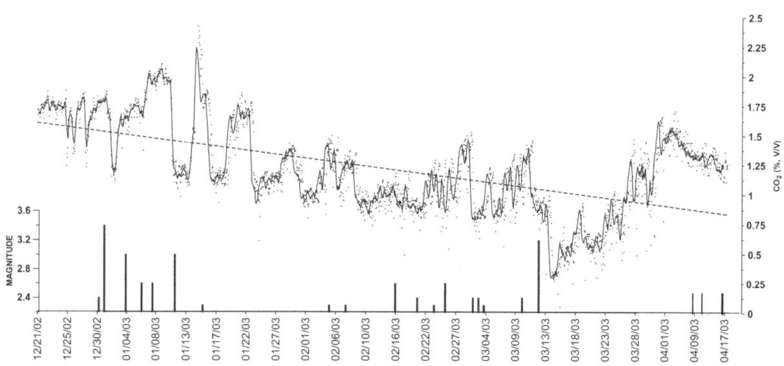

Figure 3. Dissolved CO_2 results from the monitoring station installed in Vasche (after Annunziatellis et al., 2004). Dots indicate individual values while the running average is a black line. The vertical bars are earthquakes near the station registered by the Istituto Nazionale di Geofisica e Vulcanologia in Italy.

The geochemical monitoring station began to collect data on December 20, 2002, and during the intervening period it has collected approximately 2 years worth of dissolved gas data. Some of the data collected during the monitoring period is shown in Figure 3 (note that the values are not dissolved gas concentrations but rather the concentration in the silicone tubing which has equilibrated with the groundwater). The general decreasing trend of the line chart also shows several sharp drops in the signal, which could be linked to different environmental variables (such as water and atmospheric temperatures affecting the sensors) or to the occurrence of earthquakes. The first case cannot be verified because this prototype station was not equipped with temperature sensors. The second hypothesis can be studied by comparing the dissolved gas data with the occurrence of earthquakes in an area surrounding the station site, as shown in the figure. In particular there were about 4 magnitude-3 earthquakes in the region during the monitoring period. The line chart shows abrupt decreases in correspondence with earthquakes which occurred in the Reatini Mts, the Sibillini Mts and at Gran Sasso Mt. There are, however, other CO_2 decreases which cannot be correlated with seismic events in the region and thus there may be multiple processes which contribute to the signal.

6. SUMMARY AND CONCLUSIONS

The present work presents an overview of some results obtained by the authors within the NASCENT project. This EC-funded project examined various sites throughout Europe which can be considered as natural analogues of anthropogenic geological CO_2 storage, due to the accumulation, or leakage at surface, of naturally-produced deep CO_2. The goals of this work can be subdivided into three broad fields: i) understanding why some reservoirs leak while others don't; ii) understanding the possible effects of CO_2 should it leak into the near-surface environment; and iii) using the leaking sites to develop, test and optimise various monitoring technologies.

In regards to the first point, the leaking natural analogue site at Latera provides insight into both macro- and micro-scale gas migration mechanisms, as well as spatial and temporal variations in gas behaviour. This study suggests that: i) gas emanation along "gas bearing" faults is not homogeneously distributed; ii) gas release to the atmosphere is highly localised and spatially restricted; and iii) the near-surface spatial distribution of different deep-origin gases depends on their mobility, reactivity and solubility. In regards to the non-leaking Sesta site, both soil gas and flux values highlight that the clayey cover (which is characterized by recent but sealed faults) prevents gas migration towards surface and that this site has the main features needed for a suitable CO_2 geological storage site (such as clay cover, sealed faults, the lack of seismicity, etc.). A clear comparison of the CO_2 data from various natural analogue study sites is given in Figure 4, with the leaking sites of Latera, Ciampino and San Vittorino showing a markedly different statistical distribution as compared to that for the non-leaking Sesta site.

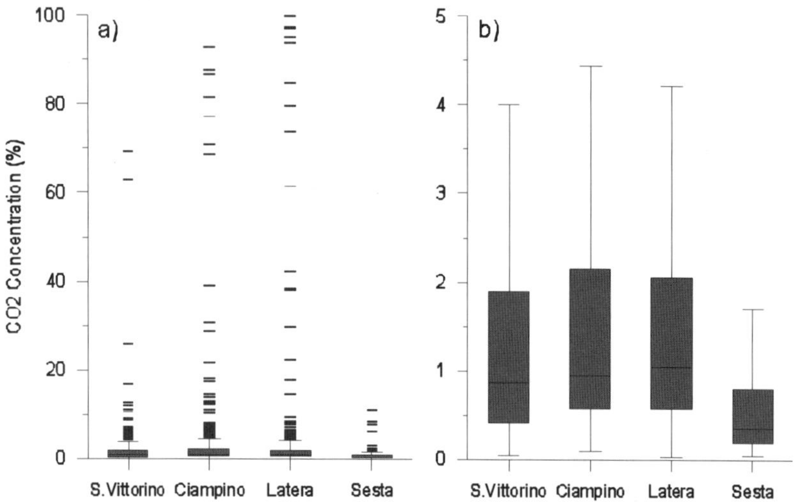

Figure 4. Box plots of soil gas CO_2 data from 4 studied sites. The boxes show the upper and lower quartiles (and median inside the box), the vertical lines show the normally-distributed minimum and maximum while the horizontal lines show outliers. The plot in (a) shows all the data while (b) is the same but without outliers. Note that the three sites showing CO_2 leakage are very similar whereas that the non-leaking Sesta site has significantly lower values.

In regards to the second point, related to near-surface effects of CO_2 leakage, much work was conducted by the authors on this topic however these results were not discussed in detail in this paper. In particular the effect on vegetation (Latera), groundwater composition and sinkhole formation (San Vittorino) and air quality (Ciampino) was observed, however in most cases the effects are highly localized and any risks can be managed with relatively simple measures.

Finally in terms of the third goal, that is the testing of various monitoring techniques on the naturally leaking sites, work during the various soil gas and gas flux surveys have shown that these methods are uniquely adapted to near-surface studies for both site assessments and eventual site monitoring. This was particularly well shown in the gas injection tests, which tested monitoring techniques as well as examined migration mechanisms. Results from the gas injection test also highlighted the importance of understanding the chemical and physical characteristics of the gases being monitored, as aqueous solubility, gas density and both water- and gas-phase diffusivities play a critical role in the travel times and mass attenuation features of migrating gases. This was clearly shown by the highly different breakthrough and migration behaviour of He and CO_2 during the test. Another technique tested during the project was geochemical continuous monitoring stations, which have the potential to monitor for leaking CO_2 in the vicinity of deep injection wells or along known tectonic structures. The development, construction, testing and application of the prototype geochemical station during this work has provided "proof of concept" and has shown that the unit is capable of monitoring dissolved CO_2 in groundwater, sensing variations in concentration and transmitting the data in real-time back to the laboratory for processing and interpretation.

The experience gained during this work on natural analogues in Italy has already been shown to be practical, economical and highly useful during its application to real-life CO_2 geological sequestration sites in Algeria (In Salah EGR project) and Canada (Weyburn EOR project).

ACKNOWLEDGEMENTS

Funding provided by the European Community (NASCENT project - ENK5-CT-2000-00303) and the Civil Protection Agency of the Region of Lazio, Italy, is gratefully acknowledged.

REFERENCES

Annunziatellis, A., Beaubien, S.E., Ciotoli, G., and Lombardi, S., 2004, Allaying public concern regarding CO_2 geological sequestration through the development of automated stations for the continuous geochemical monitoring of gases in the near surface, *GHGT7 - 7th International Conference on Greenhouse Gas Control Technologies*, Vancouver, Canada, 5-9 September 2004, 4 p.

Barberi, F., Innocenti, F., Landi, P., Rossi, U., Saitta, M., Santacroce, R., and Villa, I.M., 1984, The Evolution of Latera Caldera (Central Italy) in the Light of Subsurface Data, *Bull. Volcan.*, **47**:125-140.

Bertrami, R., Cameli, G.H. and Lovari, F., 1984, Discovery of Latera Geothermal Field: Problems of the Exploration and Research., *Seminar on Utilization of Geothermal Energy for Electric Power Production and Space Heating.*, Florence, May 14-17, pp.

Centamore, E., and Nisio, S., 2003, Effects of uplift and tilting in the Central-Northern Apennines (Italy), *Quaternary International*, **101-102**:93-101.

Ciotoli, G., Etiope, G., Guerra, M., and Lombardi, S., 1999, The detection of concealed faults in the Ofanto basin using the correlation between soil-gas fracture surveys, *Tectonophysics,* **299**:321-332.

Ciotoli, G., Guerra, M., Lombardi, S., and Vittori, E., 1998, Soil gas survey for tracing seismogenic faults: a case-study the Fucino basin (central Italy). *J. Geophys. Res.* **103B:** 23781-23794.

Ciotoli, G., Etiope, G., Guerra, M., Lombardi, S., Duddridge, G.A., and Grainger, P., Migration and behaviour of gas injected into a fault in low permeability ground. *Submitted to Quarterly Journal of Engineering Geology and Hydrogeology*.

Duchi, V., Minissale, A., Paolieri, M., Prati, F., and Valori, A., 1992, Chemical composition of natural deep and shallow hydrothermal fluids in the Larderello geothermal field, *Journal of volcanology and geochemical research*, **49**:313-328.

Gascoyne, M., Wuschke, D.M., 1997, Gas migration through water-saturated, fractured rock: results of a gas injection test, *Journal of Hydrology*, **196**:76-98.

Hutchinson, G.L., Livingston, G.P., 1993, Use of chamber systems to measure trace gas fluxes. In: *Agricultural ecosystem effects on trace gases and global climate change*, ASA Special Publication n°55.

Jacinthe, P.A., and Dick, W.A., 1996, Use of silicone tubing to sample nitrous oxide in the soil atmosphere, *Soil Biology and Biochemistry*, **26**(6):721-726.

Jacinthe, P.A., and Groffman, P.M., 2001, Silicone rubber sampler to measure dissolved gases in saturated soils and waters, *Soil Biology and Biochemistry*, **33**:907-912.

Lineham, T.R., Nash, P.J., Rodwell, W.R., Bolt, J., Watkins, V.M.B., Grainger, P., Heath, M.J., and Merefield, J.R., 1996, Gas migration in fractured rock: results and modelling of a helium gas injection experiment at the Reskajeage Farm Test Site, SW England, United Kingdom. *J. of Contaminant Hydrology*, **21**:101-113.

Pearce, J.M. (ed.), 2004, Natural analogues for the geological storage of CO_2. *Final report of the Nascent project.* British geological Survey Technical Report. 122 pages.

GEOCHEMICAL INTERACTIONS BETWEEN CO_2, PORE-WATERS AND RESERVOIR ROCKS

Lessons learned from laboratory experiments, field studies and computer simulations

I. Czernichowski-Lauriol[1], C. Rochelle[2], I. Gaus[1], M. Azaroual[1], J. Pearce[2], and P. Durst[1]

[1]*BRGM, French Geological Survey, 3 avenue Claude Guillemin, BP 6009, 45060 Orleans Cedex 2, France (Email: i.czernichowski@brgm.fr);* [2]*British Geological Survey, Kingsley Dunham Centre, Keyworth, Nottingham, United Kingdom, NG12 5GG.*

Abstract: The degree of reactivity between CO_2, pore-waters and minerals may have significant consequences on CO_2 storage capacity, the injection process, and long-term safety and stability. Geochemical reactions are highly site-specific and time-dependent. They need to be assessed on a site-to-site basis according to best practises by combining numerical modelling and observations from laboratory experiments, field monitoring, and natural analogues. A selection of lessons learned from three European projects about the reactivity of CO_2 with reservoir rocks and cap rocks is presented here for three sites: Sleipner (Norway) and Weyburn (Canada) where more than 1 Mt of CO_2 per year has been injected underground since 1996 and 2000 respectively, and Montmiral, a natural CO_2 field in France.

Key words: CO_2 geological storage, geochemical studies, reactivity, reservoir, cap rock, experiments, modelling, CO_2 trapping, Sleipner, Weyburn, Montmiral.

1. INTRODUCTION

The capture and geological storage of CO_2 is increasingly seen as a viable strategy to reduce the release of greenhouse gases to the atmosphere (IPCC, 2002; OECD/IEA, 2004). The injection of a relatively reactive substance such as CO_2 into the deep subsurface will result in chemical disequilibria and the initiation of various chemical reactions. This paper gives an overview of the potential impact of geochemical reactions, based on

S. Lombardi et al. (eds.),
Advances in the Geological Storage of Carbon Dioxide, 157–174.

work carried out jointly by BRGM and BGS in several European projects since 1993. Both institutes belong now to the CO_2GeoNet European Network of Excellence on CO_2 geological storage that was initiated in 2004 within the EC 6th Framework Programme.

2. WHY IS GEOCHEMISTRY IMPORTANT WHEN ASSESSING CO_2 STORAGE

Depending on the nature and scale of the chemical reactions, CO_2 interactions with reservoir rocks and cap rocks may have significant consequences, either beneficial or deleterious, on injectivity, CO_2 storage capacity, sealing efficiency, and long-term safety and stability (Czernichowski-Lauriol et al., 1996 a,b; Rochelle et al., 2004).

As well as being trapped as a buoyant supercritical CO_2 'bubble' (physical trapping), reaction with formation water can trap CO_2 as a dissolved phase (solubility trapping). Furthermore, reaction of this dissolved CO_2 with minerals in the host formation can result in pH buffering, enhancing solubility trapping due to the formation of dissolved bicarbonate ions and complexes (ionic trapping). Reaction of dissolved CO_2 with certain non-carbonate calcium-rich (or Fe- and Mg-rich) minerals can even trap the CO_2 as a solid carbonate precipitate (mineral trapping), essentially immobilising the CO_2 for geological time periods (Bachu et al., 1994; Gunter et al., 2004). Table 1 summarises trapping mechanisms associated with the deep underground storage of supercritical CO_2 and their relative importance with time, up to long timescales measurable in thousands of years.

Table 1. Possible trapping mechanisms associated with the deep underground storage of supercritical CO_2.

Trapping mechanisms	Increasing importance with time ↓
> **Physical trapping**	
• CO_2 'bubble': dense supercritical CO_2 phase	
> **Chemical trapping**	
• Solubility trapping: $CO_2(aq)$ or $H_2CO_3^0$	
• Ionic trapping: HCO_3^-, $CaHCO_3^+$, $MgHCO_3^+$, $NaHCO_3^0$, ...	
• Mineral trapping: $CaCO_3$ (calcite), $CaMg(CO_3)_2$ (dolomite), $MgCO_3$ (magnesite), $FeCO_3$ (siderite), $NaAlCO_3(OH)_2$ (dawsonite), ...	

All the mineralogical reactions, which are induced by the dissolution of CO_2 into the pore-waters of the reservoir rocks and cap rocks and the resulting pH decrease, are highly complex. They are the result of interdependent thermodynamic, kinetic, flow and transport processes and can occur in the bulk of the reservoir rock or cap rock, but can also occur in fractures. They result in modification of porosity and permeability, which can either hinder the actual injection of CO_2, or aid its migration out of the storage volume. For example, rapid calcite dissolution around the injection well might increase injectivity while rapid mineral precipitation might block flow pathways needed to maintain high injection rates. Dissolution of minerals in the cap rock might result in the formation of flow pathways that might aid CO_2 migration; equally, a self-sealing mechanism could occur which would aid containment. The geomechanical consequences of the chemically-induced changes in fractures and bulk rock petrophysical properties need to be assessed as they will have an effect on long term storage stability and security.

A recurring theme is that geochemical reactions are highly site specific, depending on the precise mineralogy, fluid chemistry, pressure and temperature of the host formation. They are also strongly time-dependent, due to the wide range of reaction kinetics. So it is important to recognise that geochemical investigations need to be carried out on a site-to-site basis.

3. BEST PRACTISE FOR ASSESSING THE GEOCHEMICAL EFFECTS OF CO_2 STORAGE

General recommendations for assessing the geochemical effects of CO_2 storage can be found in the SACS Best Practise Manual (IEA-GHG, 2003), with specific examples being based on experience gained during the SACS and SACS2 projects investigating CO_2 storage at Sleipner.

3.1 Determination of baseline geochemical conditions prior to CO_2 injection

A good geochemical understanding of the system will require knowledge of the 'baseline' conditions of mineralogy and fluid chemistry prior to CO_2 injection. Only with this information can changes due to the presence of CO_2 be assessed. It is important therefore, that sample acquisition be implemented prior to CO_2 injection operations.

As a minimum requirement, the first prerequisite is to have core material from a well intersecting the reservoir formation and the cap rock formation. Core and cuttings from additional wells will improve characterisation, particularly if vertical and lateral heterogeneity is suspected. The second prerequisite is to have at least one formation water sample, carefully collected from the borehole at either the surface or downhole. A special effort should be focused on obtaining both fluid and rock samples from the same location as it is essential to relate fluid chemistry to mineralogy for a good assessment of baseline conditions. Accurate measurements of reservoir temperature and pressures are also essential for accurate experimental and modelling simulations.

3.2 Determination of the geochemical impact of injected CO_2

There are a variety of approaches that can be used that combine numerical modelling and observations from laboratory experiments, field monitoring, and natural analogues. Hence observations can be made at different spatial and temporal scales: from an individual sample scale to field scale; from hours to millions of years; from direct study of the selected injection site to indirect study through natural analogues. Constrained by these three types of observations, numerical models can make predictions from shorter to longer timescales, which is essential for assessing long-term geochemical processes associated with CO_2 storage.

3.2.1 Observations from laboratory experiments

Direct observations of reactions can be achieved through well-controlled laboratory experiments reacting samples of reservoir or cap rock with CO_2 and formation pore-water under simulated reservoir conditions. Geochemical changes can be followed in detail through direct observation and careful sampling. Such investigations are particularly useful for the study of shorter-term processes. Although limited in scale and timeframe, laboratory experiments have the advantage that they can help identify the key geochemical reactions on actual rock material under simulated reservoir conditions, which is very important as such reactions are known to be highly site-specific. They are also helpful to test the ability of geochemical codes to reproduce the experimental observations before using them to make long-term predictions over timescales of up to thousands of years.

Both static batch experiments and dynamic coreflood experiments are useful. Batch experiments can highlight the potential for reaction of samples of reservoir rock or cap rock when in contact with CO_2 and formation pore-water at reservoir temperature and pressure, over different timescales. Core flood experiments are aimed at reproducing open systems where the rock is continuously flooded by pore-water rich in injected CO_2, which represents more closely actual reservoir conditions, especially during injection. They may elucidate inter-relationships between geochemical and hydrodynamic processes, and overall geochemical impact may be higher than in batch experiments.

3.2.2 Observations from field monitoring

The most obvious way to obtain direct geochemical information is by direct sampling of a CO_2 injection site. This approach requires observation boreholes with repeat fluid sampling and permanent in-situ sensors to monitor fluid chemical changes. Sidewall coring, or the drilling of boreholes through the CO_2 'bubble' would be necessary to obtain samples of rock that had been in contact with CO_2 for a variety of timescales. Such an approach is useful in providing highly relevant 'real time' information about a large-scale system.

3.2.3 Observations from natural analogues

This approach utilises relevant information from large-scale sources other than the selected injection site to generate a better understanding of the CO_2 injection system. Natural accumulations of CO_2 exist in many parts of the world and have many analogous features to a CO_2 storage operation, although these may not be directly comparable. As such, these 'natural analogues' can provide much useful information, especially for long-term processes as the CO_2 can, in many cases, be shown to have been trapped for thousands or millions of years. Study of natural accumulations of CO_2 has the advantage of similar physical size and timescale of reaction. This can build confidence in models that predict likely responses of reservoirs to geological storage.

3.2.4 Numerical modelling

Computer simulations are a very useful way to rapidly scope a range of different scenarios. They can predict the effects of adding CO_2 to formation pore-waters, and the subsequent changes in fluid chemistry and reservoir mineralogy. Some codes only deal with static (i.e. not flowing) systems, whereas others couple geochemical reactions and solute transport to produce simulations where mineral precipitation has a direct impact on fluid flow. A wide variety of different geochemical models are available, but they are generally based on the same underlying mathematical expressions. They do differ however, in their emphasis, and the way they handle input and output files. Many come with their own underlying thermodynamic data files, though kinetic data usually have to be supplied by the user. An important benefit of such codes is their ability to make predictions from shorter to longer timescales, which is essential for the long-term assessment of CO_2 storage. However, their output is crucially dependent on the reactions considered and the underlying data files, as well as on the reliability of the conceptual model chosen (which requires a good expertise in geochemical processes). Uncertainties resulting from this can be minimised by carrying out a wide range of sensitivity runs and by comparing modelling results with observations from laboratory experiments, field monitoring and natural analogues, which serve as useful validation test cases.

4. CASE STUDIES FROM THREE EUROPEAN FP5 PROJECTS

A selection of lessons learned from three European projects (5th Framework Research Programme) about the reactivity of CO_2 with reservoir rocks and cap rocks are presented for the following sites:

- Sleipner, North Sea, Norway, where 1 Mt of CO_2 per year has been injected into a deep saline aquifer since 1996 (Research carried out as part of the SACS and SACS2 projects).
- Weyburn, Saskatchewan, Canada, where 1.8 Mt of CO_2 per year has been injected into a depleted oil reservoir since 2000 (Research carried out as part of the European contribution to the IEA Weyburn CO_2 storage and monitoring project).
- Montmiral, Southeast basin, France, where a natural CO_2 field (98% CO_2) has been exploited since 1990 for CO_2 production for various industrial uses (Research carried out as part of the NASCENT project).

The main characteristics of these sites are summarized in Table 2.

Table 2. Main characteristics of the Sleipner, Weyburn and Montmiral sites.

	Depth (m)	T°C	Sal. (g/l)	Reservoir	Caprock
Sleipner	800	37	35-40	Sand	Shale
Weyburn	1400	60	35-110	Carbonate	Evaporite/shale
Montmiral	2400	100	>85	sandstone	Clay/marl

4.1 Long term predictions of clay cap rock reactivity at Sleipner

CO_2 is being injected into the Utsira sand, which is overlain by the 250 m thick Nordland shale acting as a cap rock. Due to buoyancy effects the injected CO_2 moves upward from the injection point and accumulates under the overlying cap rock - under the current conditions CO_2 is in a supercritical state. The CO_2 will partially dissolve into the cap rock formation water and may thus subsequently diffuse upward into the cap rock. It is likely that this will lead to reaction with minerals present in the cap rock. Studies of CO_2 migration in low-permeability caprocks indicate that diffusion is an extremely slow process (e.g. Hildenbrand et al., 2002)

Numerical modelling was used to predict the long term effects of geochemical reactions on the porosity of the cap rock (Gaus et al., 2005). A conceptual diagram for the modelled scenario is shown in Figure 1. The upward diffusion into the cap rock has been modelled via 1D reactive transport modelling combining reaction kinetics and diffusive transport. The modelling code used is PHREEQC (V2.6) (Parkhurst and Appelo, 1999).

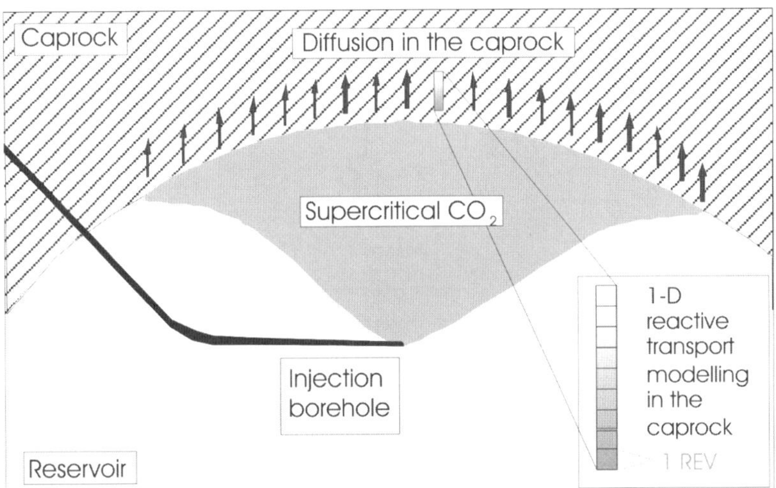

Figure 1. Conceptual model of dissolved CO_2 diffusion and reactivity into the base of the cap rock at Sleipner. "Reprinted from Chemical Geology (Vol 217, 2005, Gaus et al., p 319-337, Copyright 2005) with permission from Elsevier".

At the time of the modelling study, no cap rock core was available. The mineralogy of the Nordland shale cap rock used for the modelling was based on the composition of a selection of cutting samples taken at the base of the cap rock. As no porosity measurements were available, a porosity of 5% was assumed based on observations from the Nordland Shale in UK Quadrant 16, northern North Sea. The initial chemical compositions of the cap rock pore-waters were assessed by equilibrating the water chemistry of the Utsira formation with the caprock mineralogy by means of geochemical modelling. In the absence of more accurate data, this is a common procedure to reconstruct the pore-waters chemistry of argillaceous rocks, given the long time scale over which this equilibrium could be established.

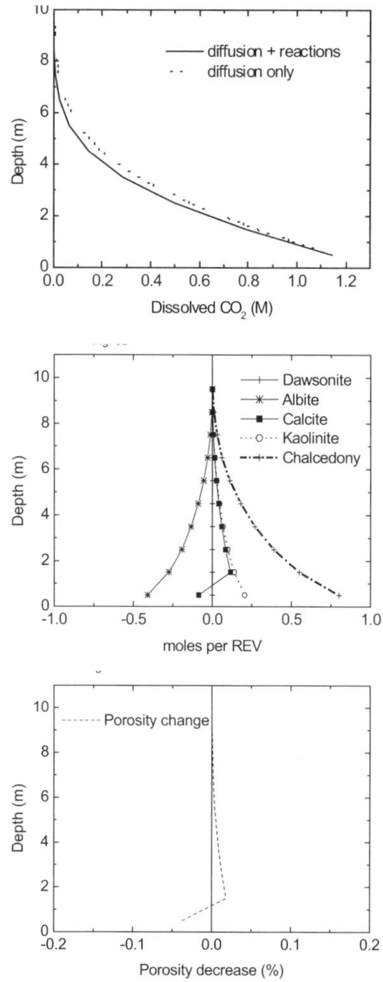

Figure 2. Profiles of impact of dissolved CO_2 in the cap rock after 3000 years at Sleipner. A) diffusion of dissolved CO_2 into the caprock. B) dissolution and precipitation of minerals. C) porosity changes "Reprinted from Chemical Geology (Vol 217, 2005, Gaus et al., p 319-337, Copyright 2005) with permission from Elsevier".

The diffusion of dissolved CO_2, the major mineralogical changes and the porosity changes after 3000 years are shown in Figure 2. The main results can be summarized as follows:

- Diffusion of CO_2 in the cap rock is predicted to be a slow process. After 3000 years, only the lower 10 meters of the 250 m thick cap rock have elevated dissolved CO_2 concentrations. Depending on the reactivity of the cap rock, diffusion can be further retarded because the dissolved CO_2 is consumed by feldspar alteration reactions.

- The calculated porosity change is small and limited to the bottom few metres of the cap rock. A decrease in porosity is predicted, with a maximum of 3% for the most reactive cap rock composition modelled. This will reduce the rate of diffusion and therefore improve the cap rock sealing capacity.
- At the very bottom of the cap rock some carbonate dissolution is expected to occur, leading to a minor porosity increase. However, this is predicted to not migrate further into the cap rock.

These conclusions are based on the assumption that the cap rock is a homogeneous medium and that diffusion is the only mass transport mechanism in the cap rock causing the displacement of dissolved CO_2.

Further investigations should aim to reduce uncertainties and strengthen confidence in the predictions. A core sample of the base of the cap rock is needed to obtain an accurate mineralogical and chemical characterisation and to initiate a series of laboratory experiments. The presence of any fractures that might lead to preferential pathways has to be studied. Information about the behaviour of similar clay cap rocks in natural CO_2 fields has to be searched for. Once information from these studies has been obtained, then numerical simulations need to be revised and a detailed sensitivity analysis should be carried out. Part of this work is currently being achieved through the CO2STORE European project (2003-2005) carried out within the 5th Framework Programme.

4.2 Reactivity of carbonate reservoir rocks at Weyburn

CO_2 is being injected into the Midale Formation carbonate reservoir as part of an enhanced oil recovery project at the Weyburn oilfield, Saskatchewan, Canada. As this is an active and mature oil field, a lot of wells are present, core samples are available and fluid samples can be taken at almost any time. A three-pronged approach has been used by the North American and European teams as part of the IEA CO_2 monitoring and storage project (Wilson and Monea, 2004) to study the impact of CO_2 upon reservoir geochemistry: monitoring changes in actual reservoir fluids from deep boreholes; laboratory experiments to simulate in-situ conditions within the reservoir; and predictive modelling of evolving conditions within the reservoir over the short to long term. A selection of results is presented below.

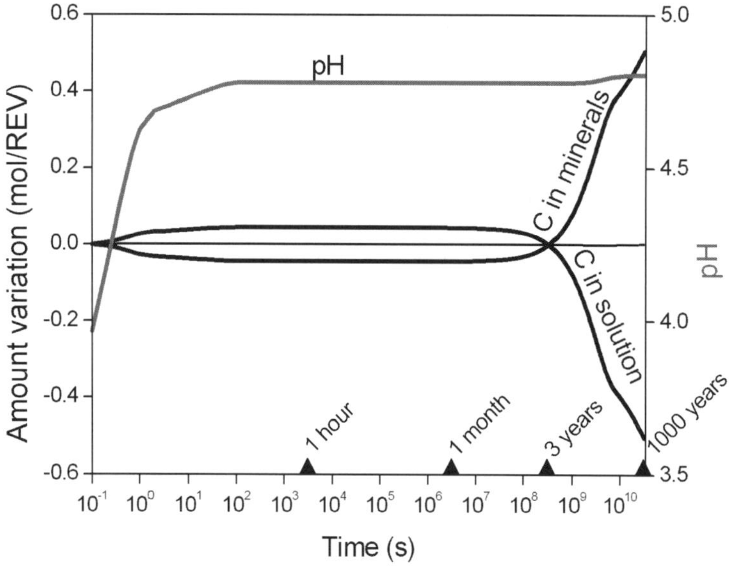

Figure 3. Batch thermo-kinetic modelling reacting a CO_2-rich Weyburn brine (1 molal CO_2) with reservoir minerals at 50°C over 1000 years: Evolution of pH and total amount of carbon in dissolved and mineral forms, expressed in mol/REV (Durst et al., 2003). A REV is a Representative Elementary Volume of 3.85 l corresponding to 1 l of fluid and 2.85 l of rock.

The potential for long term mineral trapping in the Weyburn reservoir has been assessed by numerical modelling. Figure 3 presents the results for the case where a typical Weyburn brine (50°C, salinity 81 g/l) saturated with 1 mole of dissolved CO_2 reacts with the reservoir minerals in a closed system over 1000 years. Calculations were carried out using the PHREEQC code. Key results from this modelling are as follows (Durst et al., 2003):

- A rapid thermodynamic re-equilibration of the CO_2-rich saline solution with respect to carbonates, sulphate and sulphide is predicted. The resulting quick dissolution of carbonates induces an increase in dissolved carbon, whilst the amount of carbon in mineral form decreases.
- slow dissolution of alumino-silicates (Fe-chlorite, illite and K-feldspar), allowing the precipitation of dawsonite and siderite which trap CO_2 in mineral form. After about 3 years of simulated time, the predicted carbon concentration in solution continuously decreases while more and more carbon is trapped in mineral form.

- as the system is closed, the total amount of carbon in the system is constant.
- it is predicted that 50% of the CO_2 initially dissolved in the brine will be trapped in mineral phases (siderite, dawsonite) after 1000 years,
- however, thermodynamic equilibrium is still not reached after 1000 years.
- no significant variation of porosity was predicted as the system was closed (batch calculations). However for open systems with significant advective flow, carbonate reactions could lead to noticeable porosity changes.

Extrapolation to the reservoir scale was attempted by Perkins et al. (2004). Planned CO_2 injection is on the order of 20-25 million tons. In the long term, the Weyburn Midale reservoir has the potential to store all of the injected CO_2 through solubility and mineral trapping mechanisms, but assuming that every part of the reservoir is in contact with dissolved CO_2.

A series of laboratory experiments has been undertaken where samples of well-characterised borehole material from the Midale Formation were reacted with both CO_2 and synthetic reservoir formation waters under simulated in-situ conditions (60°C, 15-25 MPa). The duration of the experiments ranged from 1 week to 6 months. Changes in fluid chemistry, mineralogy, porosity and permeability were measured. The basic layout of the batch experiments is reproduced in Figure 4. Core flooding experiments with a fluid velocity of 15–29 cm/day were also carried out. Interpreted results of these experiments, based on observed changes and numerical modelling using the SCALE 2000 and PHREEQC codes, are described below (Azaroual et al., 2004; Riding and Rochelle, 2005).

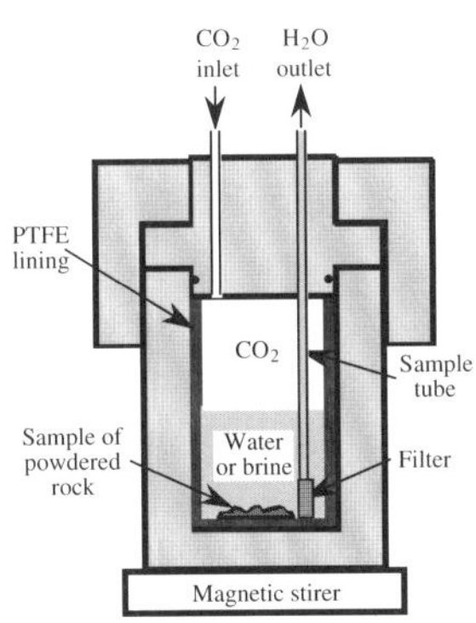

Figure 4. Layout of the batch experiments used to react rock samples with CO_2-saturated brines under reservoir temperature and pressure conditions. Vessel volume is 100-150 ml.

In the closed system batch experiments the Midale lithologies (limestones, dolostones and anhydrite) showed variable amounts of carbonate mineral dissolution, some precipitation of gypsum incorporating the Ca released from carbonate minerals (Figure 5), 'inhibition' of anhydrite dissolution, and poor reactivity of alumino-silicate minerals over the timescale of the experiments.

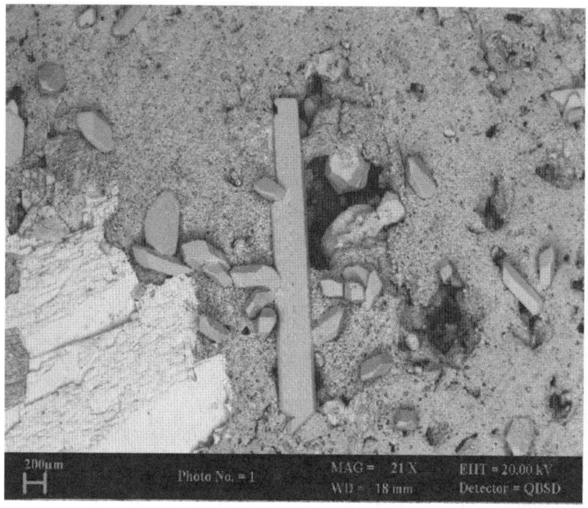

Figure 5. Gypsum crystals up to 2.5 mm long after 8 weeks reaction of Weyburn reservoir rock with CO_2 in batch experiments under reservoir conditions (60°C, 15 MPa) (Riding and Rochelle, 2005).

In the CO_2 flooding tests observed dissolution features were confined to the inlet end, where CO_2-saturated brine entered the sample (Figure 6). Calcite and dolomite showed various states of corrosion. Porosity and matrix gas permeability increased. No precipitating minerals were observed in the CO_2 flooding experiments.

The 1D PHREEQC reactive transport code was used to simulate the long-term evolution of Midale material under CO_2 injection conditions in the Midale Formation of the Weyburn reservoir (150 bar and 54°C). The effect of mineral dissolution / precipitation reaction kinetics coupled to advective / diffusive / dispersive transport processes was predicted over a simulated 10,000 year time period. Varying flow rates were considered, from the undisturbed natural aquifer flow (0.25 m/year) up to higher values close to the injection zone (50 m/year). Simulations show that main reactions occur close to the injection zone. Calcite dissolves at the very entrance of the system whereas a calcium sulphate phase (anhydrite or gypsum) precipitates. Feldspars dissolve over a long distance while dawsonite precipitates. Clays show two reaction fronts, dissolving close to the injection zone but precipitating further on. Chalcedony, a silica phase, has the opposite behaviour. Finally, as a consequence of these dissolution / precipitation reactions, the porosity is predicted to increase. In the case of a flow rate of 0.25 m/year, the initial 20% porosity is expected to increase in the first 10 meters of the 1D column by up to 20.3% after 1000 years, and by up to 25% after 10,000 years.

Figure 6. Inlet end of the Weyburn reservoir rock sample after the CO_2 flooding experiment carried out at GEUS. Numerous small pits are present in the sample surface after the experiment. They represent dissolved calcite grains. (Riding and Rochelle, 2005; photograph reproduced with kind permission of GEUS).

Further investigations will be carried out in the Weyburn Phase II project. More accurate estimates coupling the CO_2 plume displacement, the CO_2 dissolution process as well as the geochemical interactions will be necessary to investigate the ultimate fate of the injected CO_2.

4.3 Long term reactivity of sandstone reservoir rocks at Montmiral

At Montmiral, supercritical CO_2 has accumulated in Triassic sandstones below 2400m depth (Czernichowski-Lauriol et al., 2002; Pearce et al., 2004). Based upon the tectonic setting and the results of petrographic studies, this accumulation occurred in post-Pyrenean times about 15 million years ago. The reservoir is sealed by Domerian and Callovian clay and marl. The Montmiral CO_2 field (98% CO_2) was discovered during gas and oil exploration in the 1960's and is currently used for industrial CO_2 production. The availability of core material and access to reservoir fluids through the

CO_2 producing well have allowed determination of the geochemical processes occurring as a result of CO_2 emplacement. This has been achieved through diagenetic studies, fluid inclusion analyses, isotopic analyses, numerical modelling, and has also been supported by a detailed reconstruction of reservoir evolution (Gaus et al., 2004).

Observed dissolution and precipitation reactions that can be attributed to interactions with injected CO_2 are, in diagenetic order: K-feldspar dissolution; extensive anhydrite cement removal; infilling of secondary porosity in K-feldspars by kaolinite (Figure 7); barite precipitation; mica splaying in open porosity; dolomite/ankerite development as a major pore-filling cement in some horizons; and the corrosion of ankerite. All these processes created an additional secondary porosity of about 2.4%.

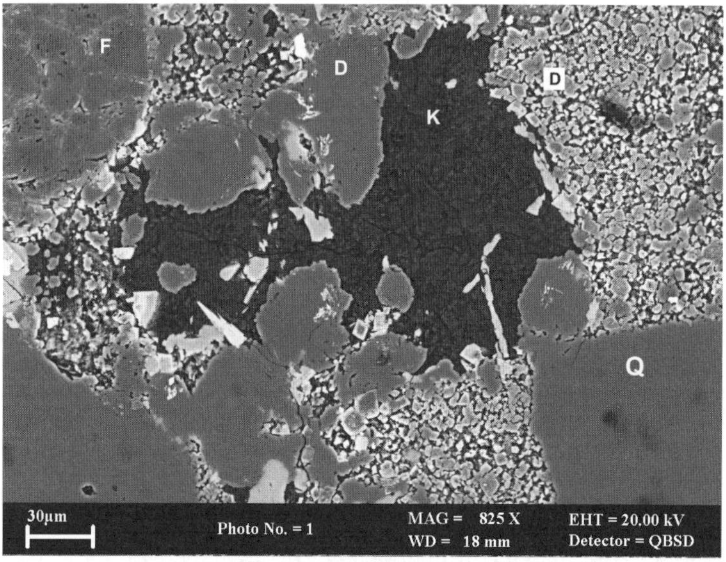

Figure 7. Observed chemical reactivity at Montmiral: large secondary pore in corroded K-feldspar that is part of a coarse lithic grain, which is infilled by fine grained moderately developed kaolinite books (Q = quartz, F = feldspar, D = dolomite, k = kaolinite, Fe = Fe oxyhydroxides, S = siderite, M = mica) (Riding and Rochelle, 2005).

Batch and flow-through geochemical modelling using PHREEQC was applied to reproduce the observed effects, to identify their driving parameters and to assess their impact in terms of potential mineral trapping and porosity changes. It has proved unnecessary to consider reaction kinetics; in the case of Montmiral, elevated temperatures and long contact times between the CO_2 and the rock allowed an assumption of local thermodynamic equilibrium to be used.

The modelling reproduced most of the observed changes that can be attributed to CO_2 ingress. However, in the batch modelling the extent to which these reactions occurred was not reproduced. The modelled reactions do not have a large impact on the mineralogy of the reservoir and an insignificant porosity increase was predicted (0.02 %). The observed reactivity implies that the sediment has been flushed intensively with CO_2-rich pore waters (open system) and that a flow regime in the reservoir must have been in place at a certain point in the reservoir's geological history. 1D reactive-transport modelling showed that an increase of secondary porosity of 2.4% requires that the pore water of the reservoir must be renewed 6000 times with CO_2-rich brine. This means that if we consider flow velocities between 0.1 and 100 meters per year, this would require between 100 and 100,000 years of flushing.

Mineral trapping can therefore be insignificant when considering certain closed systems. However, the presence of a flow regime can enhance mineral trapping, and a mineral trapping capacity of 8 kg CO_2 per m^3 of reservoir rock was modelled. In the case of Montmiral, the predicted trapping phase is dolomite. However, the precipitation of tiny amounts of dolomite is very difficult to confirm by observations. Dawsonite (another potential secondary carbonate mineral) was neither observed nor modelled at Montmiral.

5. CONCLUSIONS

Further knowledge on CO_2 reactivity with reservoir rocks and cap rocks has been gained through studies of three CO_2 storage sites, Sleipner and Weyburn ('industrial' sites), and Montmiral (a natural site).

Modelling of long timescale upward diffusion of dissolved CO_2 through the thick clay cap rock at Sleipner shows that diffusion is a very slow process, and is potentially retarded further by geochemical reactions. The overall net result is that only the bottom few meters of the cap rock adjacent to the reservoir is exposed to chemical reactions. As a consequence, the overall performance of the caprock seal is likely to be good. That said, however, migration of CO_2 may still be possible through pre-existing or newly formed fractures, or via the well bore. Indeed, for the latter, there is still a question regarding the chemical reactivity of CO_2 with borehole completions and the consequent impact on sealing.

Laboratory experiments, modelling and field monitoring at Weyburn show that the carbonate reservoir undergoes mainly dissolution reactions and an increase of porosity, especially close to the injection zone. However, mineral trapping could play a significant role in the long term due to the predicted precipitation of dawsonite. A full 3D dynamic modelling study still needs to be undertaken, where flow, transport and chemical processes are coupled in order to quantify the temporal and spatial changes in CO_2 solubility and mineral trapping, and consequent porosity changes.

Although CO_2 has accumulated over millions of years in the Montmiral sandstone reservoir, observed chemical reactivity is low. Detailed mineralogical and fluid characterisation combined with numerical modelling show that the dissolution of feldspars is the main reaction and that the porosity has increased by less than 3%. Mineral trapping is very limited and is restricted to minor precipitation of dolomite. Dawsonite was neither observed nor modelled at Montmiral, although often referred to in the literature as an important CO_2-trapping mineral.

Overall, therefore, CO_2 reactivity is highly site-specific and time dependent. Investigations need to be based on a very precise characterisation of minerals and fluids from the host formation. They also benefit from the combined outputs of laboratory investigations, numerical modelling, field monitoring and comparisons with natural analogues.

Reactivity can be minor for certain types of reservoirs or cap rocks, but significant for others. It is worth noting that chemical reactions are accelerated by high reservoir temperatures, and can be amplified in the presence of a flow regime. The impact of reactivity, either minor or significant, on the geomechanical properties of the reservoir and cap rock needs to be assessed to investigate long term storage stability and security.

ACKNOWLEDGEMENTS

This paper highlights a selection of lessons learned from several European research projects: '*The underground disposal of carbon dioxide*' (Joule II project, 3rd Framework Programme, 1993-1995), 'SACS' (4th Framework Programme, 1998-1999), 'SACS2' (5th Framework Programme, 2000-2002), NASCENT and WEYBURN (5th Framework Programme, 2001-2004). We thank the European Commission and the industrial and research consortia of the SACS, NASCENT and WEYBURN projects. We thank also the whole consortium of the international 'IEA Weyburn CO_2 Monitoring and Storage Project'.

BGS staff published with permission of the Executive Director, BGS (NERC).

REFERENCES

Azaroual, M., Kervévan, C., Durance, M.-V., and Durst, P., 2004, *SCALE2000 (V3.1) User's Manual; Software for thermodynamic and kinetic calculations*. BRGM Editions, ISBN2-7159-0939-X, 70 p.

Azaroual, M., Durst, P., Czernichowski-Lauriol, I., D., Olsen, D. N,. Springer, N., Rochelle, C. A., Pearce, J., Bateman, K., and Birchall, D., 2004, Geochemical reactions resulting from CO_2 injection into the Midale formation, Weyburn oilfield; a laboratory experimental and modelling study. *GHGT7- 7th International Conference on Greenhouse Gas Control Technologies*, Vancouver, Canada, 5-9 September 2004, 4 p.

Bachu, S., Gunter, W.D., and Perkins, E.H., 1994, Aquifer disposal of CO_2: hydrodynamic and mineral trapping. *Energy Conversion and Management*, **35**:269-279.

Czernichowski-Lauriol, I., Pauwels, H., Vigouroux, P., and Le Nindre, Y.M., 2002, The French carbogaseous province: an illustration of natural processes of CO_2 generation, migration, accumulation and leakage. *Proceedings of the 6th International Conference on Greenhouse Gas Control Technologies (GHGT-6), Vol. 1*, J. Gale and Y. Kaya (Eds), p. 411-416, Kyoto, Japan, Oct. 1st-4th 2002. Elsevier Science Ltd.

Czernichowski-Lauriol, I., Sanjuan, B., Rochelle, C., Bateman, K., Pearce, J., Blackwell, P., 1996a, Analysis of the geochemical aspects of the underground disposal of CO_2: scientific and engineering aspects. in: *Deep Injection Disposal of Hazardous and Industrial Waste*, J.A. Apps and C.F. Tsang, ed. pp. 565-583, Academic Press.

Czernichowski-Lauriol, I, Sanjuan, B, Rochelle, C, Bateman, K, Pearce, JM, and Blackwell, P., 1996b, Inorganic geochemistry. In: *Final Report of the Joule II Project No. CT92-0031: The Underground Disposal of Carbon Dioxide*, S. Holloway, ed., pp. 183-276, British Geological Survey, Keyworth, Nottingham, UK.

Durst, P., Azaroual, M., and Czernichowski-Lauriol, I., 2003, Weyburn database validation problem for predictive geochemical modelling. *Report BRGM/RP-52375-FR*, 46 p. Restricted.

Gaus, I., Le Guern, C., Pauwels, H., Girard, J.-P., Pearce, J., Shepherd, T., Hatziyannis, G., Metaxas, A., 2004, Comparison of long term geochemical interactions at two natural CO_2-analogues: Montmiral (Southeast Basin, France) and Messokampos (Florina Basin, Greece) case studies. *GHGT7 - 7th International Conference on Greenhouse Gas Control Technologies*, Vancouver, Canada, 5-9 September 2004, 9 p.

Gaus, I., Azaroual, M., and Czernichowski-Lauriol, I., 2005, Reactive transport modelling of the impact of CO_2 injection on the clayey cap rock at Sleipner (North Sea). *Chemical Geology,* **217**:319-337.

Gunter, W.D., Bachu, S., and Benson, S., 2004, The role of hydrological and geochemical trapping in sedimentary basins for secure storage of CO_2. in: *Geological Storage of Carbon Dioxide,* S.J. Baines and R.H. Worden eds., Special Publication of the Geological Society of London n°233, pp. 129-145.

Hildenbrand, A., Schloemer, S., and Krooss, B.M., 2002, N_2 and CO_2 gas breakthrough experiments on fine-grained sediments. in: *Poromechanics*, Auriault et al. (eds). Swets & Zeitlinger, Lisse, 445-450.

IEA-GHG, 2003, Saline Aquifer CO_2 Storage project (SACS) Best Practice Manual. *Report Number PH4/21*, 53 p.

IPCC, 2002, Proceedings of the IPCC Workshop on carbon dioxide capture and storage, Regina, Canada, 18-21 November 2002.

OECD/IEA, 2004, Prospects for CO_2 capture and storage. IEA Publications, *ISBN 92-64-10881-5*, 249 p.

Parkhurst, D.L., and Appelo, C.A.J., 1999, User's guide to PHREEQC (version 2) - A computer program for speciation, batch-reaction, one-dimensional transport, and inverse geochemical calculations: *U.S. Geological Survey Water-Resources Investigations Report* 99-4259, 312 p.

Pearce, J., Czernichowski-Lauriol, I., Lombardi, S., Brune, S., Nador, A., Baker, J., Pauwels, H., Hatziyannis, G., Beaubien, S.E., Faber, E., 2003, A review of natural CO_2 accumulations in Europe as analogues for geological sequestration. in: *Geological Storage of Carbon Dioxide,* S.J. Baines and R.H. Worden eds., Special Publication of the Geological Society of London n°233, pp. 29-41.

Perkins, E., Czernichowski-Lauriol, I., Azaroual, M., Durst, P., 2004, Long term predictions of CO_2 Storage by mineral and solubility trapping in the Weyburn Midale Reservoir. *GHGT7 - 7th International Conference on Greenhouse Gas Control Technologies*, Vancouver, Canada, 5-9 September 2004, 4 p.

Riding, J.B., and Rochelle, C.A., 2005, The IEA Weyburn CO_2 Monitoring and Storage project: Final report of the European research team. British Geological Survey, *Research Report RR/05/03*, 54 p.

Rochelle, C.A., Czernichowski-Lauriol, I., and Milodowski, A.E., 2004, The impact of chemical reactions on CO_2 storage in geological formations: a brief review. in: *Geological Storage of Carbon Dioxide,* S.J. Baines and R.H. Worden eds., Special Publication of the Geological Society of London n°233, pp. 87-106.

Wilson, M., and Monea, M., 2004, IEA GHG Weyburn CO_2 Monitoring & Storage Project Summary Report 2000-2004. *Proceedings of the 7th International Conference on Greenhouse Gas Control Technologies*, *Vol.3*, Vancouver, Canada, 5-9 September 2004, 273 p.

STUDY OF NATURAL CO_2 EMISSIONS IN DIFFERENT ITALIAN GEOLOGICAL SCENARIOS

A refinement of natural hazard and risk assessment

N. Voltattorni, G. Caramanna, D. Cinti, G. Galli, L. Pizzino and F. Quattrocchi
Istituto Nazionale di Geofisica e Vulcanologia, Via di Vigna Murata n° 605, 00143 Rome, Italy

Abstract: Natural gas emissions represent extremely attractive surrogates for the study of CO_2-effects both on the environment and human life. Three Italian case histories demonstrate the possible co-existence of CO_2 natural emissions and people since Roman times. The Solfatara crater (Phlegraean fields caldera, Southern Italy) is an ancient Roman spa. The area is characterized by intense and diffuse fumarole and hydrothermal activity. Soil gas flux measurements show that the entire area discharges between 1200 and 1500 tons of CO_2 a day. In proximity of Panarea island (Aeolian islands, Southern Italy), a huge submarine volcanic-hydrothermal gas burst occurred in November, 2002. The submarine gas emissions locally modified seawater pH (from 8.0 to 5.0) and Eh (from +80 mV to -200 mV), causing a strong modification of the marine ecosystem. Collected data suggest an intriguing correlation between the gas/water vent location/evolution and the main local and regional faults. CO_2 degassing also characterizes the Telese area (Southern Italy), one of the most seismically active segments of the southern Apennine belt with the occurrence of five large destructive earthquakes in the last 500 years. Geochemical surveys in this area reveal the presence of high CO_2 content in ground-water. Carbon isotopic analysis of CO_2 reveal its deep origin, probably caused by the presence of a cooling magmatic intrusion inside the carbonate basement. All the above mentioned areas are constantly monitored since they are densely populated. Although natural phenomena are not always predictable, local people have nevertheless learnt to manage and, in some case, exploit these phenomena, suggesting significant human adaptability even in extreme situations.

Key words: natural emissions; carbon dioxide; soil gas; dissolved gas; flux measurement.

S. Lombardi et al. (eds.),
Advances in the Geological Storage of Carbon Dioxide, 175–190.

1. INTRODUCTION

Natural degassing phenomena can be studied as "natural analogues" in the frame of geological storage and sequestration of anthropogenic CO_2 emissions, especially when the risk of possible leakage at surface is taken into account with potential consequences in the biosphere.

Active and quiescent volcanic areas release high amounts of CO_2 to the atmosphere (Chiodini et al., 1998) from both active craters, as plumes and fumaroles, and along the flanks of volcanic edifices as diffusive soil emanations (Allard et al., 1991a; Baubron et al., 1990).

The two basic processes which are recognised to drive soil gas movement through rocks and sediments are *diffusion* and *advection*.

If transport through a stationary medium take place by diffusion, the steady state diffusive flux Φ_d is proportional to the concentration gradient, $dC/d\lambda$, as expressed by Fick's first law:

$$\Phi_d = -\nu D(dC/d\lambda) \qquad (1)$$

where ν and D represent soil porosity (i.e., the fraction of pore volume to total soil volume) and the diffusion coefficient, respectively, and the minus sign indicates that gas moves from points of high concentration towards points of low concentration (or partial pressure). On the contrary, advection involves the movement of matter due to the action of a force, i.e., a pressure gradient $dP/d\lambda$. Advective flow Φ_a is described by the well-known Darcy's law:

$$\Phi_a = (k/\mu)(dP/d\lambda) \qquad (2)$$

where k is the specific permeability and μ is the viscosity of the fluid.

Although equation 2 was experimentally derived for the steady flow of liquids in porous media, it has also been extensively used to describe the advective flow of compressible fluids in porous media (Gurrieri and Valenza, 1988). For the sake of correctness, the steady advective flow of compressible fluids is described by comparatively complex, well-known equations (e.g. Scheidegger, 1974). Because of mathematical complexities many workers have decided to study the steady flow of gases assuming that they are incompressible. Although this assumption might seem unreasonable, it is justified when the pressure gradient is comparatively small.

Exhalation is the process that transfers gas from the soil to the atmosphere via the already-described mechanisms of diffusion and advection. Diffusive gas exhalation takes place due to the different concentrations of that gas in the soil and atmosphere. Advective exhalation exists when the gas in the soil has a higher total pressure than that in the atmospheric (eg. barometric pumping). Exhalation can be measured in terms of flux from the soil, i.e. the quantity of gas per unit area per unit time (Φ_{max} = Kg $m^{-2}s^{-1}$; Φ_{volume} = $m^3m^{-2}s^{-1}$). Generally, mean values for CO_2 exhalation are 3.7 x 10^{-7} $m^3m^{-2}s^{-1}$ (de Jong and Schappert, 1972) or 0.4 – 4 x 10^{-7} $m^3m^{-2}s^{-1}$ (Kanemasu et *al.*, 1974).

2. SAMPLING AND ANALYSIS

Many techniques are used to sample soil-gases, waters for dissolved gas determination and gas/water from submarine gas emission points. Samplings were performed using well tested methods (Lombardi & Reimer; 1990; Bertrami et al., 1990; Hutchinson and Livingston, 1993; Matthias et al., 1980).

2.1 Soil-gases

Soil gas samples are collected from shallow point sources. The soil-gas survey is generally carried out in the summer during a period of stable meteorological conditions. A 1m hollow steel probe with a conic point at the bottom and a sampling port on top is inserted to a depth of 0.5m below the ground surface. Two 50 cc samples of soil gas are extracted with a syringe for cleaning the probe, then a soil-gas sample is extracted and stored in an evacuated 25 ml steel cylinder for laboratory analysis (He, H_2, O_2, N_2, CO_2, CH_4 and H_2S) by means of a *Perkin-Elmer AutoSystem XL* gas chromatograph.

Furthermore, soil-gas flux measurements have been performed in situ by means of an accumulation chamber (volume: 50 litres) set on the ground in such a way as to eliminate the input of atmospheric air. Every five minutes a gas sample is extracted from an external septum located on the top of the chamber and analysed in loco using a portable gas chromatograph. These measurements result in an accumulation curve which can be used to calculate soil gas flux.

2.2 Dissolved gases in water

The method used is based on the partitioning equilibrium of gases between liquid and gas phases. It is possible to derive the initial concentrations of the gases dissolved in the liquid phase from concentrations in the gas phase, using the partitioning coefficients of the different species (Capasso & Inguaggiato, 1998). Samples are collected in glass flasks sealed with gas-tight rubber/teflon plugs. In the laboratory, 10 cc of pure argon are injected into the flask drawing out the equivalent water volume through a needle, thus leaving the apparatus at atmospheric pressure. Since dissolved gases can be under or over saturated in natural water samples it is necessary to restore atmospheric pressure inside the flask before extraction. This is obtained by adding or removing a few ml of water by connecting the flask upside down to a tube filled with water open to the atmosphere. After restoring the sample to standard conditions (25°C, 1 atm) and shaking it for 5 min, a portion of gas is drawn for gas chromatography analysis by means of a *Perkin-Elmer AutoSystem XL* gas chromatograph.

2.3 Submarine fluid sampling

To collect gas samples a plastic funnel (30 cm diameter) is placed (with a 12 kg ballast around the lower ring) on the gas vent. Then the funnel is connected to a glass flask filled with air at a pressure which is greater than the expected hydrostatic pressure to prevent sea water from entering the sample. This system allows one to change flasks underwater in order to collect multiple samples for each vent. The same device is also used to collect gas samples in NaOH-filled flasks. In this case the external section of the flask valve is filled with distilled water to avoid sea water contamination. A traditional glass sampler, filled with water collected close to the vent, is used for dissolved gas. In regards to hot water sampling from underwater springs, the main difficulty is due to the low water flux. In order to avoid seawater contamination, a evacuated glass flask connected to a steel hose is used. Once the hose is fixed in the spring outlet, a 60 cl syringe connected by a 3 way valve, is used to expel the seawater from the hose and then fill the evacuated flask with only the geothermal water. In this case, seawater contamination is as low as reasonably achievable.

3. GEOLOGICAL SETTINGS OF STUDIED AREAS

Three Italian sites are considered here to show different geological scenarios with natural CO_2 gas accumulations and emissions. Both the Panarea (Aeolian islands, Southern Italy) and Solfatara (Phlegraean fields, Southern Italy) areas are characterised by volcanic activity, whilst the Telese area hosts one of the most seismically active segments of the southern Apennine belt.

3.1 Panarea island (Aeolian islands, Southern Italy)

Panarea Island belongs to the Aeolian Arc, a volcanic structure extending for about 200 km along the north-western side of the Calabro-Peloritano block. This is a fragment of the Hercynian-Alpine orogenic belt (consisting of various types of metamorphic, sedimentary and intrusive rocks) that detached from the Corsica-Sardinia block and migrated south-eastward to its present position during the opening of the Tyrrhenian Sea. The exposed volcanic activity took place entirely during the Quaternary, from about 400 ka to the present (Gillot, 1987). The distribution of volcanoes is strongly controlled by regional fault systems which are oriented E-W, NW-SE and NE-SW. Seismic studies reveal a crust of about 20-25 km beneath the Aeolian arc (Gasparini et al., 1982; Barca and Ventura, 1991, and references therein) which indicates mantle upwelling.

The volcano of Panarea is a cone shaped edifice, rising from a depth of 1700-1200 m below sea level up to 421 m above sea level; its maximum diameter is about 20 km (Gabbianelli et al., 1993). The emergent rocks form the main island of Panarea and a number of islets. The submerged summit of the volcano is characterised by a broad and almost flat surface covering about 50 km^2 at an average depth of 100-150 m below sea level.

3.2 The Solfatara area (Phlegraean fields, Southern Italy)

The ancient Romans were aware of the Solfatara of Pozzuoli (located near the centre of the Phlegraean caldera) since Imperial times. Strabone (66 B.C. -24 A.C.) gives the most ancient written testimony, indicating it in his work "Strabonis geographica" with the name "*Forum Vulcani*", which means dwelling of the god Vulcano, entrance to Hades. The Solfatara is a sub-circular depression with a 12 km diameter. It originated about 35 ka BP after the eruption of the Campanian Ignimbrite (Rosi et al., 1984). The Phlegraean Fields magmatic system is still active, as the last eruption occurred in 1538 A.D. at Monte Nuovo. Faults affecting the Phlegraean Fields caldera follow two preferred strikes, NW-SE and NE-SW, that also affect the Campanian Plain and the inner sectors of the Apennine belt (Hyppolite et al., 1994; Orsi et al., 1996). The most impressive thermal manifestations (including fumaroles, mud pools and vigorously boiling pools) are located in the Solfatara area. The *Bocca Grande* (Large Mouth) is the name of the main fumarole and it has water vapour temperatures reaching approximately 160°C. Within the mouth, the vapour condensate contains salts such as realgar (AsS), cinnabar (HgS) and arsenic sulphide (As_2S_3) which give a yellow-reddish colour to the surrounding rocks. In the middle of the Solfatara area a bubbling mud pit (called "La Fangaia") is made up of rainwater and vapour condensation.

3.2.1 Bradyseism phenomenon in the Solfatara area

From 1969-72 and 1982-84 the inhabitants of the Phlegraean area, and Pozzuoli in particular, were witness to and victims of a phenomenon where the earth's surface rose; within a few months the ground level rose by 3.5 metres. This phenomenon is called *bradyseism,* which is literally a slow movement of the earth's surface as opposed to the fast movement due to an earthquake. Deformation connected with bradyseism is due to chemical and physical variations linked to the evolution of the volcanic system in the Phlegraean Fields; this produces an increase in temperature and pressure in the rock above the magma chamber. The increase in pressure translates into a variation in stresses in the rock which leads to the characteristic lifting of the ground surface. The place which shows the best evidence of Phlegraean bradyseism over the centuries is the *macellum* (a market of Roman period, better known as the *Temple of Serapide*) situated close to the port of Pozzuoli (Fig. 1). On the columns of this building there are holes made by lithodomes (sea molluscs living in coastal areas between high and low tide line) which provide evidence of the variations in ground level relative to sea level from the IV century A.D. onwards.

Figure 1. The "Macellum" (Temple of Serapide) within the Phlegrean Fields: at the left a view during the 1960's showing that the area was below sea level. At the right, the same site today. Remains of the ancient buildings provide evidence of the ground level variations due to bradyseism thanks to holes made by lithodomes, sea mollusks living between high and low tide line.

3.3 The Telese area (Southern Italy)

The studied area belongs to the Sannio-Matese zone, one of the most active seismic regions of the southern Apennines, where destructive earthquakes have occurred several times in the past. At the present, the area is characterised by low-energy sequences (M_{dmax}= 4.1) but not much is known about the tectonic structures responsible for such seismicity (Vilardo et al., 2003). The Sannio-Matese area is located in the frontal portion of the Apennine fold and thrust belts, at the junction between the northern and southern subsidiary arcs. At this location different tectonic styles are superimposed as a result of Neogene and Quaternary thrust, strike-slip and extensional tectonics (Di Bucci, 1995; Corrado et al., 1997). Two geological units characterise the Sannio-Matese area:

1. the Molise Unit having highly variable structural trends, ranging from E-W in the south to N-S in the north;
2. the Sannio Unit is interpreted as a rootless thrust sheet originally deposited in a deep basin area located internally alongside the Latium-Abruzzo carbonate platform (Patacca et al., 1992). According to unpublished surface and subsurface data (Corrado et al., 1998) the entire thickness of the Sannio Unit is estimated to be about 3500 m.

Carbonates in this area are intensely fractured and host very important aquifers. They are locally affected by huge CO_2 and H_2S fluxes that play a very important role in defining the degree of water-rock interaction and, consequently, the final water chemistry. These gases rise along well known local and regional faults that cross the area. Moreover, near the town of Telese, extensive outcropping travertine layers are found, testifying to past geochemical processes acting in the area. Moreover, the Telese area is characterised by the presence of an ancient spa / hot spring tradition. The sulphur spring formed after the 1349 earthquake causing the mineral water discharge.

4. RESULTS AND DISCUSSION

4.1 Panarea island

In proximity of Panarea island (Aeolian islands, Southern Italy) a huge submarine volcanic-hydrothermal gas burst occurred during November, 2002. The high-pressure gas release created sinkholes with the collapse of the seafloor. During the first months of Panarea "activity", some vertical logs were performed in the area characterised by gas emissions (Fig. 2) which showed the presence of convective cells containing a mixture of water and gas from the vents. Furthermore, they revealed a modification of sea water pH from 8.0 to 5.0 and Eh from +80 mV to -200 mV, probably due to the presence of H_2S in the emission points. The *STS* data logger permitted the temperatures of the gas emission points to be measured. The data acquisition interval was one measurement per hour in order to have a correlation with sea tide fluctuations. The gas emission temperatures range from 30 to 90°C, except for the "*black point*" whose mean value was around 120°C.

More than 100 samples (both water and gas) were collected from November 2002 to December 2004. Analyses were performed in order to determine major and minor elements, trace elements, dissolved gases in water and gases in free phase. Results suggest that the sea water was locally affected by the acidic, reducing and more saline fluids that circulated at high temperatures in the volcanic rocks.

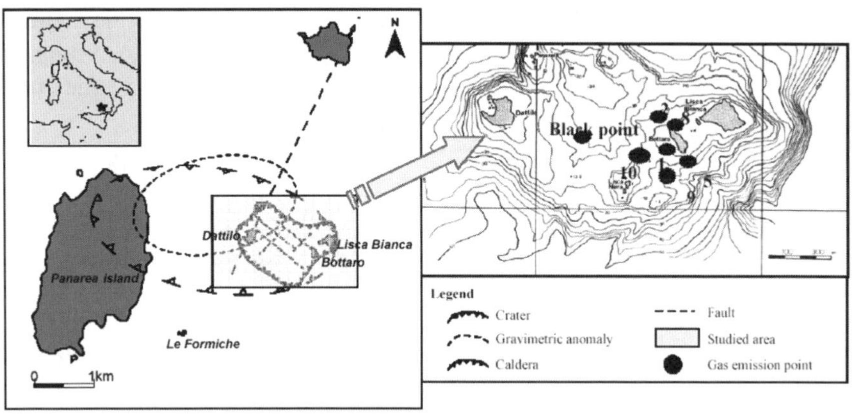

Figure 2. Panarea island location: on the right a detailed map of the studied area where gas emission points have been monitored since November 2002.

Gas samples (Tab.1) are mainly composed of carbon dioxide (98%, v/v) and methane (around 10 ppm) except for the "*black point*", whose mean value is around 600 ppm.

Figure 3 shows the direct relationship between temperatures and CH_4 content in the "*black point*", suggesting the thermogenic origin of the methane. Slightly elevated helium concentrations (mean value = 11 ppm) were found at every point. He isotopic analyses were performed on 32 samples in order to define its origin, with values from 4.1 to 4.5, suggesting a magmatic origin. Temporary variations of both H_2 and H_2S content have the same trend, inferring fractionation of gases by partial dissolution in water. The observed gaseous and chemical composition of the Panarea emission points is very different from that reported in the literature (Fig. 4), as data collected before November 2nd, 2002 have values typical of hydrothermal fields (represented by N_2), whilst samples collected after the recent gaseous emission have concentrations typical of volcanic fields (high He concentrations). This data suggests an evolution of the system controlling the Panarea emissions.

Table 1. Mean values of gases emitted at Panarea island.

Gas species	Mean value
CO_2 (vol., %)	98
CH_4 (ppm)	10
N_2 (vol. %)	0.4
He (ppm)	11
H_2 (ppm)	1100
H_2S (vol. %)	2.2

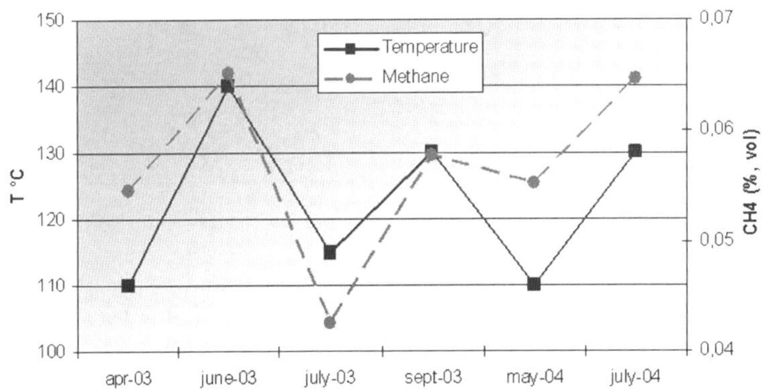

Figure 3. Comparison between methane (%, v/v) and temperature variations (°C) at the "black point" (Panarea island). The observed direct relationship infers a thermogenic origin of the methane gas.

4.2 The Solfatara area

The Solfatara area has been characterized by intense and diffuse fumarolic and hydrothermal activity since Roman times. The area was investigated by means of a detailed soil-gas survey in the inter crater sector, during which 32 soil gas flux measurements (1 sample/100 m) were carried out using an accumulation chamber and a portable gas-chromatograph. Φ_{CO2} is typically 1127.32 gr/m^2*d, although the highest flux values were found in the "La fangaia" and near the "Bocca Grande" and "Bocca Nuova" fumaroles. These fumaroles have the highest outlet temperatures (145°-165°C) amongst the several fumaroles present in the area (mean discharge temperature = 100°C). Fumarole effluents have similar chemistry, with H_2O as the main component, followed by CO_2 and H_2S (Chiodini et al., 2001).

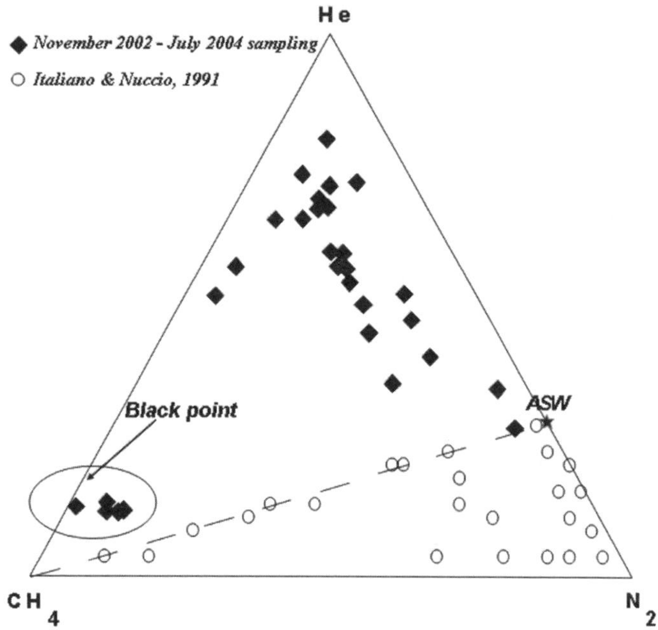

Figure 4. Ternary diagram (N_2-He-CH_4) for gas samples at Panarea island. Data from the literature (white dots, Italiano & Nuccio, 1991) have values typical of hydrothermal fields, whilst samples collected from November 2002 to December 2004 have concentrations typical of volcanic field suggesting an evolution of the system controlling Panarea emissions.

Based on their stable isotopes, the fumarole fluids were interpreted as magmatic fluids which are variably contaminated by metamorphic and meteoric components (Panichi & Volpi, 1999; Allard et al., 1991b). On the basis of outlet temperatures and the content of H_2O, CO_2 and of un-reactive gases (i.e., Ar, N_2 and He) from fumarole fluids it was possible to estimate the flux of steam and thereby the heat flux involved in the diffusive degassing process. Chiodini et al. (2001) calculated that the entire Solfatara area releases 1500 t/d of hydrothermal CO_2 through diffusive degassing. The contour map (Fig. 5) shows an area of about 0.5 km^2 of high Φ_{CO2} values representing an important diffuse degassing structure crossed by a NW-SE band of low fluxes. The highest Φ_{CO2} values overlap with faults and fractures, confirming that the degassing process is strictly related to tectonic structures. In fact, the area is characterised by a regional extensional tectonic feature and by transverse structures considered as transfer faults along which the main regional volcanoes are located (Acocella *et al.*, 1999).

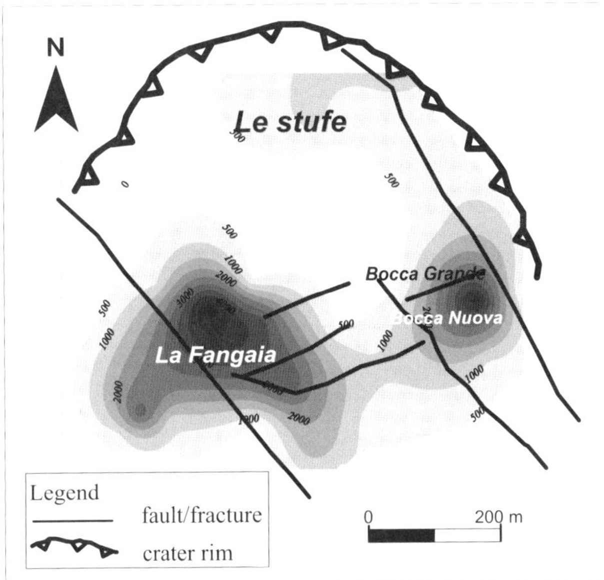

Figure 5. CO_2 flux contour map of the inter crater sector of the Solfatara area. Highest Φ_{CO2} values overlap faults and fractures confirming that the degassing process is strictly related to tectonic structures.

4.3 The Telese area

The Telese area is one of the most seismically active segments of the southern Apennine chain. Several destructive earthquakes (M = 7) struck the area and surroundings in the past centuries. The last energetic seismic sequence (2800 events) occurred in October 1997 (M_{dmax} 4.1) a few km away from the Telese area, during which a first geochemical survey was performed. The survey was repeated in May 1998 during a quiescent period. A total of 82 water samples were collected in order to understand the main water-rock interaction processes and the possible influence of the stress field on water chemistry. Results indicated two main chemical families, Na-HCO_3 and Ca-HCO_3 waters, characterised by different salinities. These types of waters reflect the complexity of the geochemical processes occurring in a geological environment that is made up mainly of outcropping Mesozoic platform carbonates (locally with dolomite), covered by sedimentary layers (clays, marls and sands) and the occurrence of a volcanic complex made up primarily by ignimbrites.

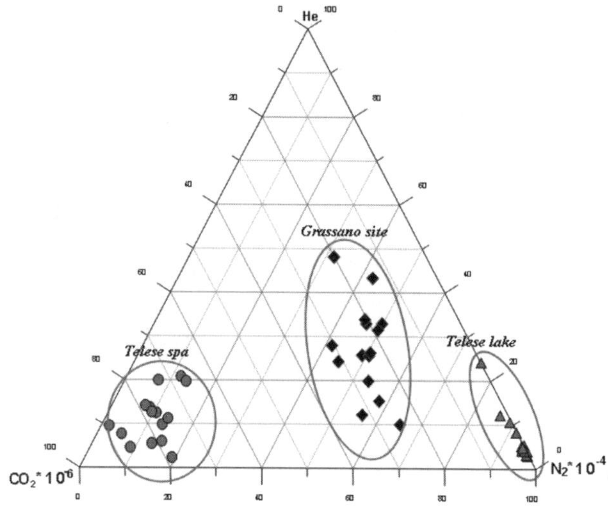

Figure 6. Ternary diagram N_2-He-CO_2. The three sampled aquifers have different chemical compositions. It is clearly evident the CO_2 enrichment of the Telese spa in comparison with the Telese lake samples (shallow aquifer composition) and those of the Grassano river (representing the intermediate aquifer).

On the basis of these first results, three sites representing three different geochemical environments (that is, three different aquifers, Fig. 6) were chosen for monthly geochemical monitoring:

- the Diana spring (Telese spa) is a mineralised (2000-2500 μS/cm), acidic (pH = 6.1) and ipo-thermal (about 20°C, constant all over the year) water having a Ca-HCO_3 chemistry. Water chemical analyses revealed the presence of a high CO_2 content (more than 90%, v/v) in the groundwater. Isotopic analysis of total carbon ($\delta^{13}C$ = +0.4‰) revealed a deep origin probably due to the presence of a cooling magmatic intrusion within the carbonate basement. Helium isotope ratios (R/Ra = 2) strongly support this hypothesis. Due to its geochemical features, the Diana spring represents the end-member of the deep aquifer.
- the Grassano spring is located a few kilometres from the town of Telese. It is a medium-mineral water (800 μS/cm) with a temperature of 11.5°C, a near-neutral pH and a Ca-HCO_3 chemistry. Its geochemical characteristics are typical of a relatively long residence time in the hydrologic circuit with non negligible water-rock interaction processes with carbonates. The discharge is fed by local meteoric waters that only partially receive the deep inputs. Due to its characteristics and considering the other two analysed springs, the Grassano spring is considered the end-member of the intermediate aquifer.

- the Telese Lake is a small, 17 m deep lake located about 1 km away from the town of Telese; the water has a medium-low value of salinity (TDS = 0.3 g/l), an alkaline pH (8.0-8.5) and a Ca-HCO_3 chemistry. Water temperature, at the surface, is linked to seasonal variations, ranging from 6°C in the winter up to 25°C or more during the summer. All reported chemical features, together with a dissolved gas content very similar to ASW (Air Saturated Water), define this water as a shallow one without deep input. For these reason, this represents the shallowest hydrologic circuit presently discharging in the studied area.

Several studies on the monitoring of gas content in waters for seismic prediction suggest that during phases preceding or following seismic energy release rocks undergo physical-structural variations which favour the mobilization of deep fluids, such as CO_2 and H_2 (Wakita et al., 1985; Honda et al., 1982; Lombardi, 1981). In view of these researches, monthly water sampling in the Telese area and surroundings have the primary goal to detect eventual relationships between seismic activity and chemical/geochemical variations in the aquifers. In figure 7 results from the Diana spring (Telese spa) monitoring (April 2003-October 2004) are reported. The graph clearly highlights the different behaviour of dissolved CO_2 in comparison with dissolved H_2 observed during the sampling.

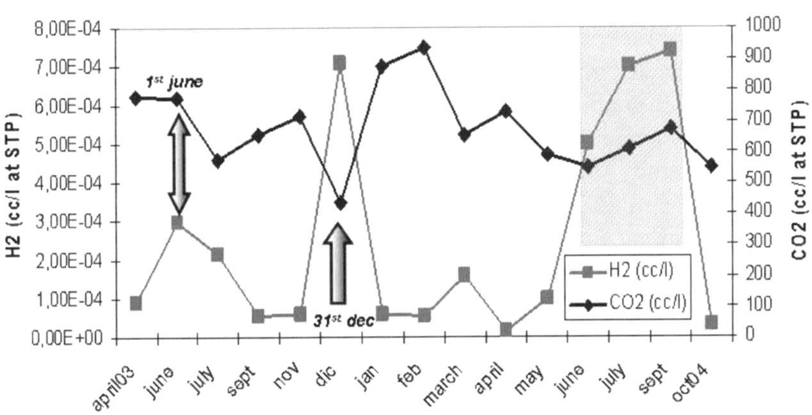

Figure 7. H_2 versus CO_2 content at the Telese spa (Diana spring). It is interesting to note that in correspondence with earthquakes (the two arrows) H_2 and CO_2 contents change. The shaded rectangle indicates a three-month period of small earthquakes during which the H_2 and CO_2 contents undergo a slight increase.

The two arrows in the graph indicate the correspondence between earthquakes and H_2 and CO_2 content trends. The June survey was performed

17 days after an earthquake (M_d = 4.1) occurred in the adjacent Molise region, at a distance of about 70 km from Telese. Another earthquake (M_d = 4.2) occurred on December, 16th, 2003 in the same area. A geochemical survey was performed 15 days before this earthquake. During June-September 2004, a sequence of small earthquakes (M_{max} = 4.1) in an area just 100 km away from the studied sites took place. During the same period an increase of both dissolved H_2 and CO_2 content was observed (Fig. 7, shaded rectangle). It is worth noting that in the period January-May 2004 no earthquakes with M≥ 4.0 occurred within 100 km from the Telese area (INGV, 2004) and, in the same period, dissolved H_2 has an almost constant trend.

5. CONCLUSIONS

The main purpose of this paper is to show some results from areas characterised by natural accumulations and emissions of CO_2. These areas represent interesting natural analogues for the study and prediction of the possible consequences of leakage from geological sequestration sites of anthropogenic CO_2 (i.e., the return to surface, potentially causing localised environmental problems).

Three Italian sites, representing different geological scenarios, were described. All these sites are linked to migration towards the surface of significant quantities of carbon dioxide. In particular, two of these sites are densely populated but people have learned to live with CO_2 gas emissions since ancient Roman times and, for this reason, they have even tried to exploit them. Nowadays the above mentioned areas are constantly monitored in order to study the temporal evolution of different phenomena. In particular, the aim of our research is to evaluate and mitigate risks for local populations, taking into account several factors controlling CO_2 emissions and studying the gas behaviour in time and space. To date, the obtained results suggest that gas uprising is generally well localised around restricted areas, often controlled by local tectonics (faults and/or fractures). This implies that, in the frame of geological CO_2 sequestration, it is necessary to carefully assess the presence of pathways (fault and/or fractures) that might allow the migration of CO_2 out of the reservoir.

REFERENCES

Acocella, V., Salvini, F., Funiciello, R., and Faccenna, C., 1999, The role of transfer structures on volcanic activity at Campi Flegrei (Southern Italy), *J. Volcan. Geotherm. Res.*, **91:**123-139.

Allard, P., Carbonelle, J., Dajlevic, D., Le Bronec, J., Morel, P., Robe, M. C., Maurenas J.M., Faivre-Pierret R., Martin D., Sabroux J.C., and Zettwoog P., 1991a, Eruptive and diffusive emissions of CO_2 from Mount Etna, *Nature*, **351:**387-391.

Allard, P., Maiorani, A., Tedesco, D., Cortecci, G., and Turi, B., 1991b, Isotopic study of the origin of sulphur and carbon in Solfatara fumaroles, Campi Flegrei caldera, *J. Volcanol. Geotherm. Res.*, **48:**139-159.

Barca, D., and Ventura, G., 1991, Evoluzione vulcano-tettonica dell'isola di Salina (Arcipelago delle Eolie), *Mem. Soc. Geol. Ital.*, **47:**401-415.

Baubron, J. C., Allard, P., and Toutain, J. P., 1990, Diffusive volcanic emissions of carbon dioxide from Vulcano Island , Italy, *Nature,* **344:**51-53.

Bertrami, R., Buonasorte, G., Ceccarelli, A., Lombardi, S., Pieri, S., and Scandiffio, G., 1990, Soil Gases in Geothermal Prospecting: Two Case Histories (Sabatini Volcanoes and Alban Hills, Latium, Central Italy), *J. Geoph. Res.*, **9:**21,475-21,481.

Capasso, G., and Inguaggiato, S., 1998, A simple method for the determination of dissolved gases in natural waters: An application to thermal waters from Volcano Island, *Applied Geochemistry*, **13**(5):631-642.

Chiodini, G., Cioni, R., Guidi, M., Raco, B., and Marini, L., 1998, Soil CO_2 flux measurements in volcanic and geothermal areas, *Applied Geochemistry*, **13**(5):543-552.

Chiodini, G., Frondini, F., Cardellini, C., Granirei, D., Marini, L., and Ventura, G., 2001, CO_2 degassing and energy release at Solfatara volcano, Campi Flegrei, Italy. *J. Geoph. Res.*, **106**(8):16216-16221.

Corrado, S., Di Bucci, D., Naso, G., and Butler, R. W. H., 1997, Thrusting and strike-slip tectonics in the Alto Molise region (Italy): implications for the Neogene-Quaternary evolution of the Central Apennine Orogenic System, *J. Geol. Soc. London*, **154:**679-688.

Corrado, S., Di Bucci, D., Naso, G., Giampaolo, C., and Adatte, T., 1998, Application of the organic matter and clay mineral studies to the tectonic history of the Abruzzo-Molise-Sannio area, Central Apennines, Italy, *Tectonophysics*, **285:**167-181.

de Jong, E., and Schappert, H. J. V., 1972, Calculation of soil respiration and activity from CO_2 profiles in the soil, *Soil Science*, pp. 328-333.

Di Bucci, D., 1995, Rapporti tra piattaforme carbonatiche e "Alloctono" lungo la media valle del Sangro, *Mem. Soc. Geol. Ital.*, **114:**443-463.

Gabbianelli, G., Romagnoli, C., Rossi, P.L., and Calanchi, N., 1993, Marine geology of the Panarea-Stromboli area (Aeolian Arcipelago, Southeastern Tyrrenhian Sea). *Acta Volcanol.,* **3:**11-20.

Gasparini, P., Iannaccone, G., Scandone, P., and Scarpa, R., 1982, The seismotectonic of the Calabrian Arc. *Tectonophysics,* **84:**267-287.

Gillot, P. Y., 1987, Histoire volcanique des Iles Eoliennes:Arc insulaire ou complexe orogenique anulaire ?, *D. T. IGAL 11*, 35-42.

Gurrieri, S., and Valenza, M., 1988, Gas transport in natural porous mediums: a method for measuring CO_2 flows from the ground in volcanic and geothermal areas, *Rend. Soc. Ital. Mineral. Petrologia,* **43:**1151-1158.

Honda, M., Kurita, K., Hamano, Y., and Ozima, M., 1982, Experimental studies of He and Ar degassing during rock fracturing, *Earth Planet. Sci. Lett.*, **51:**206-214.

Hutchinson, G. L., and Livingston, G. P., 1993, Use of chamber systems to measure trace gas fluxes, *American Society of Agronomy*, special publication, **55**.

Hyppolite, J., Angelier, J., and Roure, F., 1994, A major change revealed by quaternary stress patterns in the Southern Apennines, *Tectonophysics*, **230:**199-210.

INGV (2004) – Seismological report

Italiano, F., and Nuccio, P. M., 1991, Geochemical investigations of submarine volcanic exhalations to the east of Panarea, Aeolian Islands, Italy, *J. Volcan. Geoth. Res.*, **46**(1-2):1-185.

Kanemasu, E. T., Powers, W. L., and Sij, J. W., 1974, Field chamber measurements of CO_2 flux from soil surface, *Soil Science,* **118**(4):233-237.

Lombardi, S., 1981, Variazione del tenore in elio in alcune manifestazioni sorgive e gassose della valle del Sele nel periodi immediatamente successivo all'evento sismico del 23/11/1980, *Rend. Soc. Geol. Ital.,* **4.**

Lombardi, S., and Reimer, G. M., 1990, Radon and Helium in soil gases in the Phlegrean fields, central Italy, *Geoph. Res. Let.*, **17**(6):849-852.

Matthias, A. D., Blackmer, A. M., and Bremner, J. M., 1980, A simple chamber technique for field measurements of emissions of nitrous oxide from soils, *J. Envir. Qual.,* **9**(2): 251-256.

Orsi, G., De Vita, S., and Di Vito, M., 1996, The restless, resurgent Campi Flegrei nested caldera (Italy): constrains on its evolution and configuration, *J. Volcanol. Geotherm. Res.,* **74:**179-214.

Panichi, C., and Volpi, G., 1999, Hydrogen, oxygen and carbon isotope ratios of Solfatara fumaroles (Phlegrean Fields, Italy): further insight into source processes, *J. Volcanol. Geotherm. Res.,* **91:**321-328.

Patacca, E., Scandone, P., Bellatalla, M., Perilli, N., and Santini, U., 1992, The Numidian-sand event in the Southern Apennines, *Mem. Soc. Geol. Padova,* **43:**297-337.

Rosi, M., and Santacroce, R., 1984, Volcanic hazard assessment in the Phlegrean Fields: a contribution based on stratigraphic and hystorical data, *Bull. Volcanol.,* **47:**359-370.

Scheidegger, A. E., 1974, The physics of flow through porous media, University of Toronto Press, Toronto, 353p.

Vilardo, G., Nappi, R., Petti, P., and Ventura, G., 2003, Fault geometries from the space distribution of the 1990-1997 Sannio-Benevento earthquakes: interferences on the active deformation in Southern Apennines, *Tectonophysics,* **363:**259-271.

Wakita, H., Nakamura, Y., and Sano, Y., 1985, Groundwater radon variations reflecting changes in regional stress field, *Earthq. Predict. Res.,* **3:**545-557.

NATURAL LEAKAGE OF HELIUM FROM ITALIAN SEDIMENTARY BASINS OF THE ADRIATIC STRUCTURAL MARGIN

Perspectives for geological sequestration of carbon dioxide

G. Ciotoli[1], S. Lombardi[1] and F. Zarlenga[2]

[1]*Earth Sciences Department, University of Rome "La Sapienza", Rome, Italy;* [2]*ENEA-Cr-Casaccia- Prot/Prev, Italy*

Abstract: Soil gas surveys were performed along the Adriatic foredeep (Vasto, Ferrandina, Pisticci basins), in order to evaluate the relationship between neotectonics and gas leakage from hydrocarbon reservoirs. More than 4000 soil gas samples were collected in the area and analyzed for helium, which is a good fracture tracer due to its chemical inertness and high mobility. Furthermore, helium is enriched in more than 90% of known reservoirs, and displays many characteristics of the ideal geochemical tracer for buried faults and gas and oil reservoirs (abiogenic, non-reactive, and mobile). Statistical analysis shows an average helium concentration of 5.7 ppm in the foredeep basin areas (atmospheric value is 5.2 ppm). This value is higher than that calculated using over 30,000 soil gas samples collected throughout Italy in different geological scenarios (about 5.4 ppm). Results obtained using a geostatistical approach are consistent with the presence of high helium concentrations as linear or spot anomalies due to irregular, channelled flow along faults above hydrocarbon accumulations. Considering the Plio-Miocene age of the Adriatic foredeep reservoirs, the magnitude of these diffuse gas microseepages highlights that losses from hydrocarbon reservoirs should be low. Results indicate that despite the fact that the area is heavily faulted and that gas seepage has occurred from the reservoir over geological periods of time, no environmental effects are observed at surface. This fact lends support to the idea that geological sequestration of CO_2 within a less structurally-active area would result in the safe, long-term isolation of this green-house gas.

Key words: soil gas geochemistry, tracer, helium, safety and monitoring of CO_2 geological sequestration sites

S. Lombardi et al. (eds.),
Advances in the Geological Storage of Carbon Dioxide, 191–202.

1. INTRODUCTION

In this research helium soil gas concentrations have been considered as a tracer of gas microseepage through thousands of meters of fractured and/or faulted overburden . Helium is a good fracture tracer because it is chemically inert, physically stable, highly insoluble in water, and as such it is highly mobile (Pogorski and Quirt, 1981; Oliver et al., 1984). Helium migration, as well as other gases, occurs primarily by advection or diffusion from deep accumulations (geothermal fluids, uraniferous ore and in 90% of known hydrocarbon reservoirs) towards the surface along permeable fault and fracture systems (Eremeev et al., 1973; Lombardi and Reimer, 1990; Duddridge et al., 1991; Ciotoli et al., 1998). Its distribution in soil gas depends on the permeability of the rocks, depth of origin, the presence of intersecting structures, and the influence of near-surface conditions such as groundwater flow, weathering, etc. The concentration of helium in the atmosphere is constant (5220+/-15 ppb; Holland & Emerson, 1990) as the result of the dynamic budget between degassing helium and helium diffusion into space.

Extensive experience in soil-gas prospecting by the authors indicates that helium soil-gas anomalies generally occur as linear, fault-linked anomalies, irregularly-shaped diffuse or "halo" anomalies and irregularly-spaced plumes or "spot anomalies" (Lombardi et al., 1996; Ciotoli, 1997; Ciotoli et al., 1997, 1999). These features reflect gas migration dominated by brittle deformation, both at the macro and/or micro scale. Linear anomalies longer than several hundreds of meters are often taken as strong evidence of tectonic control of gas migration along gas-bearing faults, (Ciotoli et al., 1998). It is important to realize that most faulted areas consist of fault intersections and extensive microfractures that allow gases to escape laterally and vertically, producing "halo" or diffuse anomaly. Furthermore, recent research has demonstrated how the gas-bearing properties of faults are not necessarily continuous along a tectonic structure (Ciotoli et al., 1998). In these cases isolated points with high concentration values ("spotty anomalies") are frequently observed in soil gas distributions. When multiple "spot" anomalies occur along a linear trend, one can infer that they lie along a structural feature which has spatially discontinuous gas-transport properties (Lombardi et al., 1996; Ciotoli et al., 1998) (Fig.2).

Since helium is more soluble in oil than in water, oil-water mixing will result in the preferential partitioning of helium into oil, thus the surface distribution of helium (similarly to those for hydrocarbon gases) can indicate the presence of underground helium reservoirs in which petroleum may also be found (Zartman et al., 1961; Cunningham, 1988). For example, the helium distribution above the Eldingen oil field in Germany is thought to be controlled by a combination of structural and hydrogeological conditions (Roberts, 1981; Van den Boom, 1987).

A further aspect in the interpretation of helium values considers their magnitude, which can be reduced to the following main populations:

- negative or near zero concentrations that indicate zones of weathering processes (i.e. high moisture, good soil aeration, or zones where the fractures are water-conducting);
- normal soil-air content, which ranges between about –100 to +100 ppb in Italy (determined statistically using more than 32,000 data values collected in central and southern Italy by the authors since 1970);
- anomalous values that represent the subsurface presence of a deep helium source.

2. GAS MICROSEEP DIFFERENCES BETWEEN THE THYRRENIAN AND ADRIATIC COASTS

From a geological point of view Italy is characterized by active and recent volcanoes (from the Pliocene to Present) and geothermal fields along the Thyrrenian margin. Two mountain chains (the Alps and the Apennines) transect the country and sedimentary basins (the Padanian Foredeep, Adriatic sea and other small peri-thyrrenian basins) host oil and natural gas reservoirs and saline aquifers. Therefore there is a very strong geological complexity which controls the gas leakage from deep reservoirs (Ciotoli et al., 1997; 1998; 1999; 2002). Within this framework various CO_2 reservoirs are located throughout the different geological domains, and many are considered as natural analogues of the geological storage of anthropogenic CO_2. These are characterized by different pressures, temperatures and leakage rates, and thus represent a large range of situations which may eventually occur in an engineered system. This combination of great depths and long time periods mean that the natural analogue sites can help us better understand eventual system evolution, which is difficult to model or to represent experimentally.

The experience accumulated in Italy has shown widespread macro-scale (gas vents, fumaroles, mineralised aquifers) and micro-scale (micro seeps) gas leakage from oil, natural gas and geothermal reservoirs along active faults (Ciotoli et al., 1998; Etiope and Lombardi, 1994). In active volcanic areas the CO_2 flux from degassing activity ranges from 25 at Etna, to 1-2 at Stromboli to 0.066 Mton y^{-1} at Vulcano (Morner and Etiope, 2002). In the sedimentary basins along the Tyrrhenian coast, where geothermal reservoirs also occur, the flux of CO_2 derives from diffuse soil degassing and from gas vents. The output of diffuse degassing ranges from 7300 ton $y^{-1}km^{-2}$ (Cava deSelci, Alban Hills Volcanic District), to > 73,000 ton $y^{-1}km^{-2}$ (Latera, Sabatinian Volcanic District), to 0.83-1,123 kg m^{-2} y^{-1} (in the sedimentary basins of the Siena Graben). In the case of gas vents the measured output ranges from 1,800 ton y^{-1} (at Massa Martana and Montecastello Vibio) to 310,000 ton y^{-1} (Mefite d'Ansanto). The regionally evaluated output of the central Apennine chain is 4.4-13.2 x10^6 ton y^{-1} (over 12,564 km^2) (Chiodini et al., 2000).

3. SOIL GAS DATA FROM THE SEDIMENTARY BASINS OF THE ADRIATIC FOREDEEP

Soil gas surveys were carried out in different sedimentary basins located along the Adriatic foredeep (Vasto, Ferrandina, Pisticci basins) in order to evaluate the relationship between neotectonics and gas leakage from hydrocarbon reservoirs. In these areas about 5,000 soil gas samples were collected and analysed. Statistical analysis of soil gas data collected in the foredeep basins (FB) shows a mean helium concentration of 5.7 ppm (atmospheric value is 5.2 ppm; Holland and Emerson, 1990). This value is higher than the mean value (about 5.4 ppm) calculated using over 30000 samples collected throughout Italy in different geological settings (Tab. 1). Furthermore, it is worth noting that intermontane basins (IB) show the highest maximum value (315.22 ppm) and have a higher mean value (5.55 ppm) than that of volcanic areas (VA) and neogenic basins (NB) (Tyrrhenian sedimentary basins). This characteristic is probably linked to the presence in these zones of active tectonics and seismogenic faults that provide a vertical pathway for gas migration towards the surface.

Elevated outlier values shift the data distribution toward higher concentrations, as suggested by the high skewness values (FB, VA and IB). In contrast, NB shows a more symmetrical data distribution, as suggested by the low values of standard deviation (0.31) and skewness (0.52), and by the similar values of the mean (5.39 ppm) and the median (5.37 ppm). The lower quartile values indicate atmospheric content, and the upper quartile values generally correspond to the anomaly threshold. Considering the similar values of the interquartile range (IQR) (50% of data) it is possible to conclude that in Italy He concentrations higher than about 0.25 -0.30 ppm can be considered as anomalies.

Figure 1 compares helium concentrations in the Italian peri-Adriatic basins (Vasto, Pisticci, Agri, Pescara and Ferrandina basins) with data from the NB and VA. The box plot highlights that the high helium values observed in the investigated areas are probably linked to deep hydrocarbon reservoirs, as suggested by the similar mean helium values for the various Vasto and Agri basin sectors. It is worth noting that high helium concentrations were measured in the Agri basin before the discovery of the oil field in 1986.

Table 1. Main statistical parameters of helium concentrations in soil gas from different investigated geological scenarios.

	Foredeep Basins (FB)	Volcanic areas (VA)	Neogenic Basins (NB)	Intermontane Basins (IB)
N	4893	13255	6597	4338
Mean	5.74	5.42	5.39	5.55
Median	5.31	5.34	5.37	5.26
Min	3.82	2.12	1.76	1.2
Max	146.81	127.52	8.82	315.22
LQ	5.22	5.22	5.22	5.15
UQ	5.48	5.5	5.51	5.45
IQR	0.26	0.28	0.29	0.29
Std Dev	5.13	1.51	0.31	5.8
Skewness	20.05	50.83	0.52	42.29

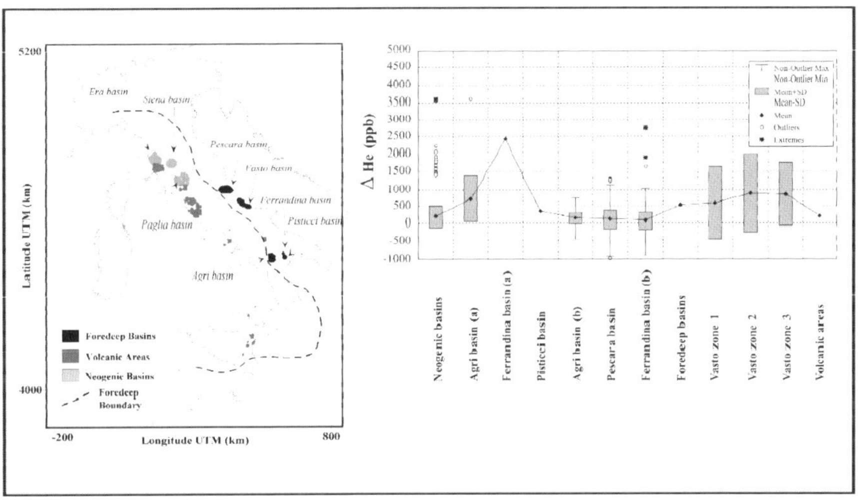

Figure 1. The figure to the right makes a statistical comparison of helium data from different surveys conducted in Italy by the authors since the 1970's; survey locations are shown to the left (modified after Ciotoli et al., 2004). The high helium concentrations detected in the Vasto basin are comparable to those measured in the Agri basin, in which hydrocarbon accumulations occur. Note that the mean helium values at Vasto are higher than those measured in volcanic areas where helium migration and high concentrations are well known in the literature.

Finally, a comparison of helium anomaly magnitudes with the different thicknesses of the sedimentary covers occurring in the central and southern sectors shows that the soil-gas distribution does not seem to be affected by lithology changes and/or thickness. These results could, however, be affected by the different permeability of the fracture systems due to different activation age and/or self-sealing phenomena, and by different soil moisture content. In conclusion, the overall spatial association of geochemical data with geomorphological and structural features suggests the presence of zones of enhanced gas permeability (with gas microseeps) resulting from neotectonic activity in the area.

4. HELIUM GAS MICROSEEPS IN THE ADRIATIC FOREDEEP: THE VASTO BASIN

The Vasto basin is located between the Sangro river to the north and the Biferno river to the south, and covers part of the Adriatic foredeep which constitutes the extensive NW-SE trending structural and morphological depression along the eastern side of the Italian peninsula (Fig 2). This is the only domain of the Apennines where folding and thrusting are still active, associated with progressive eastward shifting of the Adriatic foredeep and migration of the foreland peripheral bulge at rates of 1.5-3 cm y^{-1} (Patacca and Scandone, 1989).

The outcropping formations consist of more than 2,500 m (8,300 ft) of Plio-Pleistocene sandy-clays and a pre-Pliocene substratum of Lower and Middle Miocene biogenic limestones which are transgressive over Cretaceous carbonates.

The structural evolution of the area involved different phases. Middle Pliocene compressional tectonics caused a translation of the Apennine chain towards the east, which formed thrust faults (characterized by strike-slip movement) which have been active since the Plio-Pleistocene. In the northern-central sector of the studied area Pleistocene extensional tectonics caused the formation of structural highs (e.g. the Cupello anticline) that are often characterized by the presence of hydrocarbon reservoirs. About 100 small hydrocarbon fields have been discovered in the entire foredeep region since 1950. In 1955 the largest gas field, with an area of 10 km^2, was discovered SW of Vasto ("Cupello-S Salvo"). This is mainly a gas province where most of the hydrocarbon reservoirs occur in a belt that parallels the contact between the outer thrust front and the Adriatic foredeep (Carlin and Dainelli, 1998). Hydrocarbons are generally trapped at the top of the Middle Miocene and Mesozoic carbonates, or in structural (thrust-faulted anticlines) and stratigraphic (pinch-outs) traps located within different Pliocene sand bodies. The youngest reservoirs are confined to the outermost position, in Middle-Late Pliocene-Pleistocene sands (Casnedi, 1983).

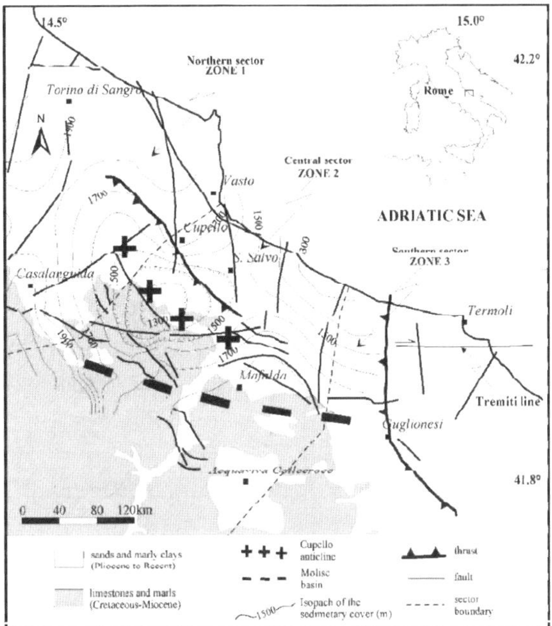

Figure 2. Geological map of the investigated area. Figure also shows the boundaries of the three sectors (zone 1 - northern sector; zone 2 - central sector; zone 3 - southern sector) defined to better interpret the helium distribution.

Drilling in the hydrocarbon fields indicates the occurrence of overpressures up to lithostatic in compartmentalized domains, which are laterally bounded by thrust faults and vertically trapped by low-permeability beds. This closed circulation in low-permeability domains is indicative of thrust faults acting as barriers to meteoric and deep fluids (and to hydrocarbons) (Ghisetti et al., 2000; Ghisetti and Vezzani, 2002). Progressive eastward fluid expulsion, connected with the eastward migration of shortening, is suggested by the west to east migration of gas reservoirs, and by geochemical indicators like helium.

5. RELATIONSHIPS BETWEEN TECTONICS AND REGIONAL HELIUM DISTRIBUTION

Helium trend surfaces were calculated (using a 2nd-degree polynomial regression) in order to highlight large-scale trends and patterns related to regional tectonics, such as basins and/or structural highs (Fig 3). The map shows a clear anisotropic helium distribution, consisting of:

- mainly NE-SW-oriented anomalies in the northern sector;
- N-S-oriented anomalies in the central sector ;
- E-W-oriented anomalies in the southern sector.

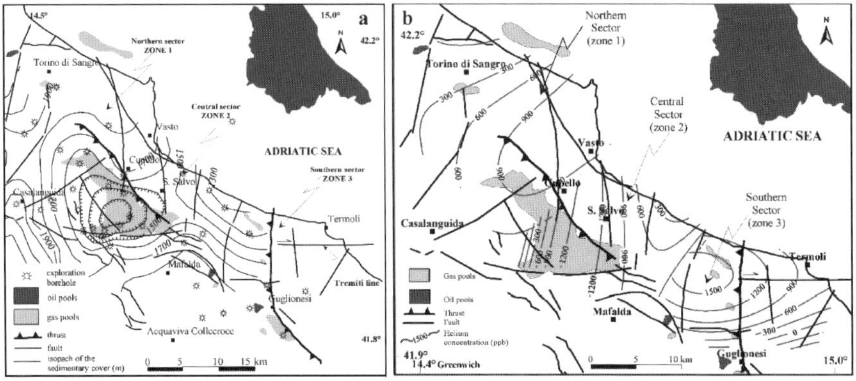

Figure 3. Location map of the hydrocarbon reservoirs and some of the producing and / or exploration wells (a). The map also shows the thickness of the sedimentary cover, as isopach lines, calculated using borehole data. The figure also shows soil-gas helium-trend surfaces (b) that may be related to regional tectonic features. The figure shows a clear anisotropic helium distribution in the central (N-S oriented) and southern (E-W oriented) sectors probably related to the Cupello anticline and the Molise Basin (modified after Ciotoli et al., 2004).

The anisotropic spatial behavior of regional helium anomalies are probably linked to the different orientations of the main fault systems which characterize these sectors. In particular, in the northern sector the regional helium distribution is influenced by the NE-SW trending faults related to the extensional tectonics phase, which controls the direction and abrupt deviation of the major rivers in this zone. In the central sector the N-S trend of the regional helium anomalies may be related to the most recent fault systems which defined the morphology of the coastline, as well as the orientation of some river courses. In the southern sector the regional helium anomalies agree well with the right-lateral transfer zone of the Tremiti fault line (Doglioni et al., 1994), which is also suggested by the orientation of the isopach lines (Fig. 3a). Furthermore, a comparison of helium trends versus sedimentary cover thickness (Fig. 3b) shows high values even where the carbonate substratum is structurally lowered and overlain by a clay-rich sedimentary cover that is up to 1500 m thick in the southern sector. In these sectors structural hydrocarbon traps also occur at different depths. This fact suggests a possible deep source for the helium in the hydrocarbon reservoirs, with leakage and distribution influenced by the presence of more permeable pathways, such as regional faults and associated fractures.

The map shows E-W contour lines which reflect the general helium trend in this sector, as reported in Figure 3. The eastward shifting and the N-S distribution of the high helium residual spots away from the location of the recognized small gas fields is consistent both with the N-S reservoir distribution and with the idea of gas migration along more recent and permeable fault systems (i.e. low-angle N-S thrust faults).

The map of residual anomalies (Figure 4a), calculated from the helium values measured in the southern sector of the investigated area, highlights high fault-related helium anomalies probably associated with the hydrocarbon reservoirs as shown in Figure 4b. The map shows east-west contour lines that reflect the general helium trend in this sector, as reported in Figure 3. The eastward shifting and the north-south distribution of the high helium residual spots from the location of the recognized small gas fields is consistent with both the north-south distribution of the reservoirs and with the idea of gas migration along more recent and permeable fault systems (i.e., low-angle north-south thrust faults).

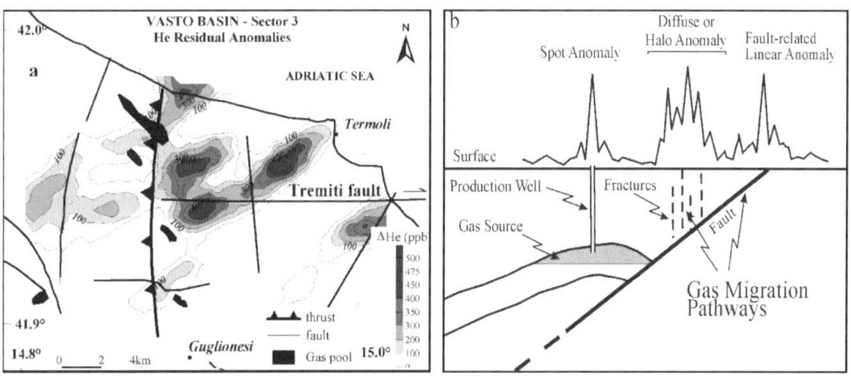

Figure 4. Residual anomalies map for the southern sector (a). This map shows the N-S distribution of the spotty residual anomalies (above 300 ppb), which is congruent with N-S thrust fault systems in the sector. A further spotty anomaly near the village of Termoli may indicate the existence of another pool. The slight eastward shifting of the residual anomalies seem to be well correlated with the location of the pools to the west, as illustrated in the schematic diagram (b) (modified after Ciotoli et al., 2004).

6. IMPLICATIONS OF GAS MICROSEEPS FOR THE GEOLOGICAL SEQUESTRATION OF CO_2

Soil gas surveys and flux measurements carried out in Italy in different geological settings show that gas leakage occurs mainly along recently-activated faults and fractures. Even in this case it only occurs along preferred pathways and is strongly localized (channelling effects) (Ciotoli et al., 1997, 2002; Lombardi et al., 1996; Lombardi et al., 1993). Even if the quantitative study of flux across faults is not complete, research performed in clayey basins, characterized by both extensional and compressive tectonics, suggests only minor gas escape along the above-mentioned channels. This hypothesis is also supported by a limited number of gas injection tests carried out in shallow, faulted and unfaulted clayey sequences (Etiope et al., 1995; Lombardi et al., 1996; Ciotoli et al., 2003). Considering these results discussed above the following work is needed to better apply this method for the geological sequestration of CO_2:

- define background values of the natural soil gas emissions, in case the area is eventually used for CO_2 storage;
- develop new monitoring systems to detect in real time increases in CO_2 escaping along vertical migration pathways

7. CONCLUSIONS

Results outlined in the present work show how the integration of near-surface gas geochemistry with lithology, structural geology and geomorphology data can be used to better understand the subsurface distribution of oil and gas reservoirs, even in areas where thick sedimentary sequences cover and isolate them. However the very fact that gas anomalies originating from great depths are observed in the shallow soil horizon implies that, although the structural and stratigraphic traps have resulted in hydrocarbon accumulation, small scale leakage does occur along the numerous faults and fractures in this tectonically active and complex area. That said, these leaks have been found to be spatially restricted and to result in relatively low levels of gas release to the atmosphere. In regards to the field of CO_2 geological storage these results have three very important implications. First, leakage is slow and isolated even in a highly faulted area, giving confidence to the idea that sequestration within un-faulted areas will be even more safe. Second the soil gas method has been shown to be well adapted to the twin tasks of site-assessment prior to installation of a CO_2 sequestration site (to look for gas permeable structures) and for eventual monitoring of the site once injection begins (for carbon credit auditing and public assurance). Finally although the Italian foredeep region is clearly not ideal, these results indicate that CO_2 sequestration in the aging hydrocarbon reservoirs of this region (due to various economic and logistical advantages) may still be feasible if eventual small leakage rates are acceptable. This conclusion does not, however, take into account the potential for an earthquake in this region which might result in a larger release, although the long term isolation of the hydrocarbons does not support this idea.

REFERENCES

Carlin, S., and Dainelli, J., 1998, Pressure regimes and pressure systems in the Adriatic foredeep (Italy), *AAPG Memoire* **70**: 145-160.

Casnedi, R., 1983, Hydrocarbon-bearing submarine fan-system of Cellino formation, central Italy: *AAPG Bulletin*, **67**:359-370.

Chiodini, G., Frondini, F., Cardellini, C., Parello, F., Peruzzi, L., 2000, Rate of diffuse carbon dioxide Earth degassing estimated from carbon balance of regional aquifers: the case of central Apennine, Italy. *J. Geophys. Res.* **105**(B4): 8423–8434.

Ciotoli, G., 1997, Introduzione di metodi geostatistici nell'interpretazione della distribuzione dei gas endogeni nei suoli quale contributo alle indagini strutturali: la Piana del Fucino: Tesi di dottorato, Dipartimento di Scienze della Terra, Università di Roma "La Sapienza", Roma, Italy, 226 p.

Ciotoli, G., Lombardi, S., Serafini, S., and Zarlenga, F., 1997, The refinement of geomorphological and geochemical statistical techniques in the study of clay-basin tectonics: the Era basin (Central Italy): *Italian Journal of Quaternary Sciences*, **10**(2):231-246.

Ciotoli, G., Guerra, M., Lombardi, S., and Vittori, E., 1998, Soil gas survey for tracing seismogenic faults: a case-study the Fucino basin (central Italy): *Journal of Geophysical Research*, **103**(B10):23,781-23,794.

Ciotoli, G., Etiope, G., Guerra, M., and Lombardi, S., 1999, The detection of concealed faults in the Ofanto basin using the correlation between soil-gas fracture surveys: *Tectonophysics*, **299**(3-4):321-332.

Ciotoli G., Lombardi S., Morandi S., and Zarlenga F.(2004) – A Multidisciplinary Statistical Approach to Study the Relationships between Helium Leakage and Neo-tectonic Activity in a Gas Province: The Vasto Basin, Abruzzo-Molise (Central Italy). *AAPG Bull.*, 88/3, 355-372.

Doglioni, C., Mongelli, F., and Piero, P., 1994, The Puglia uplift (SE Italy): an anomaly in the foreland of the Apenninic subduction due to buckling of a thick continental lithosphere: *Tectonics*, **13**(5):1309-1321.

Duddridge, G. A., Grainger, P., and Durrance, E. M., 1991, Fault detection using soil gas geochemistry: *Quatern. Journal of Engineering Geology*, **24**:427-435.

Eremeev, A. N., Sokolov, V. A., and Solovov, A. P., 1973, Application of helium surveying to structural mapping and ore deposit forecasting: *London Symposium Volume,* p. 183-192

Etiope, G., and Klusman, R. W., 2002, Geologic emission of methane to the atmosphere: *Chemosphere*, **49**:777-789.

Etiope, G, and Martinelli, G., 2002, Migration of carrier and trace gases in the geosphere: an overview: *Physics Earth Planetary Interior*, **129**:185-204.

Ghisetti, F., and Vezzani, L., 2002, Normal faulting, transcrustal permeability and seismogenesis in the Apennines (Italy), *Tectonophysics*, **348**:155-168.

Ghisetti, F., Kirschner, D., and Vezzani, L., 2000, Tectonic controls on large scale fluid circulation in the Apennines (Italy): *Journal Geochemical Exploration*, **69/70**:533-537.

Lombardi, S., and Reimer, G. M., 1990, Radon and helium in soil gases in the Phlaegrean Fields, central Italy: *Geophysical Research Letters*, **17**:849-852.

Lombardi, S., Etiope, G., Guerra, M., Ciotoli, G., Grainger, P., Duddridge, G. A., Gera, F., Chiantore, V., Pensieri, R., Grindrod, P., and Impey, M., 1996, The refinement of soil gas analysis as a geological investigative technique: Final Report. CEC, Programme on Radioactive Waste Management (1990-1994) Part A, Task 4, Eur 16929 En, 194 p.

Oliver, B. M., Bradley, J. G., and Farrar IV, H., 1984, Helium concentration in the Earth's lower atmosphere: *Geochimica et Cosmochimica Acta*, **48**:1759-1469.

Patacca, E., and Scandone, P., 1989, Post-Tortonian mountain building in the Apennines. The role of the passive sinking of a relic lithospheric slab: In *The Lithosphere in Italy*, eds. A. Boriani, M. Bonafede, G.B. Piccardo, and G.B. Vai, Accademia Nazionale dei Lincei, v. 80, p. 157-176.

Pogorsky, L. A., and Quirt, G. S., 1981, Helium emanometry in exploring for hydrocarbons: Part I. In: *Unconventional Methods in Exploration for Petroleum and Natural Gas II*. Southern Methodist University Press, p. 124-135.

Roberts, A.A., 1981, Helium emanometry in exploring for hydrocarbons: Part II. In: *Unconventional Methods in Exploration for Petroleum and Natural Gas II*. Southern Methodist University Press, p. 135-149.

Van Den Boom, G.P., 1987, Helium distribution pattern of measured and corrected data around the "Eldingen" Oil Field, Nw Germany. *J. Geophys.Res.*, **92b:**12547-12555.

TECTONICALLY CONTROLLED METHANE ESCAPE IN LAKE BAIKAL

J. Klerkx[1], M. De Batist[2], J. Poort[2], R. Hus[2], P. Van Rensbergen[2], O. Khlystov[3] and N. Granin[3]

[1] *International Bureau for Environmental Studies, Brussels, Belgium;* [2]*Renard Centre of Marine Geology, University of Ghent, Belgium;* [3]*Limnological Instute, Siberian Branch of the Russian Academy of Sciences, Irkutsk, Russia*

Abstract: Methane, which is at least partly stored in the bottom sediments of Lake Baikal as gas hydrates, is released on the lake floor in the deeper parts of the basin along major faults, forming venting structures similar to small mud volcanoes. The CH_4 venting structures are considered to be the surface expression of escape pathways for excess CH_4 generated by the dissociation of pre-existing hydrates. The existence of a local heat flow anomaly associated with the seep area is most likely due to a heat pulse causing the dissociation of the underlying gas hydrates. The heat pulse may be caused by upward flow of geothermal fluids along segments of active faults, possibly accelerated by seismic pumping. It is assumed that this fluid flow is tectonically triggered, considering that left-lateral strike-slip movements along the border faults act as a major factor in fluid accumulation: even a reduced lateral displacement is able to generate fluid flow in the compressional direction, resulting in fluid escape along faults directed along the main direction of extension. The tectonic effect may be coupled to the sediment compaction due to a high sedimentation rate in the area of mud volcanism. Both processes may generate a large-scale convective fluid loop within the basin-fill sediments which advects deeper gases and fluids to the shallow sub-surface. Even in the extensional tectonic environment of Lake Baikal, local compressional forces related to a strike-slip component, may play a role in fluid flow, accumulation and gas escape along active faults. The mechanisms that result in the expulsion of the CH_4 in the Lake Baikal sediments are considered as an analogue of what could happen during CO_2 sequestration in a similar tectonic environment.

Key words: Methane escape, mud volcanism, gas hydrates dissociation.

S. Lombardi et al. (eds.),
Advances in the Geological Storage of Carbon Dioxide, 203–219.

1. INTRODUCTION

The Baikal rift basin is commonly associated with active extensional tectonics (Logatchev, 1993; Zonenshain et al., 1992; Mats, 1993). The present-day tectonic activity is confirmed by GPS-measurements: the basin undergoes active extension along a WNW-ESE direction with a velocity of up to 4-5 mm/year (Calais et al., 1998; Petit et al., 1997, 1998). Active subsidence is testified by the three large water-filled sedimentary basins comprised by Lake Baikal. Within these basins, active fluid flow results in local heat-flow anomalies (Poort and Klerkx, 2004) and fluid escape at the lake floor (Granin and Granina, 2002; De Batist et al., 2002; Van Rensbergen et al., 2002; 2003). Gas venting has been shown during recent years to be a common phenomenon (Fig. 1).

Methane gas, which has accumulated in the sedimentary infill due to microbial decomposition of organic matter and which is -at least partly-stored as gas hydrates, is released on the lake floor along major fault zones. Reflection seismic profiles have also revealed that the bottom-simulating reflector (i.e. BSR; commonly considered to represent the base of the gas-hydrate stability zone (GHSZ), or -more precisely- the transition between gas-hydrate bearing sediments above and free-gas containing sediments below) has an irregular behaviour, instead of displaying a regular pattern and mimicking the lake floor as it is usually the case. Locally, the BSR is disrupted along faults and vertical fluid-migration conduits that develop along these faults propagate into the GHSZ and extend up to the lake floor. Morphological evidence of fluid escape at the lake floor is given by venting structures, similar to small mud volcanoes. These mud volcanoes appear to be sourced from the base of the GHSZ in areas where the BSR displays an irregular behaviour (De Batist et al., 2002; Van Rensbergen et al., 2002; 2003). Such venting structures occur in several places throughout the lake and their present activity is testified by hydroacoustically observed "plumes" above several of them and an increased concentration of dissolved methane in the bottom water in their vicinity.

Fluid flow in the sediments of Lake Baikal is thus believed to result in the local accumulation of gas (i.e. at the base of the GHSZ) and its subsequent release along zones of weakness. This fluid flow is assumed to result -at least partly- from the tectonic processes that act within the basin. But what are exactly the processes responsible for the destabilisation of gas hydrates and the generation of free methane? What is the tectonic control on methane escape from the lake floor in a tectonic basin under extension? The mechanisms that result in the expulsion of methane present in the Lake Baikal sediments are considered as a natural analogue of what could happen during CO_2 geological sequestration in a similar tectonic environment.

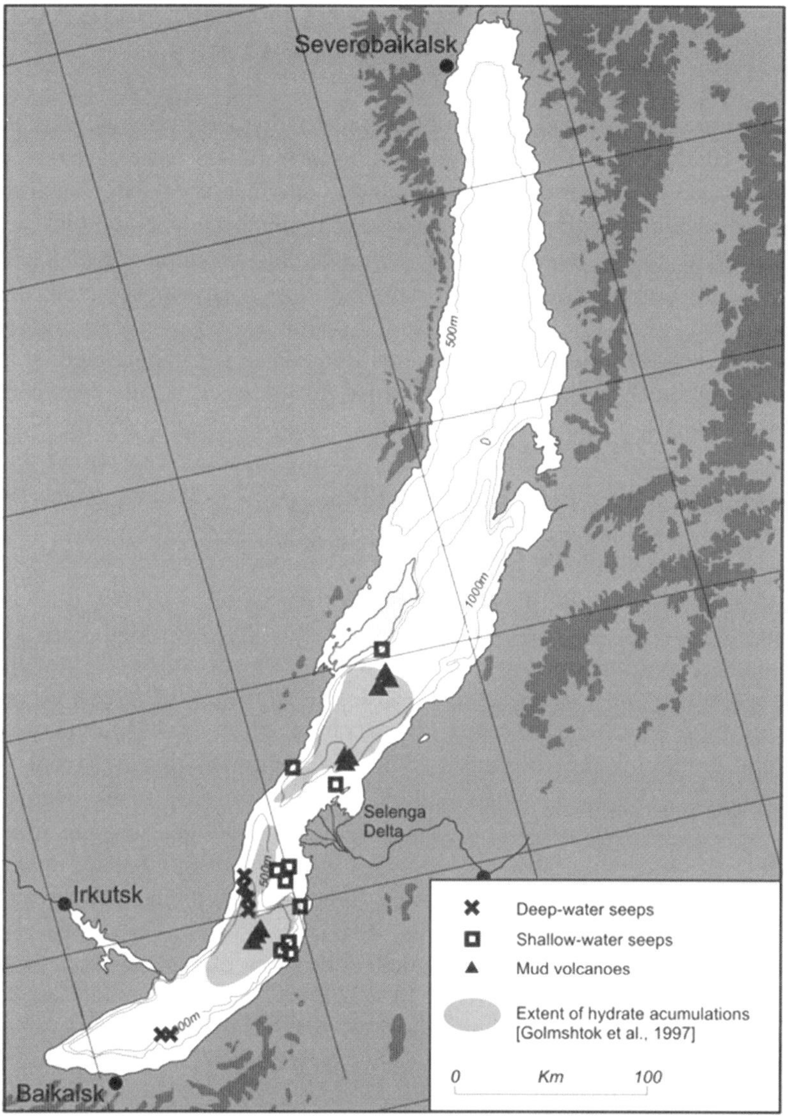

Figure 1. Overview map of Lake Baikal, with indication of the extent of the hydrate province and of the location of mud volcanoes, and of active deep-water and shallow-water seeps.

2. STRUCTURAL OUTLINE OF LAKE BAIKAL AND THE VENTING STRUCTURES

Lake Baikal consists of three distinct bathymetric basins: the South Baikal (SBB), Central Baikal (CBB) and North Baikal basins (NBB). Each of these also corresponds to a well-defined sedimentary basin, bounded by steep border faults, which are best expressed along the northwestern side of the rift. The three basins are at least moderately asymmetric and tend to adopt a half-graben shape (Zonenshain et al., 1992 ; Hutchinson et al., 1992 ; Scholz et al., 1993). They are separated by active high-relief morphostructural ridges, or accomodation zones s.l.. The Selenga Delta Accomodation Zone consists of a broad bathymetric saddle between the SBB and CBB. Although the saddle is largely obliterated by the Selenga delta sediments, it is underlain by several elevated basement blocks (Hutchinson et al., 1992 ; Zonenshain et al., 1992 ; Colman et al., 1993 ; Scholz et al., 1993) (Fig. 1).

The area south of the Selenga delta is complicated in its central part by the Posolsky Bank, a morphological ridge, striking SW-NE, transverse to the basin shape (Fig. 2). It is assumed to be a large back-tilted fault block, belonging to the Selenga Delta Accomodation Zone (Scholz and Hutchinson, 2000) and separated at its south-eastern side from the SBB by a relatively narrow fault zone with a vertical displacement of about 900 m. The south-eastern limit of the Posolsky Bank is delimited by a steep slope corresponding to the Posolsky fault transiting abruptly to the deep-water SBB at depths of about 1100 m. The basin floor gradually deepens to about 1300 m when reaching the steep slope of the main eastern border fault. The sediments of this deep basin are not affected by major faults (Logatchev, 1993 ; Zonenshain et al., 1992 ; Mats, 1993).

The venting area is located immediately south of the structural high of Posolsky Bank. It is delimited at its northern side by a small, secondary fault, antithetic to the Posolsky fault (Fig. 3) (Van Rensbergen et al., 2002). The venting structures are aligned almost parallel to the small antithetic fault (Fig. 2), which offsets the lake floor by about 20 m. Some of the venting structures themselves are also faulted.

The present or recent activity of the venting structures is testified by the freshness of their morphology as observed on side-scan sonar mosaics (Fig. 2), and by the presence of 10-25 m high acoustically non-transparent plumes in the water column above them or along their rims. The venting structures coincide with "gas chimneys"-or fluid-flow conduits- (Figs. 3 and 4) in the sub-surface that are observed on several seismic profiles (e.g. Van Rensbergen et al., 2002; 2003).

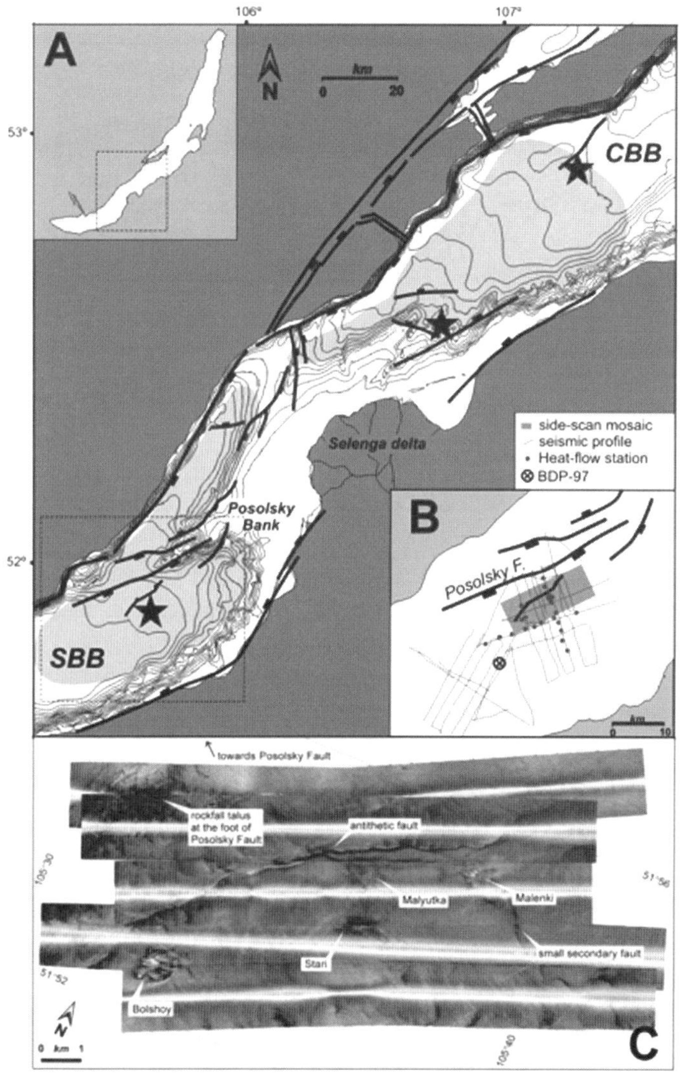

Figure 2. Location map of the study area within the SBB near the Selenga River delta, showing the bathymetry and the main faults (Scholz & Hutchinson, 2000; Mats et al., 2000). The area with inferred hydrate occurrences in the SBB and CBB (Golmshtok et al., 1997) is indicated in grey. Stars indicate areas with anomalously shallow and disrupted BSR and/or with mud volcanoes. A: General location of the study area within Lake Baikal. B: Available data from the Posolsky Bank study area. C. Side-scan sonar mosaic of the Pososlky Bank venting area, with clear indication of the venting structures (i.e. Bolshoy, Stari, Malyutka and Malenki).The venting area is located immediatelhy south of the structural high of Posolsky Bank. It is delimited at its northern side by a small, secondary fault, antithetic to the Posolsky fault (Fig. 3) (Van Rensbergen et al., 2002). The venting structures are aligned almost parallel to the small antithetic fault, which offsets the lake floor by about 20 m. Some of the venting structures themselves are also faulted.

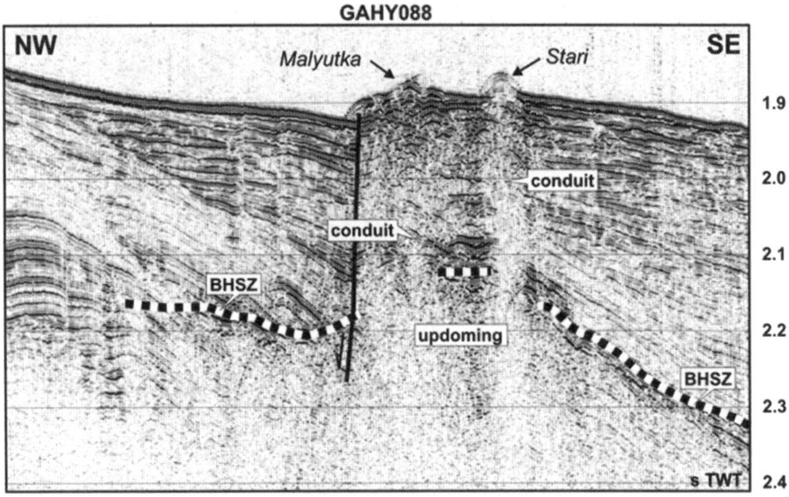

Figure 3. Seismic profile GAHY088 crossing the Stari and Malyutka venting structures, showing the irregular behaviour of the "BSR" and the presence of fluid-flow conduits adjacent to the small fault.

3. METHANE ESCAPE AND ITS RELATION TO BSR AND HEAT FLOW ANOMALIES

A distinct BSR is evidenced in the area south of the Selenga delta on single-channel high-resolution seismic reflection profiles (Vanneste et al., 2000; De Batist et al., 2002). Outside the venting area, the BSR runs regularly and smoothly at a predictable sub-bottom depth, in agreement with the gradually deepening lake floor. In the venting area, the BSR no longer mimics the lake floor, but it becomes highly disrupted and rises to a very shallow sub-bottom depth (figs. 3 and 4) (De Batist et al., 2002).

Intra-basin faults in Lake Baikal generally do not affect the continuity of the BSR or the thickness of the hydrate-bearing layer: the BSR simply cuts across the folded and faulted stratigraphic reflections (Golmshtok et al., 1997). The irregular pattern of the BSR and its disruption suggest the presence of very localized, drastic variations in one or more of the variables controlling the depth of the hydrate phase boundary: i.e. temperature (higher than average), pressure (lower than average) or gas composition. Pressure as a controlling factor can be ruled out as it is difficult to invoke a mechanism by which sub-bottom pressures would be generated that are lower than the background hydrostatic pressure. Heat-flow values in the studied area, determined using marine thermoprobe measurements of the temperature gradient and conductivity in surface sediments, are generally slightly higher (between 55 and 90 mW/m^2, locally up to 165 mW/m^2) than the average heat-flow values for the Baikal Basin which are about 50-70 mW/m^2 (Duchkov et al., 1999). This indicates that the irregular morphology of the BHSZ is predominantly temperature-controlled and directly related to the existence of a local anomaly of high heat flow.

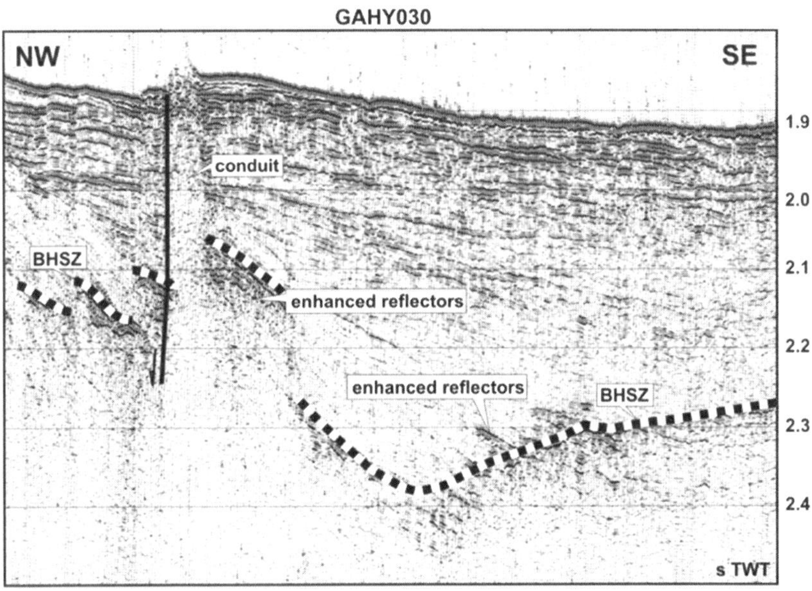

Figure 4. Seismic profile GAHY030 across the small antithetic fault and the adjacent Malenki mud venting structure, showing the strongly disrupted "BSR" and the acoustically chaotic zone extending to the lake floor and affecting the lake-floor morphology.

As the lake-floor venting structures in the deeper parts of Lake Baikal only occur in areas of anomalously shallow and disrupted BSR, characterised by an anomaly of high heat flow, it is concluded that they are the surface expression of escape pathways for excess methane generated by the dissociation of pre-existing hydrates. Moreover, as the gas-venting structures are spatially associated with active faults, it is postulated that gas escape is favoured by the presence of these faults. The fact that the gas-venting structures appear to be confined to particular segments of active faults, suggests that upward fluid flow along the fault plane may have been facilitated by recent dilatation ("leakage") along that segment. Given the fact that the hydrate province in Lake Baikal is located in an active tectonic region, such a process of tectonically driven upward fluid flow and temperature-induced destabilization of hydrates may occur quite regularly, but along different segments of the various active faults.

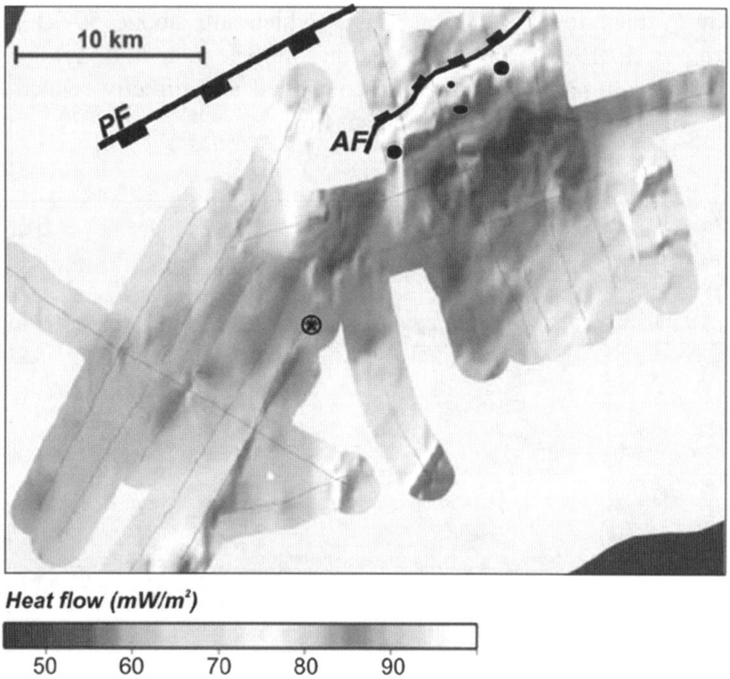

Figure 5. Regional contour map of the BSR-derived heat flow for the Posolksy Bank area (after Vanneste et al., 2003), showing that the localized heat-flow high in the venting area is flanked by a prominent heat-flow low towards the south.

As the position of the BSR, which in general mimics the lake-floor topography, is primarily controlled by temperature, it can be used to estimate the local geothermal field, and consequently to construct a regional contour map of the heat flow (Fig. 5) (Vanneste et al., 2003). This map shows the irregular pattern of the heat flow, with in particular positive anomalies over the venting area, flanked to the south by an area of very low heat flow. However, when comparing the heat flow inferred from the BSR with direct heat-flow measurements, it appears that deviations between measured and inferred heat flow are up to 40 %. The measured heat flow ranges from 56 up to 165 mW/m^2 and is consistently higher than background values.

Maxima are located on the venting sites. The concentric shape of the thermal anomaly with heat-flow values of twice the background heat flow in the venting sites is typical for the focused upward flow of warm fluids (Fig. 6).

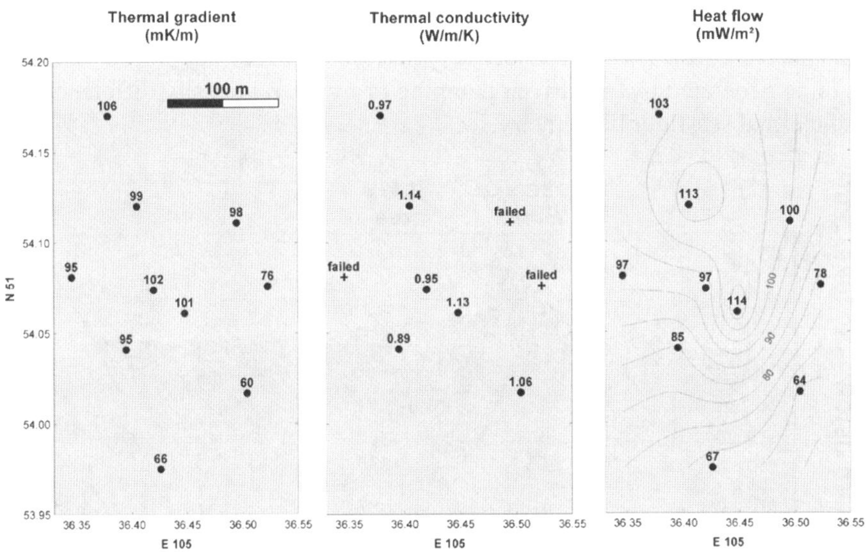

Figure 6. Lateral distribution of the thermal gradient, thermal conductivity and heat flow in and around the Stari mud volcano.

4. INTERPRETATION ON THE ORIGIN OF THE METHANE ESCAPE STRUCTURES

4.1 Presence of a convective fluid flow system

The discrepancy that is locally observed between measured and inferred heat flow implies that the conditions at the surface or/and at the BHSZ are in a non-equilibrium state. Only heat transport by fluid flow is believed to be strong enough to create heat-flow differences of 20-50 % between surface and BSR-depth. In addition, fluid flow is the only process, which can ultimately explain a strongly reduced heat flow at the BSR-depth. This fluid flow has to be downward and active only at BSR-depth. In general, in sedimentary basins fluids are expelled upwards, driven by compaction. Therefore, downward flow demands a driving force which is opposite the compaction and stronger. This force can be delivered by density differences in the fluids and will result in local convection loops.

A conceptual model for a convective fluid loop that accounts for both the positive heat-flow anomalies near the venting sites and the negative heat-flow anomalies zone south of the venting sites, has been proposed (Fig. 7) (Vanneste et al., 2003). The density differences necessary for such a loop can be generated by gas-driven pumping or by lateral density differences by isothermal sea-floor bathymetry.

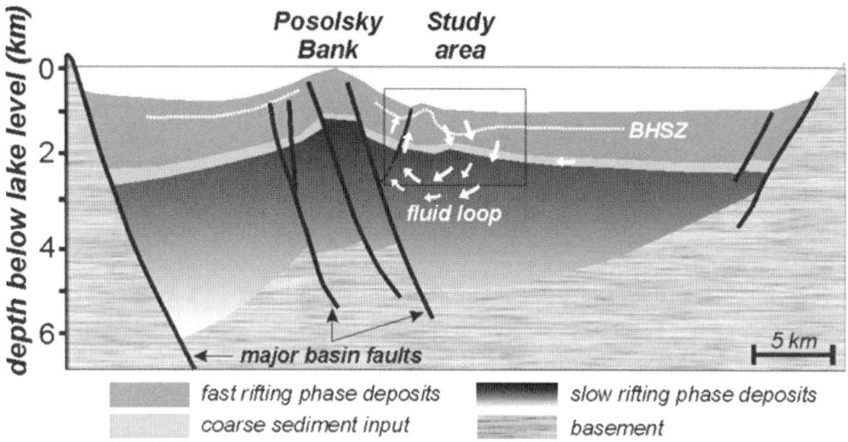

Figure 7. Conceptual fluid-loop model for explaining the characteristic heat-flow signature in the Posolsky Bank study area (after Vanneste et al., 2003).

4.2 Destabilisation of gas hydrates

The cold methane seeps in Lake Baikal are associated with expulsion of sediment which is the result of increase of pore-fluid pressure in the sub-surface to lithostatic values, initiating hydraulic fracturing of the overburden and fluid expulsion. The origin of gas and pore-fluid overpressure appears to be the rapid increase of pore-fluid volume caused by the disintegration of gas hydrates.

Methane seeps are often interpreted to result from gas- and fluid-release by gas-hydrate destabilization in response to various processes. However, gas-hydrate dissociation alone is usually not sufficient to explain methane seepage, as has been shown for example, at the Blake Ridge gas-hydrate area (Paull et al., 1995): gradual decomposition of hydrates at the base of the hydrate layer is insufficient to account for the estimated upward fluid flow (Egeberg, 2000) and, consequently, for the methane released at the sea floor.

In Lake Baikal, the existence of a local anomaly of high heat flow associated with the seep area is most likely due to a heat pulse causing the dissociation of the underlying gas hydrates. The heat pulse can be caused by upward flow of geothermal fluids along active segments of the Posolsky fault, maybe accelerated by seismic pumping. The elevated heat flow is probably transient as it is related to fault activity that may jump to other segments of the fault, possibly causing destabilization of hydrates, seeps, and mud volcanoes elsewhere in the basin (Van Rensbergen et al., 2002).

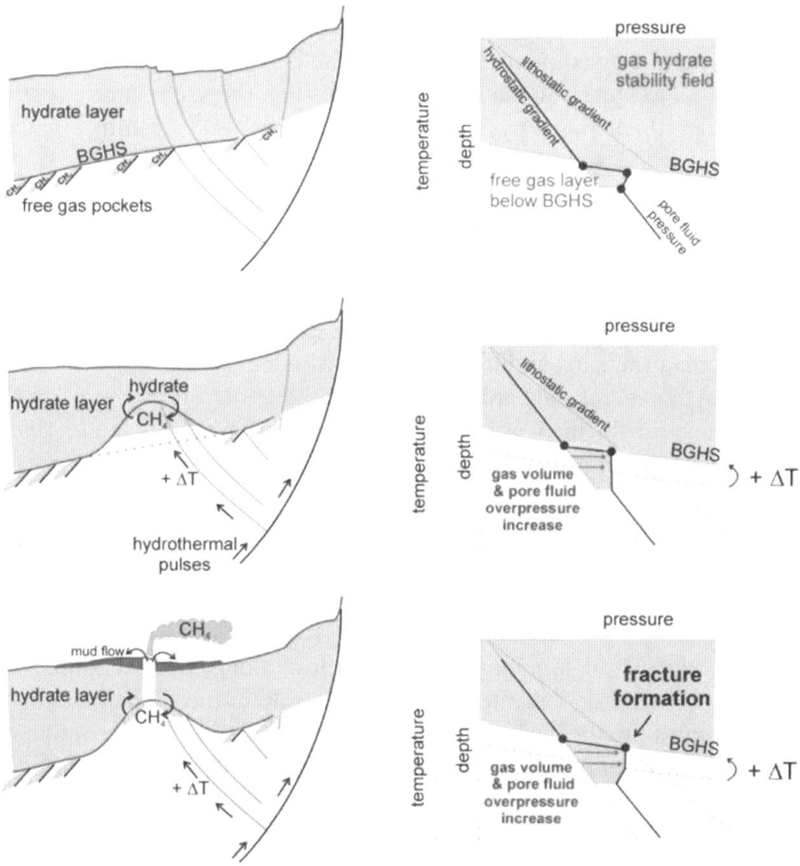

Figure 8. Schematic model explaining the localised destabilisation of gas hydrates in response to a structurally guided upward flow of fluids advecting heat to the BHSZ, creating an overpressure compartment and leading to the expulsion of remobilised sediment and fluids at the lake floor as gas seeps and mud volcanoes (after Van Rensbergen et al., 2003).

4.3 A tectonic driving force

Among the processes that may be involved in the shallowing of the base of the GHSZ, the destabilization of the gas hydrates and the upward channeling of fluids (Dillon et al., 1998; Henriet and Mienert, 1998; Tréhu et al., 1999), only two are relevant to the specific sedimentary and tectonic environment of Lake Baikal: compaction-driven fluid flow and tectonically driven fluid flow.

- *Compaction-driven fluid flow* may indeed play a role, as the area south of Selenga delta is the site of major sediment input in the lake. On the Selenga delta the sedimentation rate for the last 30,000 years is between 15 and 19 cm/kyr (Colman et al., 1996). When the sedimentary section is rapidly thickening by sedimentation, the GHSZ shifts upward continuously destabilizing gas hydrates at its base. Free gas consequently accumulates from the dissociation of the gas hydrates (von Huene and Pecher, 1999). When a fault reaching the base of the GHSZ is reactivated, this fault can tap the area of fluid accumulation underneath the BSR allowing the fluids to discharge along the fault and to reach the lake floor. Fluids ascending from below the BSR are warmer than the surrounding sediments, which results in destabilization of gas hydrates and upward migration of the BSR. Such a process has been proposed for the Cascadia convergent margin (Suess et al., 1999). The presence of mud volcanoes in the Black Sea (Limonov et al., 1997), under an extensional regime, also is interpreted by fast subsidence and burial of mobile source beds resulting in overpressure. But the high rapid sediment accumulation doesn't account for the irregular behavior (i.e. the combined occurrence of areas with shallower-than-background and deeper-than-background BSR) of the BSR.
- *Tectonically driven fluid flow,* by fluid flow along fault conduits, has been sufficiently documented for playing a role in the destabilization of gas hydrates in the compressional environment of active continental margins (a.o. Booth et al., 1998; von Huene and Pecher, 1999; Suess et al., 1999).

Even in the extensional tectonic environment of Lake Baikal, local compressional structures have been observed. Levi et al. (1997) emphasizes folding structures affecting the sedimentary section in the western part of the SBB, consisting of NW-SE trending folds with up to 1 km amplitude and 7-8 km wavelength. The folded structures are attributed either to a phase of horizontal principal compression in a NE-SW direction, parallel to the longitudinal axis of the basin, or -more probably- to strike-slip movements. Folding of the sedimentary section also has been evidenced on many seismic sections obtained in the SBB and CBB by Golmshstok et al. (2000), who relate these folding structures to movements along strike-slip faults. Their axes are considered to be almost orthogonal to the axis of the basin. Also a prominent roll-over structure is evidenced in the eastern part of the central basin (Moore et al., 1997). This roll-over structure is considered to have developed at the western side of a strike-slip fault zone which extends over more than 50 km in the CBB.It is interesting to note that the fold axes have a WNW trend. On the contrary, the axes of depocentres for several depositional stages trend NNE, oblique to the main fault structures.

Evidences for a left-lateral strike-slip component in the main extensional kinematics of the Baikal basin consequently exist, the axes of compression varying between WNW in the CBB and NW in the extreme south of the SBB. Although the significance of the strike-slip movement for the tectonic evolution of the Baikal basin is not fully understood, it is evident that it plays a role in the deformational pattern of the sedimentary fill.

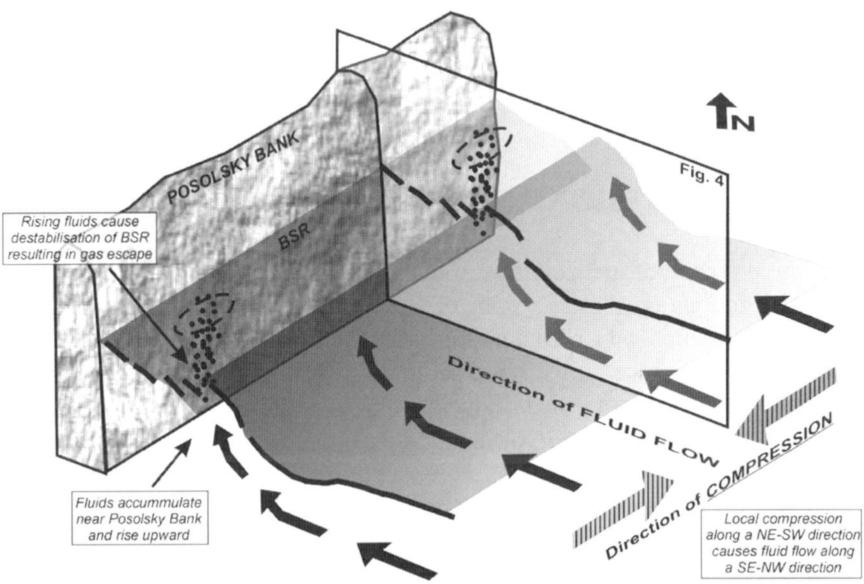

Figure 9. Schematic diagram showing the fluid flow that is generated along the NW-SE trending compression related to NE trending strike-slip movement. Fluids accumulate against Posolskii Bank, resulting in updoming of the BSR and CH_4 escape along NE trending extensional faults.

The left-lateral strike-slip movement that has been evidenced results in shortening of the sedimentary layers and dewatering of the sediments, and consequently generates a fluid displacement. The direction of fluid transport should be NW, orthogonal to the direction of principal compression which is

NE-SW. As this direction corresponds to the progressive shallowing of the BSR, we propose that fluids are moving northwestwards along the sedimentary layering in the fluid-enriched zone underneath the base of the GHSZ. The basement block of Posolsky Bank forms a barrier for the moving fluids. When approaching this barrier, the amount of fluids progressively increases and their heat supply results in upward movement of the BSR. Close to Posolsky Bank, fluids will consequently concentrate, creating overpressure that results in thermal up-doming and finally disruption and destabilisation of the gas-hydrate layer. When a fault oriented along the NE-SW major direction of extension will be activated, the BSR is disrupted and fluids erupt along this fault. Also in the other areas of known mud volcanism in Lake Baikal, the mud volcanoes and methane escape occur along major extensional intra-basin faults trending NE-SW.

Although the rapid sediment accumulation in the area of the Selenga delta may play a role in the process of fluid accumulation, it is assumed that fluid flow induced by left-lateral strike-slip movements along the border faults is a major factor for the fluid accumulation, which is responsible for up-doming and destabilization of the base of the GHSZ, and consequently generation of methane and its escape on the lake floor.

5. DISCUSSION AND CONCLUSIONS

Active escape of methane and/or other fluids - i.e. cold seeps – occurs at several locations on the lake floor of Lake Baikal, forming mud volcanoes which involve not only methane/fluid seepage into the lake water, but also mobilize and expulse sub-surface sediment. In all cases, methane venting occurs along well-constrained segments of active, first-order rift-basin faults. Moreover, the areas of mud volcanism are always underlain by gas hydrates in the sub-bottom sediments.

A tentative model is proposed which relates the methane released at the lake floor in the seep areas, as well as the associated expulsion of re-mobilized sub-surface sediments, to the destabilization of gas hydrates at the base of the hydrate-bearing sediment section. This destabilization - involving the disruption of the BSR, a localized shallowing of the BHSZ, and the generation of an overpressure compartment - must have occurred in response to an upward flow of geothermal fluids, advecting heat to the BHSZ. It is assumed that this fluid flow is tectonically triggered, considering that even in an extensive tectonic environment, a reduced lateral displacement component is able to generate a fluid flow in the compressional direction, resulting in fluid escape along faults directed along the main direction of extension. The tectonic effect may be coupled to the sediment compaction due to a high sedimentation rate. Both processes may generate a large-scale convective fluid loop within the basin-fill sediments which advects gases and fluids of deeper origin to the shallow sub-surface within the upward-flowing limb of this fluid loop.

Is methane escape on the lake floor of Lake Baikal presenting analogies with gas release in the process of CO_2 sequestration? There are major differences between the natural occurrence of methane venting in Lake Baikal, related to destabilisation of gas hydrates, and the processes involved in CO_2 sequestration: in Lake Baikal, only methane is released on the lake floor, and moreover, strong overpressure is created when huge amounts of methane are liberated when gas hydrates dissociate. Nevertheless, the tectonic processes resulting from local compression that are supposed to play a role in fluid-flow generation in Lake Baikal, as well as the role of sediment compaction due to a high sedimentation rates, may also be effective in the processes of CO_2 sequestration.

Heat flow anomalies may be an indicator of such processes. It is clearly evidenced in the area of mud volcanism in Lake Baikal that the gases ascending from depths are accompanied by fluids that are warmer than the environment where they escape, and that they create local heat flow anomalies. These heat flow anomalies are consequently an indicator of intensive, even cold gas and fluid seepages when the fluids generate at a sufficient depth.

References

Booth, J.S., Winters, W.J., Dillon, W.P., Clennell, M.B. and Rowe, M.M., 1998, Major occurrences and reservoir concept of marine clathrate hydrates : implications of field evidence. In: *Gas Hydrates: Relevance to World Margin Stability and Climate Change,* J.P. Henriet and J. Mienert, eds, Geological Society, London, Special Publications, **137**: 113-127.

Brown, K.M., 1990, The nature and hydrological significance of mud diapirs and diatremes for accretionary systems, *Journal of Geophysical Research*, **95**:8969-8982.

Calais, E., Lesne, O., Deverchere, J., San'kov, V., Lukhnev, A., Miroshnitchenko, A., Buddo, V., Levi, K., Zalutzky, V., Bashkuev, Y., 1998, Crustal deformation in the Baikal rift from GPS measurements, *Geophysical Research Letters,* **25**: 4003-4006.

Colman, S.M., Jones, G.A., Rubin, M., King, J.W., Peck, J.A, and Orem, W.H., 1996, AMS radiocarbon analyses from Lake Baikal, Siberia: Challenges of dating sediments from a large, oligotrophic lake, *Quaternary Geochronology*, **15**:669-684.

De Batist, M., Klerkx, J., Van Rensbergen, P., Vanneste, M., Poort, J., Golmshtok, A., Kremlev, A., Khlystov, O., and Krinitsky, P., 2002, Active Hydrate Destabilization in Lake Baikal, Siberia ? *Terra Nova,* **14**(6):436-442.

Dillon, W.P., Danforth, W.W., Hutchinson, D.R., Drury, R.M., Taylor, M.H., and Booth, J.S., 1998, Evidence for faulting related to dissociation of gas hydrate and release of methane off the southeastern United States, In: *Gas Hydrates: Relevance to World Margin Stability and Climate Change,* J.P. Henriet and J. Mienert, eds,. Geological Society, London, Special Publications, **137**, pp. 293-302.

Duchkov, A.D., Lysak, S.V., Golubev, V.A., Dorofeeva, R.P., and Sokolova, L.S., 1999, Heat flow and geothermal field of the Baikal region, *Geologiya i Geofizika,* **40**:287-303.

Ginsburg, G.D., 1998. Gas hydrate accumulation in deep-water marine sediments. In: *Gas Hydrates : Relevance to World Margin Stability and Climate Change,* J.P. Henriet and J. Mienert, eds. Geological Society, London, Special Publications, **137**, pp. 51-62.

Ginsburg, G.D. and Soloviev, V.A., 1997, Methane migration within the submarine gas-hydrate stability zone under deep-water conditions, *Marine Geology*, **137**:49-57.

Golmshstok, A.Y., Duchkov, A.D., Hutchinson, D.R., Khanukaev, S.B., and El'nikov, A.I., 1997, Estimations of heat flow in Lake Baikal by seismic data on the lower boundary of gas hydrate layer, *Russian Geology and Geophysics*, **38**(10):1714-1727.

Golmshstok, A.Ya., Duchkov, A.D., Hutchinson, D.R., and Khanukaev, S.B., Heat flow and gas hydrates of Baikal Riftt Zone, *Geol. Rundschau*, in press.

Granin, N.G. and Granina, L.Z., 2002, Gas hydrates and gas venting in Lake Baikal. Russian Geology and Geophysics, **43**(7):589-597.

Henriet, J.P. and Mienert, J. (eds), 1998. *Gas Hydrates: Relevance to world margin stability and climatic change.* Geological Society, London, Special Publications, **137**.

Hutchinson, D.H., Golmshtok, A.J., Zonenshain, L.P., Moore, T.C., Scholz, C.A., and Klitgord, K.D., 1992, Depositional and tectonic framework of the rift basins of Lake Baikal from multichannel seismic data, *Geology*, **20**:589-592.

Hyndman, R.D. and Davis, E.E., 1992, A mechanism for the formation of methane hydrate and seafloor bottom-simulating reflectors by vertical fluid expulsion, *J. Geophys. Res.*, **97**(5B):7025-7041.

Kennett, J.P. and Fackler-Adams, B.N., 2000, Relationship of clathrate instability to sediment deformation in the upper Neogene of California, *Geology*, **28**(3):215-218.

Kuzmin, M.I., Kalmychkov, G.V., Geletii, V.F., 1998, First find of gas hydrates in sediments of Lake Baikal. *Doklady Akademii Nauk*, **362**(4):541-543 (in Russian).

Kuzmin, M.I., Kalmychkov, G.V., Duchkov, A.D., Getetii,V.F., Golmshstok, A.Y., Karabanov, E.B., Khakhaev, B.N., Pevzner, L.A., Ioshida, H., Bazhin, N.M., Dyadin, Y.A., Larionov, E.G., Manakov, A.Y., Mandel'baum, M.M., and Vashchenko, I.F., 2000, Methane hydrates in sediments of Lake Baikal. *Geology of Ore Deposits*, **42** (1):25-37 (in Russian).

Kvenvolden, K.A., 1988, Methane hydrate - a major reservoir of carbon in the shallow geosphere ? *Chemical Geology,* **71**:41-51.

Kvenvolden, K.A., 1993, Gas hydrates – Geological perspective and global change, *Reviews of Geophysics*, **31**:173-187.

Levi, K.G., Miroshnichenko, A.I., Sankov, V.A., Babushkin, S.M., Larkin, G.V., Badardinov, A.A., Wong, H.K., Colman, S., and Delvaux, D., 1997, Active faults of the Baikal depression, *Bull. Centre Rech. Elf Explor.-Prod.*, **21**(2):399-434.

Limonov, A.F., Van Weering, Tj.C.F., Kenyon, N.H., Ivanov, M.K., and Meisner, L.B., 1997, Seabed morphology and gas venting in the Black Sea mudvolcano area : Observations with the MAK-1 deep-tow sidescan sonar and bottom profiler, *Marine Geology*, **137**:121-136.

Logatchev, N.A., 1993, History and geodynamics of the Lake Baikal rift in the context of the Eastern Siberia rift system: a review, *Bull. Centres Rech. Explor-Prod. Elf Aquitaine*, **17**(2):353-370.

Mats, V.D., 1993, The structure and development of the Baikal rift depression, *Earth Science Reviews*, **34**:81-118.

Mats, V.D., Khlystov, O.M., De Batist, M., Ceramicola, S., Lomonosova, T.K., and Klimansky, A., 2000, Evolution of the Academician Ridge Accommodation Zone in the central part of the Baikal Rift, from high-resolution reflection seismic profiling and geological field investigations, *International Journal of Earth Sciences,* **89**(2):229-250.

Moore, T.C., Klitgord, K.D., Golmshstok, A.J., and Weber, E., 1997, Sedimentation and subsidence patterns in the central and north basins of Lake Baikal from seismic stratigraphy, *Geol. Soc. America Bull.*, **109**(6):746-766.

Paull, C.K., Ussler, W., Borowski, W.S., and Spiess, F.N., 1995, Methane-rich plumes on the Carolina continental rise - associations with gas hydrates, *Geology*, **23**(1):89-92.

Petit, C., Burov, E., Deverchere, J., 1997, On the structure and mechanical behaviour of the extending lithosphere in the Baikal rift from gravity modeling, *Earth and Planetary Science Letters,* **149** (1-4): 29-42.

Petit, C., Koulakov, I., Deverchere, J., 1998, Velocity structure around the Baikal rift zone from teleseismic and local earthquake traveltimes and geodynamic implications, *Tectonophysics,* **296** (1-2): 125-144.

Scholz, C.A.,and Hutchinson, D.R., 2000, Stratigraphic and structural evolution of the Selenga Delta Accommodation Zone, Lake Baikal Rift, Siberia, *International Journal of Earth Sciences,* **89**:212-228.

Soloviev, V.A., and Ginsburg, G.D., 1997, Water segregation in the course of gas hydrate formation and accumulation in submarine gas-seepage fields, *Marine Geology*, **137**:59-68.

Suess, E., Torres, M.E., Bohrmann, G., Collier, R.W., Greinert, J., Linke, P., Rehder, G., Trehu, A., Wallmann, K., Winckler,G., and Zuleger, E., 1999, Gas hydrate destabilization: enhanced dewatering, benthic material turnover and large methane plumes at the Cascadia convergent margin, *Earth and Planetary Science Letters*, **170**:1-15.

Trehu, A.M., Torres, M.E., Moore, G.F., Suess, E., and Bohrmann, G., 1999, Temporal and spatial evolution of a gas hydrate-bearing accretionary ridge on the Oregon continental margin, *Geology*, **27**(10):939-942.

Vanneste, M., Poort, J., De Batist, M., and Klerkx, J., 2003, Atypical heat flow near gas hydrate irregularities and cold seeps in the Baikal Rift Zone, *Marine and Petroleum Geology,* **19**:1257-1274.

Van Rensbergen, P., De Batist, M., Klerkx, J., Hus, R., Poort, J., Vanneste, M., Granin, N., Khlystov, O., and Krinitsky, P., 2002, Sublacustrine mud volcanoes and methane seeps caused by dissociation of gas hydrates in Lake Baikal, *Geology*, **30**(7):631-634.

Van Rensbergen, P., Poort, J., Kipfer, R., De Batist, M., Vanneste, M., Klerkx, J., Granin, N., Khlystov, O., and Krinitsky, P., 2003, Evidence of near-surface sediment mobilization and methane venting in relation to hydrate dissociation in Southern Lake Baikal, Siberia. In: *Subsurface sediment mobilization.* P. Van Rensbergen, R.R.Hillis, A.J. Maltman and C.K. Morley, eds., Geological Society Special Publication, **216**:207-221.

Von Huene, R., and Pecher, I.A., 1999, Vertical tectonics and the origin of BSRs along the Peru margin, *Earth and Planetary Science Letters*, **166**:47-55.

Von Rad, U., Berner, U., Delisle, G., Doose-Rolinski, H., Fechner, N., Linke, P., Lückge, A., Roeser, H.A., Schmaljohann, R., and Wiedicke, M., 2000, Gas and fluid venting at the Makran accretionary wedge off Pakistan, *Geo-Marine Letters*, **20**:10-19.

Zonenshain, L.P., 1992, Baikal rift structure, *Geotectonics*, **26**(5):396-407.

PART IV

ACTIVE CO_2 INJECTION SITES

Monitoring Strategies and System Evolution of Active, Man-Made Carbon Dioxide Storage Sites

THE IEA WEYBURN CO_2 MONITORING AND STORAGE PROJECT

Integrated results from Europe

J.B. Riding

British Geological Survey, Keyworth, Nottingham NG12 5GG, UK

Abstract: The IEA Weyburn CO_2 Monitoring and Storage Project has analysed the effects of a miscible CO_2 flood into a carbonate reservoir rock at an onshore Canadian oilfield. Anthropogenic CO_2 is being injected as part of an enhanced oil recovery operation. The European research was aimed at analysing long-term migration pathways of CO_2 and the effects of CO_2 on the hydrochemical and mineralogical properties of the reservoir rock. The long term safety and performance of CO_2 storage was assessed by the construction of a Features, Events and Processes (FEP) database which provides a comprehensive knowledge base for the geological storage of CO_2. The pre-CO_2 injection hydrogeological, hydrochemical and petrographical conditions in the reservoir were investigated in order to recognise changes caused by the CO_2 flood and to assess the fate of the CO_2. The Mississippian aquifer has a salinity gradient in the Weyburn area, where flows are oriented SW-NE. The baseline gas fluxes and CO_2 concentrations in groundwater and soil were also studied. The dissolved gas in the reservoir waters has allowed potential transport pathways to be identified. Experimental studies of CO_2-porewater-rock interactions in the Midale Marly unit have indicated slight dissolution of carbonate and silicate minerals, but relatively rapid saturation with respect to carbonate minerals. Equivalent studies on the overlying and underlying units show similar reaction processes, but secondary gypsum precipitation was also observed. Carbon dioxide flooding experiments on samples of the Midale Marly unit demonstrated that porosity and gas permeability increased significantly and calcite and dolomite were shown to have undergone corrosion. Hydrogeological modelling indicates that if any dissolved CO_2 entered the main aquifers, it would be moved away from Weyburn in an E-NE direction at a rate of *c.* 0.2 metres per year due to regional groundwater flow. Analysis of reservoir fluids proved that dissolved CO_2 and CH_4 increased significantly in the injection area between 2002 and 2003 and that solubility trapping accounts for the majority of the injected CO_2, with little apparent mineral trapping. Twelve microseismic events were recorded and these are

S. Lombardi et al. (eds.),
Advances in the Geological Storage of Carbon Dioxide, 223–230.

provisionally interpreted as possibly being related to small fractures formed by injection-driven fluid migration within the reservoir. Pre- and post-injection soil gas data are consistent with a shallow biological origin for the measured CO_2. Isotopic ($\delta^{13}C$) data values are higher than in the injected CO_2, and confirm this interpretation. No evidence for leakage of the injected CO_2 to ground level has so far been detected.

Key words: CO_2 injection; geological storage of greenhouse gases; enhanced oil recovery; geoscientific monitoring; safety assessment studies

1. INTRODUCTION

The IEA Weyburn CO_2 Monitoring and Storage Project was a collaborative initiative involving workers from North America and Europe and was co-ordinated by the Petroleum Technology Research Centre (PTRC) in Regina, Canada. The project has studied the geological sequestration of CO_2 during a commercial enhanced oil recovery (EOR) operation at the Weyburn oilfield, Saskatchewan, Western Canada. EnCana Resources (formerly PanCanadian Resources) began injecting a 95% pure CO_2 stream into the oil reservoir, the Midale Beds, at Weyburn during September 2000. At the conclusion of this EOR operation, it is anticipated that over 20 million tonnes of anthropogenic CO_2 will have been permanently and safely sequestered in these Mississippian strata, which are *c.* 1.4 km underground. Hence, greenhouse gas emissions will have been significantly reduced as part of a cost-effective industrial operation. The principal objective of the monitoring project was to enhance scientific knowledge pertaining to the underground sequestration of CO_2 associated with EOR using several different techniques in geoscientific monitoring. Another major strand to this work has been the promotion of international collaboration on carbon management research between scientists in Canada, the USA and Europe. The results from this project will help guide policy makers on carbon management issues. The European project team comprises nine organisations and was co-funded by the European Commission (EC).

2. THE GEOLOGY OF THE WEYBURN OILFIELD AND THE CO_2-EOR OPERATION

The Weyburn oilfield is in south-east Saskatchewan, Canada (Figure 1a) and lies in the north-eastern part of the Williston Basin. It was discovered in 1954, is now operated by EnCana Resources and covers some 180 square kilometres of prairie. Medium gravity sour crude oil (25 to 34 degree API) is produced from the uppermost Midale Beds (Charles Formation), which represent a transgressive-regressive cycle and comprise a succession of upwards shoaling, shallow marine carbonate-evaporite sediments of Mississippian (Early Carboniferous) age. The Midale Vuggy unit is a succession of highly fractured and permeable vuggy, heterolithic limestones that were deposited in relatively deep water. This unit is overlain by the cryptocrystalline dolomites of the Midale Marly unit, which represent shallow water conditions (Wegelin, 1987). The Midale Marly unit contains the majority of the remaining oil reserves and is the target for the miscible CO_2 flood (Figure 1b).

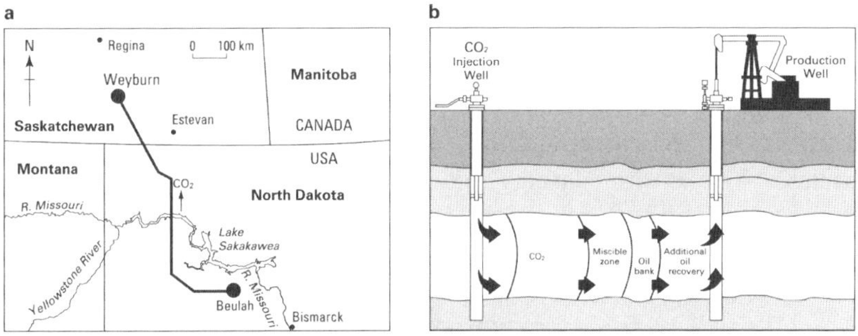

Figure 1. (a) location of the Weyburn oilfield and the route of the CO_2 pipeline. (b) how a miscible CO_2-EOR flood produces incremental oil; at the Weyburn unit, the depth to the reservoir unit is *c.* 1400 m.

Since 1964, water injection has been used as a secondary oil recovery strategy and has efficiently swept incremental oil from the Midale Vuggy Beds due to the high permeability of this unit. The water flood helped the field to achieve its peak production in 1965. The field was revitalized in 1991 by the drilling of horizontal wells. The Midale Marly Beds are less permeable and CO_2 injection is an effective method of extracting oil from this unit. This is because CO_2 is an excellent solvent and dissolves into the oil, thereby simultaneously reducing viscosity and increasing volume. The reduced viscosity makes the oil flow more easily and the swelling of the CO_2-rich oil enhances reservoir pressure. It is anticipated that this CO_2-EOR operation will extend the life of the Weyburn oilfield by the production of 130 million barrels of incremental oil (Figure 1b). The injection of CO_2 was started in September 2000 as Phase 1A and the initial injection rate was 5000 tonnes per day. Injection was originally in 18 patterns of nine wells, each at the west end of the oilfield. The CO_2 flood has been extended in a south-easterly direction and the ultimate aim is to flood 75 patterns in phases over the next 15 years. The CO_2 used is a purchased by-product of coal gasification and supplied to Weyburn through a 320 km long pipeline from the Great Plains Synfuels Plant in Beulah, North Dakota, USA that is operated by the Dakota Gasification Company (Figure 1a) (Stelter, 2001).

3. THE EUROPEAN PART OF THE IEA WEYBURN CO_2 MONITORING AND STORAGE PROJECT

The scientific tasks of the European research on the Weyburn oilfield are divided into three Work Packages (Riding et al., 2003). The scope and principal conclusions of each of these Work Packages are given in this section. This research was aimed principally at studying the potential long-term migration pathways of CO_2, and the effects of CO_2 on the hydrochemical and mineralogical properties of the reservoir rock at Weyburn. These are largely controlled by the regional geology, hydrodynamics and hydrochemistry of the aquifer systems.

3.1 Work Package 1 – Long-Term Safety and Performance of CO_2 Storage

The aim of this Work Package was to use systems analysis to provide a framework into which data can be integrated to give an assessment of the safety and reservoir integrity of CO_2 injection at the Weyburn oilfield. Features, events and processes (FEPs) describing the behaviour of CO_2 in the subsurface were used to describe and evaluate the storage concept and as a method of assessing all potentially significant factors. A generic FEP database for the geological storage of CO_2 has been developed. Natural analogue data obtained through the EC co-funded NASCENT project were also incorporated. The selected FEPs were included for their relevance to the long-term safety and performance of the reservoir, after the injection of CO_2 has been completed and the injection wells have been sealed. However, some FEPs pertaining to the injection phase have been included. The database contains some 200 FEPs arranged in a hierarchical structure that has eight categories. Each has a text description with references to the literature and discussion of its relevance to long-term performance and safety. The database is internet-enabled and incorporates hyperlinks to many other relevant information sources and is fully searchable. The generic (i.e. non Weyburn-specific) FEP database is available at http://www.quintessa-online.com/CO2/. It is therefore a comprehensive knowledge base for the geological storage of CO_2. Potentially important scenarios for the future evolution of a geological storage system were also considered. These scenarios need to be addressed in system-level models for the assessment of performance and safety. The use of the FEP database as an audit tool to evaluate the completeness of such models has been demonstrated. Deterministic and probabilistic performance assessment methodologies have been investigated to determine the long-term fate of CO_2 in the geosphere and biosphere.

3.2 Work Package 2 – Definition of Baseline Hydrogeological, Hydrochemical and Petrographical Conditions

The aim of this Work Package was to define the pre-CO_2 injection hydrogeological, hydrochemical and petrographical conditions in the Weyburn reservoir at both local and regional scales (Riding et al., 2003). A full understanding of these baseline conditions has allowed the recognition of changes resulting from the CO_2 flood and assessing the fate of the injected CO_2. To place the Weyburn oilfield in its regional context, the geological structure, hydrogeology and water chemistry were first compiled at the basin scale. The study then focused on a smaller area, centred on Weyburn. The geological structure of the area is a southward/south-westward-trending monocline. The Mississippian aquifer has a steep salinity gradient in the Weyburn area, where brines are concentrated in the south and are associated with Devonian evaporites. Salinity values in the north reflect previous dissolution of Devonian evaporites. The overall up-dip flows are dominantly oriented south-west to north-east. This is an important factor when investigating potential upward CO_2 migration and chemistry of water mixing by flooding.

The baseline chemistry of the aquifer systems was assessed at a regional scale and included analytical data from the Weyburn oilfield. Geochemical modelling has indicated the main *in-situ* chemical characteristics of the Mississippian reservoir fluids at Weyburn. Dissolved organic acid anions do not have a significant effect on the alkalinity; by contrast, sulphide concentrations may represent >60% of the total alkalinity. The reservoir waters are largely in thermodynamic equilibrium with respect to anhydrite, barite, calcite, dolomite and a silica phase. Dissolved aluminium may be controlled by clay minerals and/or K-feldspar.

Baseline mineralogical data have also been analysed to assess the initial chemical environment and to identify core material within the CO_2 flood area for use in hydrothermal experiments. Experimental geochemical studies on core samples from the first CO_2 flooding area have been carried out to ascertain key fluid properties. A further aspect of the baseline studies was to determine the pre-injection regional gas fluxes and concentrations in both groundwater and soil. The former has improved understanding of fluid-flow pathways throughout the Weyburn oilfield. A variety of dissolved and free gases and elements have been analysed. These help to determine the baseline water-rock interactions and allow potential rapid transport pathways to be identified.

Part of this task was to investigate whether soil gases and groundwater analyses can be used to identify the position of near surface features that may connect with the reservoir at depth. Seismic profiles were examined for faults, which may conduct gases and liquids above the reservoir, however none, which outcrop, were found. Consistent with this, the soil gas anomalies measured at Weyburn do not follow linear trends. The CO_2 distributions reflect both the origins and natural reactions that typify modern biogenic CO_2. The majority of the minor CO_2 anomalies may therefore be explained by anthropogenic or near surface phenomena. There is no correlation between these CO_2 anomalies with the injection wells or CO_2 pipelines. Furthermore, the expected seasonal variations in CO_2 flux data have been discerned. For example, measurements taken in September 2001 proved lower than those measured in July 2001 due to seasonal variations in soil humidity, vegetation and agricultural activity.

3.3 Work Package 3 – Define Changes to Baseline Hydrochemical, Hydrogeological and Petrographical Conditions Resulting from CO_2 Injection

Work Package 3 sought to assess the impact of injected CO_2 on the baseline conditions, especially reservoir geochemistry, mineralogy, fracture generation and fluid flow pathways (Riding et al., 2003). Experimental geochemical studies of CO_2-porewater-rock interactions investigated dissolution/precipitation reactions between CO_2, Midale reservoir rocks, the caprock, borehole cement and porewaters under reservoir conditions. The data were used in the geochemical modelling to show how mineral saturation changes when CO_2 is injected into the Midale Marly unit. It was found that the effects of CO_2 injection into Midale beds gave rapid reactions with variable amounts of carbonate mineral dissolution, some precipitation of gypsum and slow reactivity of aluminosilicate minerals. The Midale Vuggy samples proved the most reactive with etching of surfaces giving minor secondary porosity and secondary gypsum. By contrast, few observable mineralogical changes were observed in the Midale Marly samples. The Midale Evaporite caprock samples were altered slightly more than the Midale Marly lithologies and include some gypsum precipitation. The injection of CO_2 appears to have had little impact on the strength of the two cement types tested. Three CO_2 flooding experiments were completed on Midale Marly samples. Sample porosity and gas permeability increased significantly. Calcite and dolomite both underwent significant corrosion and some disintegration was observed. Anhydrite and gypsum were also corroded, whereas alkali feldspar, celestine and fluorite appeared to be unaffected. Mass balance calculations indicate a consistency between sample weight loss and the amounts of dissolved Ca^{2+} and Mg^{2+}.

The geochemical evolution of the Weyburn site has been achieved in an integrated study comprising field sampling, experimental studies and predictive geochemical modelling. It was determined that dissolved CO_2 in reservoir fluids increased 10-fold in the phase A1 injection area between 2002 and 2003; dissolved methane (CH_4) also increased significantly. Solubility trapping accounts for the majority of the injected CO_2, with little apparent mineral trapping. Regional scale hydrogeological modelling indicates that subsurface water flow is oriented south-west to north-east. Some contrasts in density and aquifer dips in the Weyburn area however, cause modifications to this principal direction. Should CO_2-charged groundwaters migrate from the reservoir succession, and the material were incorporated into the main aquifers, it is predicted that dissolved gas would only be moved 25 kilometres away from Weyburn in an east to north-east direction after 100,000 years.

Microseismic monitoring was undertaken to map hydraulic-induced fractures that have been stimulated by injection/EOR activities. Twelve microseismic events with magnitudes of -2.3 to -1.7 were recorded at 300 to 400 m from the observation well. Operations in a nearby well were shut down during ten of these events, hence this microseismicity is interpreted as possibly being related to the formation of small fractures produced by injection-driven fluid migration within the reservoir. The results of comprehensive soil gas surveys between 2001 and 2003 indicate shallow biological origins for the measured CO_2. Seasonal variability in the soil gas values and limited $\delta^{13}C$ isotopic data support this interpretation. There is no evidence so far of any leaks of injected CO_2 at the surface.

4. CONCLUSION

The Weyburn oilfield has proven to be an outstanding natural laboratory for the study of CO_2 storage; this is based on its comprehensive well/field data, the extensive core material and the accessibility of the site. The European part of the IEA Weyburn CO_2 Monitoring and Storage Project has derived significant scientific knowledge related to a commercial miscible CO_2-EOR operation. The structure of the research is intended to serve as a model, which can be applied to other CO_2 storage operations. Furthermore, the results from this project should help guide policy development on the abatement of greenhouse emissions from energy generation using underground storage. This project has also been notable for the many successful international collaborations which have taken place within the large project team. These have been both intra-European and transatlantic collaborations.

ACKNOWLEDGEMENTS

This contribution is published with the permission of the Executive Director, British Geological Survey (NERC). The European Commission is gratefully acknowledged for its generous funding of this work, which is EC project number NNE5-2000-00096 (contract ENK5-CT-2000-00304). The European research was undertaken by a collaboration of five partners; these are the British Geological Survey (BGS), the Bureau de Recherches Géologiques et Minières (BRGM), the Geological Survey of Denmark and Greenland (GEUS), the Istituto Nazionale di Geofisica e Vulcanologia (INGV) and Quintessa Limited. The Università di Roma 'La Sapienza' (URS) and the Institut Français du Pétrole (IFP) were research subcontractors to INGV and BRGM, respectively. The British Geological Survey co-ordinated the European project and the International Energy Agency Greenhouse Gas R&D Programme and the Petroleum Technology Research Centre Inc. (PTRC) facilitated linkages and integration.

REFERENCES

Riding, J. B., Czernichowski-Lauriol, I., Lombardi, S., Quattrocchi, F., Rochelle, C. A., Savage, D. and Springer, N., 2003, The IEA Weyburn CO2 Monitoring and Storage Project - the European dimension. In: Gale, J. and Kaya, Y. (eds) Greenhouse Gas Control Technologies, Volume II. Oxford: Elsevier Science Limited, 1629-1632.

Stelter, S., 2001, The New Synfuels Energy Pioneers. Bismarck, North Dakota: Dakota Gasification Company.

Wegelin, A., 1987, Reservoir characteristics of the Weyburn Field, southeastern Saskatchewan. Journal of Canadian Petroleum Technology, **26(**4):60-66.

ASSESSMENT OF THE LONG-TERM FATE OF CO_2 INJECTED INTO THE WEYBURN FIELD

System-Level Modeling of CO_2 Migration and Potential Impacts

M.J. Stenhouse, W. Zhou and R. Arthur
Monitor Scientific LLC, Suite #555, 3900 S. Wadsworth Blvd., Denver CO 80235, U.S.A.

Abstract: CO_2 is being injected into a 1450-m deep oil reservoir located in Weyburn, Saskatchewan, Canada, for enhanced oil recovery. To complement this commercial activity, a major research project to study geological sequestration and storage of CO_2, known as the International Energy Agency (IEA) Weyburn CO_2 Monitoring and Storage Project, was launched in July 2000. Phase 1 of this project was completed in 2004. This paper discusses the long-term assessment of the fate of CO_2 in the IEA Weyburn Project, describing the underlying methodology as well as the modeling approach and the results obtained. The conclusion from the modeling predictions is that if the Weyburn CO_2 storage system evolves as expected, long-term geological storage of greenhouse gas CO_2 will be achieved.

Key words: Greenhouse gas mitigation, geological storage of CO_2, CO_2 migration, system-level modeling, long-term assessment, risk assessment, abandoned wells, environmental impacts.

1. INTRODUCTION

CO_2 from the North Dakota Gasification plant is being transported and injected into a 1450-m deep oil reservoir located in Weyburn, south Saskatchewan, Canada (Figure 1) for enhanced oil recovery (EOR). The operator, EnCana Resources of Calgary, Alberta, designed a total of 75 patterns (each pattern is ~1 km^2) for this operation that is intended to last for approximately 34 years. To complement this commercial activity, a major research project to study geological sequestration and storage of CO_2, known as the International Energy Agency (IEA) Weyburn CO_2 Monitoring and Storage Project (henceforth called the IEA Weyburn Project), was launched

S. Lombardi et al. (eds.),
Advances in the Geological Storage of Carbon Dioxide, 231–242.

in July 2000. The first phase of this project (2000-2004) was sponsored by a number of governments and industrial sponsors from North America, Europe and Japan, including Natural Resources Canada (NRCan) and the U.S. Department of Energy. Planning for Phase 2 is already underway.

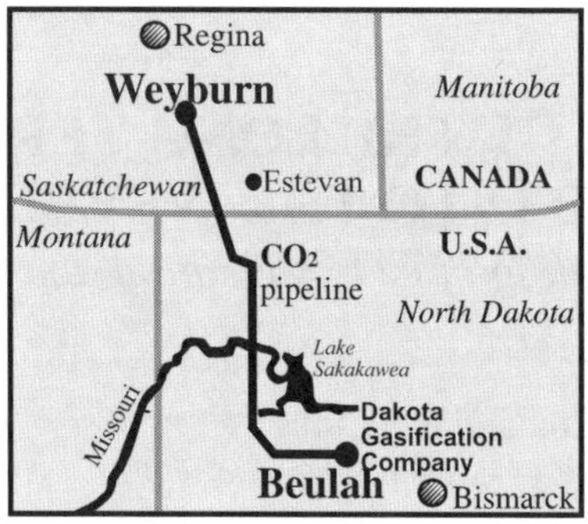

Figure 1. Location of the Weyburn field and the CO_2 pipeline from Beulah, North Dakota.

One of the objectives of the IEA Weyburn Project was to determine the long-term fate of CO_2 injected into the reservoir. Such a determination involves an evaluation of the potential for CO_2 to migrate to the environment via both natural and man-made (wellbore) pathways.

CO_2 storage projects are relatively few and so assessments associated with CO_2 storage are in their infancy. In the particular case of the IEA Weyburn Project, long-term (geological) storage of CO_2 is an additional benefit of EOR. However, safety studies for such storage of CO_2 are unusual in that they need to consider the evolution of natural systems over timeframes considerably in excess of those considered in industrial or engineering projects - typically several decades and rarely hundreds of years.

Many of the advances made in the last twenty years in the field of safety assessments for the geological disposal of radioactive wastes can also be applied to CO_2 storage (Savage, 1995). Similar to CO_2 sequestration, the final storage of nuclear waste requires an understanding of complex coupled physical-chemical-mechanical processes occurring over hundreds to tens of thousands of years. The spatial scales of both types of project are also similar.

This paper discusses the methodology applied to the long-term assessment of the fate of CO_2 in the IEA Weyburn Project, the modeling approach and the results obtained. A preliminary evaluation of the potential environmental impacts from releases of CO_2 to the biosphere is also described.

2. METHODOLOGY – SYSTEMS ANALYSIS FRAMEWORK

The Systems Analysis framework that was applied to the long-term assessment for the IEA Weyburn Project is based on the well-established methodology that has been, and still is, used widely and successfully in the field of radioactive waste management and disposal (Cranwell et al., 1982; Goodwin, 1991; Stenhouse et al., 2001). The basis of this approach is to develop a transparent and systematic treatment of the 'system' being assessed, as well as generating documentation that is auditable, should the need arise.

One of the first steps in the methodology is to define what is meant by the 'System' to be assessed. Figure 2 provides a schematic diagram of the basic components of a CO_2 storage system, *viz:*

- The **CO_2 storage reservoir** situated at depth, in this case the Weyburn field, within the
- **Geosphere**, which consists of a number of geological and hydrogeological units above and below the reservoir (units are not shown explicitly in Figure 2); and the
- **Biosphere**, the name given to the surface, or near-surface environment, where potential environmental impacts are normally evaluated.

The arrows shown in Figure 2 represent potential CO_2 migration pathways away from the reservoir. Two abandoned wells are also shown in this diagram, representing wellbores as potential (fast) pathways for reservoir CO_2 to migrate to the surface or near-surface. One of the major tasks of the IEA Weyburn Project was the detailed geological and hydrogeological characterization of the region around the Weyburn field, so that the main features of the geosphere, principally those representing potential pathways or sinks for CO_2, could be identified and incorporated in the migration modeling (see Section 3).

Having defined the 'System', the next stage of the process involved the identification of *scenarios*. Scenarios are defined as plausible and credible ways in which the Weyburn CO_2 storage system might evolve over decades to hundreds to thousands of years. One scenario was central to the modeling effort in Phase 1 - the Base Scenario, defined as "the expected evolution of the Weyburn CO_2 storage system". Table 1 provides a description of the elements that constitute the Base Scenario. Other (alternative) scenarios were also identified, though not discussed here. The documentation of scenarios occurs primarily through the listing of relevant features, events and processes (FEPs) - those factors that can affect the CO_2 storage system.

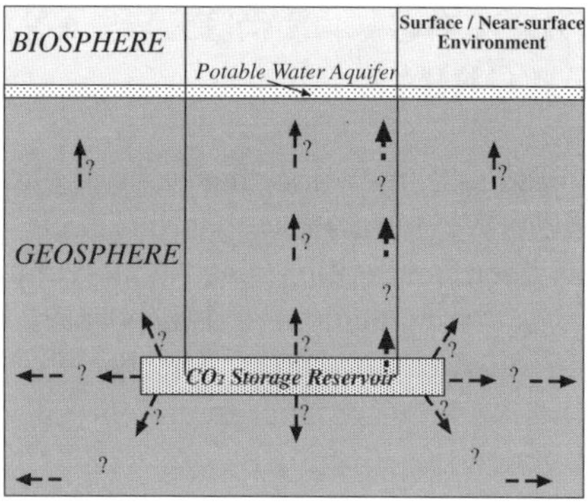

Figure 2. Schematic diagram of CO_2 storage system (cf. the more realistic representation shown in Figure 3).

Table 1. Key elements of the Base Scenario for the Weyburn CO_2 storage system.

Base Scenario

- Modeling domain: the Weyburn field (75 patterns) and the surrounding 10-km zone.
- Time frame: From the end of CO_2 injection to 5,000 years, or at which time 50% of the CO_2 in-place has migrated from the reservoir.
- Caprock: Based on existing information, the caprock integrity is not impaired; any fractures or discontinuities that exist are all isolated or sealed (see also abandoned wells).
- Geosphere: A series of aquitards/aquifers exist above and below the reservoir horizon; these formations may contain fractures and fissures.
- The System Model takes into account physical trapping features that have contained the oil/gas in the reservoir.
- Geochemical processes - *e.g.*, water chemistry changes, CO_2 removal (by solubility and ionic trapping), precipitation of carbonate minerals - occur in the aqueous phase of all aquifers.
- The Biosphere exists from the deepest potable aquifer, *i.e.*, extends to ~ 300 m below surface. It includes soil, surface water and the atmosphere as well as the flora and fauna that are found within these areas.
- Abandoned wells: All wells within the modeling domain are included and are assumed to have been abandoned according to current field abandonment procedures. This assumption includes wells that may have been abandoned previously according to different abandonment practices.
- Well seals will slowly degrade after abandonment. These seals consist primarily of the cement used to fill the annulus between casing and borehole, cement and metal plugs used to fill the casing bore, and the cap welded onto the casing ~ 4 m below ground surface.
- The metal casing will also degrade with time.
- FEPs that address storage and migration of CO_2 include hydrodynamics, buoyancy, geochemistry (precipitation / dissolution of minerals) and density-driven flow, dissolution of CO_2 in water and residual oil, and pressure-temperature changes occurring in the geological formations.

3. MODELING OF CO_2 MIGRATION FROM THE WEYBURN RESERVOIR

The modeling of CO_2 migration away from the storage reservoir was carried out in two stages, each of which is discussed separately below:

- Modeling of CO_2 migration away from the reservoir via natural pathways, i.e., through the geosphere; and
- Modeling of CO_2 migration through abandoned wells.

3.1 Modeling of CO_2 Migration Through the Geosphere

Figure 3 shows the model used as the basis for the long-term assessment of CO_2 migration in the geosphere. This geosphere model comprises a series of formations classified as either aquifers (exhibiting relatively high permeability) or aquitards (low permeability), and extends to ~10 km beyond the EOR region. The various 'compartments' of the model were identified as part of the detailed site characterization program. The geosphere also includes the reservoir (Midale Marly and Vuggy formations identified in Figure 3) beyond the EOR region.

ECLIPSE 300 (E300), a compositional simulator developed by GeoQuest/Schlumberger, was used as the modeling tool. Although not developed to specifically model CO_2 geological sequestration, one of the routines of the code is to simulate EOR employing a CO_2 flood. Key features of the code that are relevant to the modeling of CO_2 migration include equation-of-state capability to track the thermodynamic state of CO_2, which has low critical pressure and temperature, as well as the effect of the density gradient due to CO_2 dissolved in water.

The starting point for the long-term assessment modeling was the projected end of EOR activities, i.e., at 2034. Thus, results obtained from the reservoir simulation predictions for the end of the EOR period - in particular, the CO_2 in place, and pressure and fluid/component distributions within the field - were used as starting conditions for modeling CO_2 migration away from the storage reservoir. In order to avoid up-scaling problems between the reservoir simulation and long-term assessment models, the central area (75 patterns) of the Weyburn field had the same resolution in both models. Furthermore, pore volumes for the migration model were made identical to those of the reservoir simulation model. In the central area of the geosphere model (EOR region), the properties were replaced with the history-matched reservoir properties used in the reservoir simulation model.

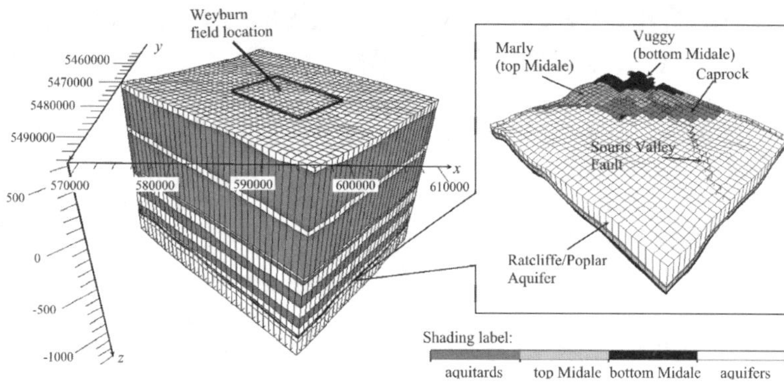

Figure 3. The model for assessing CO_2 migration within the geosphere. The x and y axes are UTM coordinates. The z axis is depth with respect to sea level in meters. Note that the vertical scale in the figure has been exaggerated.

Additional key input data were relative permeability and capillary pressure data for aquifers, aquitards, and reservoir rocks, obtained from a variety of sources. EnCana, the field operator, was the source of field data.

Key assumptions of the model included treating the fractured reservoir rock as an equivalent porous medium, with permeability values provided from field data as well as from history-matching results. This particular assumption may over-estimate mass transfer by ignoring diffusion into the rock matrix. In some cases, this is an important mass-transfer retardation process. Further details of the modeling assumptions, as well as input data and output results, can be found in Zhou et al. (2005).

Figure 4 displays the key results obtained from the long-term migration modeling. In this diagram, results are presented as the mass transfer rate of CO_2 (kg/day) out of the reservoir and into various regions of the geosphere, from 2034 up to 5,000 years after the end of EOR. Also shown in this diagram is the average oil phase pressure inside the EOR area, which has a major influence on CO_2 migration. The initially high pressure gradient across the boundary of the EOR region causes relatively fast mass transfer of CO_2, which subsequently slows down as the pressure decreases. The slight change in rate, reflected in an increase in lateral CO_2 migration at 1,000 years, is due to restoration of ambient pressure/flow field conditions.

To summarize the significance of the results shown in Figure 4, the cumulative amount of CO_2 removed from the EOR region is 26.8% of the initial CO_2-in-place at the end of EOR (~ 21 MT): 18.2% migrates into the geosphere *below* the reservoir, 8.6% migrates *laterally* (within the Midale formations), and 0.02% migrates to the geosphere *above*. No CO_2 reaches the biosphere, i.e., enters the lowest potable aquifer or any layer above, including the surface, over the 5,000-year period.

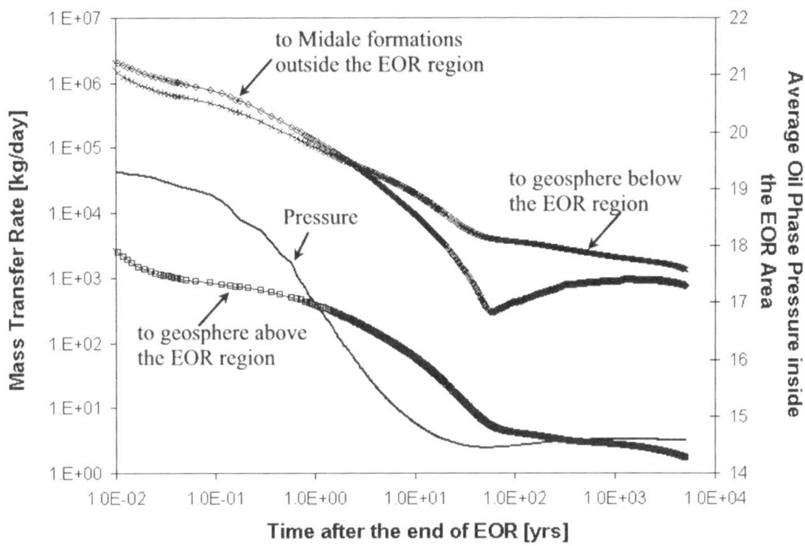

Figure 4. Time variation of CO_2 mass transfer rates (kg/day) into the geosphere, and oil phase pressure inside the EOR region (75 patterns) averaged over hydrocarbon pore volumes.

The substantial thickness of aquitards immediately above the reservoir (see Figure 3) is largely responsible for restricting the vertical movement of CO_2. In addition, sensitivity studies have demonstrated that key parameters affecting CO_2 movement include the caprock permeability and the 'entry pressure' (for entry of fluid into the caprock).

3.2 CO_2 Leakage Through Abandoned Wells

Thousands of wells exist within the study area, i.e., within the lateral extent of the system model. Thus, while CO_2 migration through the geosphere was treated deterministically, modeling of CO_2 leakage through abandoned wells was treated stochastically in order to address the substantial variability of these wells. This variability is reflected primarily in the heterogeneity of both the CO_2 distribution and the rock properties within the reservoir.

Stochastic treatment of CO_2 leakage through wells was achieved using a 'Unit-Cell' model (Figure 5). Using this model, the key assumptions were:

- Cement seal degradation corresponding to an increase in permeability from 0.001 mD initially to 1 mD at 100 years (cf. predictions from a supporting sub-task indicating such an increase in permeability but over a time period of 1,000 years rather than 100 years);

- No loss of CO_2 to fluid flow within the reservoir as well as within the formations surrounding the wellbore;
- Fast transport of CO_2 once it enters the open borehole, i.e., rapid ascent of CO_2 to the surface as gas bubbles.

Given the variability and uncertainty of key parameters used in the 'Unit-Cell' model, the above assumptions were considered to result in a conservative treatment, i.e., over-estimating CO_2 leakage.

Existing data (Yu, 1993) for fresh concrete used in low-level radioactive waste disposal systems show that this material is extremely restrictive to flow (the water phase becomes mobile at near unity saturation). The exceedingly high capillary pressure (>20 MPa at near unity water saturation) effectively blocks intrusion of non-aqueous phase fluids. While these properties are valuable to sealing performance, they may not be realistic for aged cement materials in wells. In the absence of corresponding data for aged cement, we used an estimated capillary pressure obtained by estimating the capillary pressure drop due to a permeability increase using existing data for fresh cement together with the permeability-porosity relationship found in Scheidegger (1974).

The 75-pattern EOR region was represented by 6,075 grids in the geosphere model (Figure 3). Of these grids, 287 were selected with a bias towards existing well locations, high CO_2-in-place, and high permeability values. Together, these selections correspond to a conservative treatment.

The 'Unit-Cell' model was activated for 287 simulations in which saturation, component concentrations, and permeability values in each of the 287 grids were used as input for individual runs. The results show a large range in leakage rates with the *maximum* leakage rate ranging from 0 to 0.016 kg/day. The time at which this maximum leakage rate occurred also varied significantly. Figure 6 shows the mean, 95%, and maximum leakage rates as a function of time obtained from these (287) runs, along with leakage rates *vs*. time from selected runs.

The variability in well leakage is also shown in Figure 7, which displays cumulative leakage of CO_2 via abandoned wells as a function of time. In general, the stochastic nature of reservoir conditions dictates stochastic behavior of leakage through abandoned wells across the study area.

However, when the cement permeability is low ($\leq$ 1 mD), the ultimate control of leakage lies in cement permeability. When the permeability of cement is higher (e.g., > 10 mD), reservoir properties govern the rate and amount of CO_2 leakage. To confirm this conclusion, a few 'Unit Cell' runs were carried out using an increased cement permeability (10 mD) (Millidarcy: approximately equal to 10^{-8} m/s). In these cases, the results indicated an increase in both the leakage rates and cumulative leakage, but only by a factor of about two.

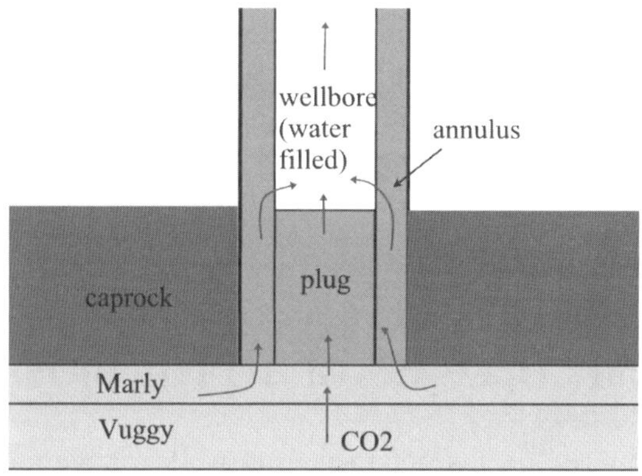

Figure 5. 'Unit-Cell' model for assessing the leakage of CO_2 via abandoned wells.

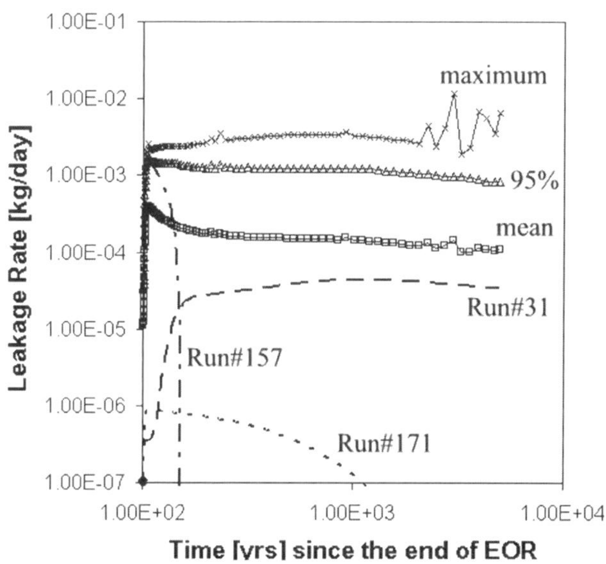

Figure 6. Stochastic results of CO_2 leakage rate as a function of time, predicted using the 'Unit Cell' model combined with a stochastic treatment. The diagram shows the results of a few selected runs. Note that the 5% result is off (below) the scale of Figure 6.

The combination of the *maximum* CO_2 flux through a wellbore (0.016 kg/day) with an estimated 1,000 wells over the 75-pattern area (currently ~824 wells), yields a total cumulative leakage of CO_2 of ~0.03 MT over

5,000 years. This total amount represents ~0.14% of the total CO_2-in-place (21 MT) at the end of EOR. This value is a highly conservative upper estimate, however, as it assumes that the maximum flux is maintained throughout the 5,000-year period for all wells, which is not supported by the observed results. Thus, a more representative value is the *mean cumulative leakage*, corresponding to less than 0.001% of the CO_2-in-place at the end of EOR.

4. POTENTIAL ENVIRONMENTAL IMPACTS FROM CO_2 LEAKAGE TO THE BIOSPHERE

In order to put the results obtained from well leakage modeling in some perspective, scoping calculations were carried out to address two potential environmental impacts:

- *Indoor air CO_2 concentration*: CO_2 leakage from an abandoned well (point source) directly into the basement or lower floor of a dwelling through cracks in the foundation. The maximum acceptable CO_2 flux from the wellbore was determined by the limiting CO_2 concentration in air (Health Canada, 1989).

- *Mobilization of trace metals into drinking water*: CO_2 leakage from an abandoned well directly into a potable groundwater, changing the water chemistry, which in turn promotes the release of a toxic trace metal (lead) into the water supply. The maximum acceptable CO_2 flux in this case was determined via geochemical modeling, taking into account an interim drinking water standard for dissolved lead (Langmuir, 1997).

Evaluation of the above cases indicated that the more restrictive example is CO_2 leakage into a potable aquifer, yielding a limiting CO_2 leakage rate of similar magnitude to the mean cumulative leakage described in the previous section.

Given that only scoping calculations were carried out, more detailed treatments of these environmental impacts are required, particularly in the case of trace metal release into drinking water, before making a definitive statement on an acceptable CO_2 leakage rate to the biosphere.

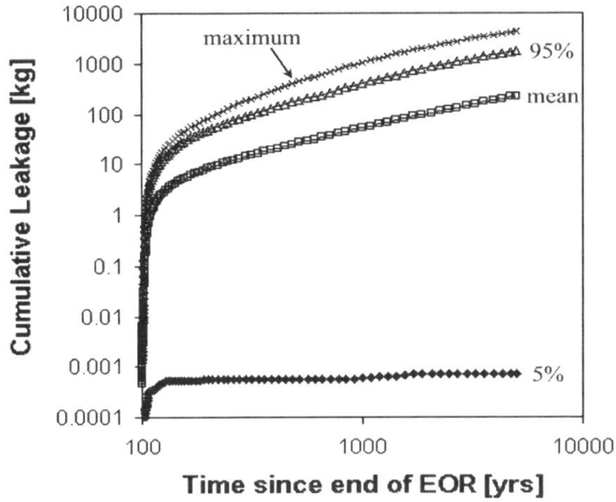

Figure 7. Stochastic results of cumulative leakages of CO_2 via abandoned wells.

5. SUMMARY

For the Base Scenario identified for the IEA Weyburn Project, modeling of CO_2 migration within the geosphere indicated that up to 75% of the initial CO_2-in-place (at the end of EOR) remained within, and in the vicinity of, the EOR region. No CO_2 enters the biosphere within the simulation period of 5,000 years.

Modeling of leakage via abandoned wells focused on the performance of degraded cement sealing combined with the variability of properties and CO_2 distribution in the reservoir. The conservative treatments incorporated in stochastic modeling provided upper bounding results indicating that the *maximum* possible leakage after 5,000 years is approximately 0.14% of the initial CO_2-in-place. However, the *likely cumulative leakage* after 5,000 years, however, is less than 0.001% of the initial CO_2-in-place.

These results mean that if the Weyburn CO_2 storage system evolves as expected, the goal of storing greenhouse gas CO_2 will be achieved. Future assessments should focus on alternative scenarios, including seismic activity, open wellbores, and human intrusions.

ACKNOWLEDGEMENTS

The modeling work described here could not have been accomplished without the technical input from project sub-tasks involving many disciplines, including geology and hydrogeology, geochemistry, geomechanics, reservoir simulation, seismic and wellbore technology. Without naming individual participants, we gratefully acknowledge their contributions as well as their willingness to discuss and review results.

REFERENCES

Cranwell, R.M., Guzowski, R.V., Campbell, J.E., and Ortiz, N.R., 1982, Risk methodology for geologic disposal of radioactive waste: scenario selection procedure, Sandia National Laboratory Report **SAND80-1429**, NUREG/CR-1667, Albuquerque, New Mexico.

Goodwin, B.W., Editor, 1991, Scenario analysis for the post-closure assessment of the Canadian concept for nuclear fuel waste disposal, AECL Report **AECL-10717, COG-93-7**, Pinawa, Canada.

Health Canada, 1989, Exposure Guidelines for Residential Indoor Air Quality, Report of the Federal-Provincial Advisory Committee on Environmental and Occupational Health. Report first published April 1987, revised July 1989. Ministry of Supply and Services, Canada.

Langmuir, D., 1997, *Aqueous Environmental Chemistry* (Prentice Hall, Upper Saddle River, New Jersey), 600 pp.

Savage, D., Editor, 1995, *The Scientific and Regulatory Basis for the Geological Disposal of Radioactive Waste,* John Wiley & Sons, Chichester, UK, 437 pp.

Scheidegger, A. E., 1974, *The Physics of Flow through Porous Media*, University of Toronto Press, Toronto, Canada.

Stenhouse, M.J., Miller, W.M., and Chapman, N.A., 2001, System studies in PA: Development of Process Influence Diagram (PID) for SFR-1 Repository: near-field + far-field, SKI Report **01:30**. Swedish Nuclear Power Inspectorate, Stockholm, Sweden.

Yu, A.D., Langton, C.A., and Serrgato, M.G., 1993, Physical Properties Measurement, Savannah River Company Report WSRC-RP-93-894, Westinghouse, Savannah River Company.

Zhou, W., Stenhouse, M.J., Arthur, R., Whittaker, S., Law, D.H.-S., Chalaturnyk, R., and Jazrawi, W., 2005, The IEA Weyburn CO_2 Monitoring and Storage Project – Modeling of the long-term migration of CO_2 from Weyburn, accepted for publication in the Proceedings of the GHGT7 Conference (2005). Prior to publication, this paper can be accessed at: <http://uregina.ca/ghgt7/PDF/papers/peer/582.pdf>.

STRONTIUM ISOTOPE ($^{87}SR/^{86}SR$) CHEMISTRY IN PRODUCED OIL FIELD WATERS: THE IEA CO_2 MONITORING AND STORAGE PROJECT

F. Quattrocchi[1], M. Barbieri[2], R. Bencini[1], D. Cinti[1], K. Durocher[4], G. Galli[1], L. Pizzino[1], M. Shevalier[3] and N. Voltattorni[1]

[1]INGV Section Roma, Fluid Geoch. Lab., Via di Vigna Murata 605, 00143, Rome, Italy [2]IGAG-CNR, c/o University "La Sapienza", P.le Aldo Moro 5, 00185, Rome, Italy; [3]Dept. of Geology & Geophysics, 2500 University Dr. NW, Calgary, AB, T2N 1N4 Canada; [4]Alberta Research Council, 3608-33rd Street, NW, Calgary, AB, T2L 2A6, Canada. (E-mail: quattrocchi@ingv.it)

Abstract: EnCana's CO_2 injection EOR project at Weyburn Saskatchewan (Canada) is the focal point of a multi-faceted research program, sponsored by the IEA GHG R&D and numerous international industrial and government partners. More than yearly strontium isotope, trace element and dissolved gas surveys were conducted by INGV in conjunction with the thrice yearly borehole fluid sampling trips performed by the Canadian partners. This paper focuses on the Sr isotope monitoring. Approximately 25 samples were collected over three years for $^{87}Sr/^{86}Sr$ analyses. At Weyburn, a water-alternating-gas (WAG) EOR technique is used to inject water and CO_2 into the Mississippian Midale reservoir. $^{87}Sr/^{86}Sr$ ratios for produced fluids fall between 0.7077 and 0.7082, consistent with published values for Mississippian fluids and carbonate minerals. A small $^{87}Sr/^{86}Sr$ component of this produced fluid is derived from waters of the Cretaceous Mannville aquifer, which has been used for water-flooding EOR since 1959. The progressively more positive Sr isotope trend from 2001 to 2003 may be due to: 1) a smaller Mannville aquifer component in the water flooding process; and/or 2) the dissolution of Mississippian host rocks during the ongoing CO_2 injection. Evidence that $^{87}Sr/^{86}Sr$ values are approaching those of Mississippian host-rock values may point towards zones of carbonate dissolution as a result of continuing CO_2 injection. This hypothesis is strengthened by i) $\delta^{13}C$ data; ii) preliminary "gross composition" of dissolved gases (H_2S, CO_2, CH_4, He, H_2) and iii) by trace elements data.

Key words: CO_2 geological storage monitoring, EOR Weyburn oil field brines, Sr isotopes.

S. Lombardi et al. (eds.),
Advances in the Geological Storage of Carbon Dioxide, 243–259.

1. INTRODUCTION

The Weyburn oil field, presently owned and operated by EnCana Resources, is located approximately 120 km SE of Regina (Canada). After the discovery of the field in 1954 and its primary depletion in middle 1964, other tools were exploited, such as horizontal wells and water flooding. Since August 2000, CO_2-Enhanced Oil Recovery (EOR) has also been performed, with around 5000 tons/day of CO_2 being injected within the Mississippian Midale Beds oil reservoir at a depth of 1300-1500 m. The "Phase A1" early injection area currently comprises around 90 oil producers, 30 water injectors and 30 CO_2 injection wells (Wilson & Monea, 2004; Tian et al., 2004). Recent studies (Tian et al., 2004) stated that after 3 years of CO_2 injection oil production increases have been relatively slow, with responses only in 30 wells and increases in the average oil production per well from around 6 to 11 m3/day and an increase in the Gas/Oil Ratio of the Phase A1 area from around 20 to 33 m^3/m^3. The CO_2-storage efficiency decreases with an increase in the amount of CO_2 produced together with oil and water. Apart from industrial exploitation, the EnCana CO_2 injection EOR project at Weyburn is the focal point of a multi-faceted research program to "co-optimise" CO_2-EOR production and CO_2-geological storage within the framework of the Kyoto Protocol. This work is sponsored by the IEA GHG R&D and numerous international industrial and government partners, including the European Community (Riding & Rochelle, 2005).

Together with reflection seismic monitoring of the evolution of CO_2 distribution at depth (recent papers on http://www.uregina.ca/ghgt7/), the geochemical monitoring tools are mainly: i) soil gas geochemistry, to discover and monitor possible CO_2 surface leakage (Jones et al., 2005) - this tool is particularly important in geo-dynamic areas where the risk assessment of CO_2 geological storage sites must be very accurate (Quattrocchi, 2003, 2004); and ii) deep-reservoir fluid geochemistry (Gunter et al., 2000; Quattrocchi et al., 2004; Perkins et al., 2004, Shevelier et al. 2004). In the Weyburn case, we are dealing with a 3-phase reservoir (water-gas-oil) requiring complex chemical equilibria, diffusion and thermodynamic models (Nitao, 1996; Czernichowski-Lauriol et al., 2001; Quattrocchi et al., 2003; Chapoy et al., 2004; Tian et al., 2004, Le Nidre and Gaus, 2004).

INGV has conducted geochemical monitoring experiments from pre-injection ("Baseline", B0, August 2000) to the present day (September, 2004, see Appendix). The purpose of produced fluid and gas monitoring was to identify and ultimately quantify water-gas-rock reactions in the reservoir; in this paper we are focusing on the Sr isotope results. These data may help in reconstructing the evolution of the fluid reservoir prior to and during the injections stages (after Sunwall and Pushkar, 1979; Smalley et al., 1988). At

Weyburn, the situation of the reservoir composition is complicated by the water-alternating-gas (WAG) EOR technique, which has been used to inject water into the Midale beds over a forty year period. Unfortunately strontium isotopic compositions of the produced reservoir oil brines prior to upper-water flooding (i.e. prior to 1959) are not available, therefore we may only record the evolution of the oil waters since the beginning of CO_2-injection. Clearly, therefore, the isotopic signature of the Mississippian Midale reservoir waters had already been modified by previous mixing with waters from the overlying, hydraulically-separated, Cretaceous, Mannville aquifer. CO_2 geological storage is mainly limited within the oil reservoir stratum (around 20-30 m thick), considering that diffusion within, and reactivity of, the cap-rock is limited, with extremely low migration rates even in the highest impact scenario of CO_2 diffusivity / advection to the surface (Czernichowski-Lauriol et al., 2001; Le Nidre & Gaus, 2004).

In mature reservoirs, Sr isotopes can be used to differentiate original formation water from injected water for water-flood surveillance (Smalley et al., 1988; Bernaby et al., 2004 and references herein). In the present work Sr isotope data (partially discussed in Quattrocchi et al., 2003) are also supported by: i) major, minor and trace element chemistry; ii) the preliminary "gross composition" of dissolved gases; and iii) carbon isotopic ratios (after Hutcheon et al., 2003).

The use of Sr isotopic ratios for this application is possible by modeling two-component mixing in the framework of Sr isotope time-tuning curves (Veizel and Compston, 1974; Burke et al., 1982; Popp et al., 1986; McNutt et al., 1990; Jones et al., 1994; Denison et al., 1994; Bruckschen et al., 1995; Romer et al., 2003; Barnaby et al., 2004). The Sr isotopic composition of formation waters could yield a significant contribution for the characterization of reservoir properties, such as connectivity between strata and reservoir heterogeneity, and thus to evaluate regional fluid flow and to interpret the evolution of fresh and saline groundwaters. Data on the Sr isotopic compositions of oil field waters from a variety of reservoir rocks are abundant in a number of recent publications (Starinsky et al., 1983; Stueber et al. 1987; Russel et al., 1988; Franklyn et al., 1991; Chaudhuri and Clauer, 1993; Russell et al., 1988; Notsu et al., 1988; Nakano et al., 1989).

Strontium isotopes thus provide a potential tracer for both static (fluid flow on geological scale) and dynamic (fluid flow on production timescale) reservoir characterization in conjunction with conventional studies using seismic, log, core, well test and production data. Strontium isotopes are especially sound to better understand formation waters patterns because the Sr isotope composition of water samples collected at the well head is known to be identical to that of the subsurface fluid (Quattrocchi et al., 2003; Bernaby et al., 2004). By using a geochemical, multi-component approach to better understand the ongoing formation oil/brine-related processes (Fisher et al., 1987; McNutt et al., 1987; Connolly et al., 1990), we may fully and better predict the long-term storage of CO_2 in the subsurface at the Weyburn oil-field, the most studied and multi-faced CO_2-EOR project in the world at present (Wilson & Monea, 2004).

2. METHODS

The Appendix summarises the INGV geochemical surveys during the Weyburn IEA-EC Project, from August 2000 to September 2004, for strontium isotope, major, minor and trace element and dissolved gas analyses. The laboratory analytical methods, applied to the liquid phase, were used after sampling the oil waters at the oil or gas well-head within collapsible, approx. 1-gal LDPE containers and allowing them to sit for a 1-2 minutes to achieve gravity separation of water and hydrocarbons.

- $^{87}Sr/^{86}Sr$ isotopic ratios were analysed from Sr dissolved in oil-water using standard mass spectrometry methods coupled with the following sample preparation technique (modified after Jones et al., 1994; Quattrocchi et al., 2003): 100 mL of sample is added to 50 mg of Na_2CO_3 to transform the sulphates ($SrSO_4$, $BaSO_4$) into carbonates. The final product of this reaction is filtered and washed to completely eliminate the sulphate ions. The residual on the filter was dissolved with 2.5 N HCl and the obtained solution was evaporated. The dry residue was dissolved with 2.5 N HCl and the strontium was separated from the other elements using a cation exchange resin (DOWEX resin type). This element is placed (as $Sr(NO_3)_2$) on a tantalium filament and the $^{87}Sr/^{86}Sr$ ratio is measured by a VG 54E mass spectrometer (University of Rome). The $^{87}Sr/^{86}Sr$ values were then normalised to the standard ratio of 0.1194. The analytical measurements are affected by an error expressed as 2 times the standard deviation (± 0,00002).
- The dissolved gas analytical method is based upon the partitioning of gases between the liquid and gas phase in the head-space of the sampler (Capasso & Inguaggiato, 1998). A *Chrompack*™ CP2002 portable gas-chromatograph is used in the laboratory to analyse CO_2, CH_4, H_2S, H_2, N_2, O_2, and light hydrocarbons (C_2-C_4), while He is also analysed with an ALCATEL mass spectrometer. As the dissolved gases were not sampled at borehole P,T,[X] conditions, it is necessary to correct the analytical data to the reservoir pressure of 16 MPa (1300-1500 m), considering also kinetic effects. The degassing during the rise of fluids from 16.5 MPa (reservoir pressure) to 0.1 MPa (atmospheric pressure) can be calculated following certain principles (Chapoy et al., 2004; Tian

et al., 2004; Renè Perez, personnel communication, work in progress). This can be done for each dissolved species, assuming a constant pump rate, sampling time and stripping effects before the immediate sampling of the oil waters at surface. At present we will only discuss the data gathered by sampling at 0.1 MPa pressure, as the "gross composition" of the reservoir evolution.

- Alkalinity is measured (U of C partner) by an ORION 960 titrometer (with 0.16M H_2SO_4) soon after the oil water sampling, considering $HCO_3 + CO_3$ and oily alkali.
- trace elements (partially discussed in this paper) were analysed by multi-component ELAN™ 6000 ICP for the following elements:, Be, B, Al, Si, Fe, Ni, Pb, Mn, Rb, Sr, Ag, Ba, Br, Cd, Cs, Cr, Cu, As, and Zn. Measurements are affected by an error expressed as 2 times the standard deviation.
- Major elements (partially discussed in this paper) are analysed by DIONEX HPLC; a charge imbalance of less than ± 5-10 % is considered acceptable.
- Geochemical Modeling, partially discussed in this paper (Cantucci B, PhD thesis in progress, INGV-University of Florence Italy 2004-2006), is accomplished by PHREEQC version 2.10 (Parkhurst and Appelo, 1999). The oil water composition is imposed both during kinetic evolution and in final equilibrium with the Vuggy and Marly rock compositions.
- Contour maps are created with the kriging technique within the program Surfer, while the "gross composition" of dissolved gases are discussed using normal probability plots (Sinclair, 1991).

3. RESULTS AND DISCUSSION

Approximately 25 samples were collected during the M 1 (March 2001), M 5 (July 2002), M 7 (March 2003), and M 9 (September 2003 partially analysed, not discussed) surveys for $^{87}Sr/^{86}Sr$ analyses. $^{87}Sr/^{86}Sr$ ratios found in the produced fluids from the Mississippian Midale beds fall between 0.7077 and 0.7082, consistent with published values for Mississippian fluids and carbonate minerals (approximately from 0.7076 - corresponding to 330 Ma ago - to 0.7082 corresponding to 360 Ma ago - Bruckschen et al., 1995; see Fig.1, Table 1).

Most of the sedimentary formation waters migrated over long distances prior to entrapment in the reservoir rocks from which the waters are sampled. Nearly two-thirds of the formation brines have higher $^{87}Sr/^{86}Sr$

ratios than marine Phanerozoic seawaters, whose values range between 0.7068 and 0.7062 (Veizer and Compston, 1974; Notsu et al., 1988).

Table 1. Sr isotope data for the M1, M5, M7 Weyburn oil-waters.

		Monitor 1	Monitor 5	Monitor 7
$^{87}Sr/^{86}Sr$	Min	0.70775	0.70800	0.70797
	Max	0.70825	0.70818	0.70818
	Mean	0.70798	0.70804	0.70807
$\delta^{87}Sr$	Min	-0.63	-0.28	-0.32
	Max	0.08	-0.03	-0.03
	Mean	-0.31	-0.23	-0.19
	Manville	Mississipian	1st survey	3rd survey
$^{87}Sr/^{86}Sr$	0.7073	0.7082	0.70798	0.70807

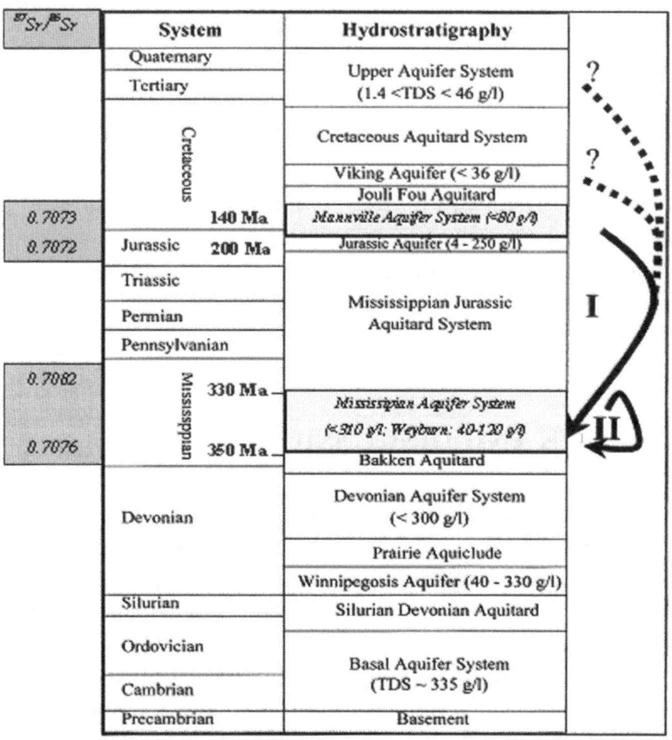

Figure 1. Williston Basin hydro-stratigraphy and nomenclature (modified from Bachu and Hicheon, 1996). Arrow (**I**) indicates that water from the Mannville Aquifer System is used for water-flooding. The symbols (?) are added to leave in question the use of upper aquifers for water-flooding. Arrow (**II**) indicates that water produced with oil is continuously re-injected.

The Pennsylvanian seawater (over the Mississippian ones) has $^{87}Sr/^{86}Sr$ values around 0.7083 (normalised to NBS 987 = 0.710255; Popp et al., 1986; Denison et al., 1994; Bernaby et al., 2004).

The Mannville aquifer, used since 1959 for water flooding of the Weyburn oil-field, is a Cretaceous arenite, with $^{87}Sr/^{86}Sr$ values between 0.7072 and 0.7073 which corresponds to an age of 140 Ma BP (Jones et al., 1994). A small component of the recycled fluid is derived from the Mannville aquifer and is re-injected into the Midale beds via a water-alternating-gas (WAG) EOR technique, as mentioned. Strontium isotopic compositions of the produced water prior to water flooding (i.e., prior to 1959) are not available. Other case histories describe the use of Sr isotopes to discriminate water-flooding injection components from the formation water (i.e., Smalley et al., 1988), mostly when the injected water has a Sr content and $^{87}Sr/^{86}Sr$ ratio which is different from the in-situ formation water. The lowest values of the Sr isotopic ratio found in the oil waters could represent a higher degree of mixing between the two aquifers (Fig. 2). We started to analyse the oil water Sr isotopes in 2001, around 40 years after the beginning of water flooding. Despite this fact we found Sr values of the oil water in the range of the Mississippian host rock (Midale reservoir), although a certain percentage of the Sr isotopic ratio could be referable to the Mannville component. Average Sr ratios and mass balance calculations suggest that as much as 25% of the produced fluid in 2001 was derived from the Mannville aquifer, decreasing to 15% in 2003.

The progressively more positive (i.e. higher) trend in the Sr isotopes from 2001 to 2003 may be due to: i) a smaller Mannville aquifer component in the water flood over the last three years and/or ii) the enhanced dissolution of Mississippian host rock during progressive CO_2 injection.

If the leaching of Mississippian host rocks is increased (as a consequence of CO_2 injection) it could shift the values towards the pre-water-flooding Sr isotopic "baseline" values, affected by 40 years of water-flooding which added a 25% Mannville component. In general terms, the lowest Sr isotopic ratio values found thus far may represent a higher "contamination" of Mississippian Midale fluids by re-injected Mannville flooding water. It may involve progressively lower Sr isotopic values which may coincide with the highest injection volumes of Cretaceous waters, as there is no natural communication or mixing between the two aquifers (no fluid pathways have yet been found in the highly impermeable Jurassic-Triassic Watrous Formation). Considering this general process, we observed instead a progressive decrease of the Mannville "contamination" from 2001 to 2003. Therefore, the more probable scenario is that the $^{87}Sr/^{86}Sr$ values are progressively approaching the Mississippian host-rock values and may point towards zones of carbonate dissolution resulting from continued CO_2

injection. Carbonate minerals within the Midale beds, which were precipitated from waters during the Mississippian, would have Sr ratios consistent with what would be considered a "baseline" isotopic composition. The leaching of strontium during dissolution of these minerals would drive the Sr composition of the produced waters to heavier values, thereby masking the Mannville aquifer contribution.

In summary, the sectors (Fig. 2) characterised by the lowest Sr isotopic ratio values represent "contamination zones" of Mississippian Midale fluids by re-injected Mannville flood water. These zones exhibit lower Sr isotopic values that may coincide with the highest injection volumes of Cretaceous water. Concurrently, the average field-wide $^{87}Sr/^{86}Sr$ values are approaching Mississippian host-rock values and may point towards sectors of the field (Fig. 2) characterised by higher carbonate dissolution, as a result of continued CO_2 injection.

This hypothesis is strengthened by the $\delta^{13}C$ data available up to September 2004, as well as by other chemical data of previous surveys (Hutcheon et al., 2003 and papers in http://www.uregina.ca/ghgt7/). Using $\delta^{13}C$ values of produced bicarbonates and CO_2, the University of Calgary has outlined both injected CO_2 and carbonate mineral dissolution in the reservoir.

The available Sr-C isotopic ratios in the produced fluids (up to September 2004) corroborate the hypothesis of chemical and isotopic input from dissolved carbonates, although further surveys could modify this statement relative to early stages of the water-rock interactions in the oil reservoir. Initially, the injected CO_2 had a distinctive $\delta^{13}C$ signature of –35‰., however since 2002 CO_2 recycling and re-injection has changed this signature to between –20 to –25‰. Therefore the second strontium isotope scenario outlined above fits well with the $\delta^{13}C$ data. The Sr isotopic ratio in produced fluids tends to increase with time, suggesting input from dissolved strontium bearing carbonates.

Sr isotopic ratios of minerals (anhydrite, dolomite, calcite of the Midale and Vuggy Formations) from the Weyburn reservoir Phase A1 area are available from Queen's University. These data are consistent with the produced Mississippian brine because they range from 0.70567 to 0.708995, with an average value of 0.708134 ± 0.0001563. It is possible to assume that the dissolution of these minerals would push the Sr isotopic ratio of the oil waters from values closer to 0.7073 (Cretaceous Mannville aquifer) towards the reservoir values of 0.7076-0.7082 (Mississippian Midale-Vuggy), again strengthening the second Sr isotope theory mentioned above.

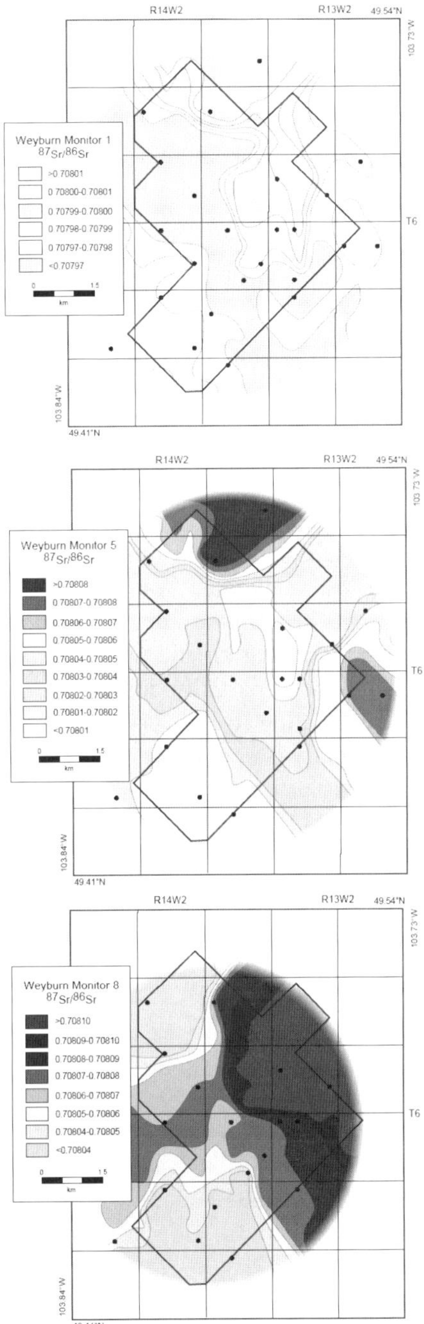

Figure 2. Contoured Sr isotopic ratio maps (a_1, b_1, c_1) from 2001, to 2002, up to 2003.

If a number of sources exist, evolving from varied mixed compositions, it could be represented by binary diagrams as 1/Sr versus $^{87}Sr/^{86}Sr$ (Fig. 3 a), where coexistent linear trends are often found (Chaudhuri and Clauer, 1993). An isotopic trend with a very high positive slope could be explained in terms of the mixing of Sr derived from two sources which are markedly different in both their isotopic composition and Sr contents. On the other hand a low positive slope in the isotopic trend could be explained by either mixing between two sources of Sr or dilution of a single source of Sr. In the Weyburn case, considering only the first two analysed surveys (Sr available only for M1-2001 and M5-2002; see Fig. 3 a), it is clear that a unique "gross" population effectively changed in terms of mixing from 2001 (lower $^{87}Sr/^{86}Sr$ ratios and higher Sr concentrations closer to the Mannville component) to 2002 (higher $^{87}Sr/^{86}Sr$ ratios and lower Sr concentrations closer to the Mississippian component). A small group forms an anomaly of relatively low Sr values (WEY 35, 42 and 45 - wells corresponding to the vertical EnCana wells 12-25 (6-14), 14-23 (6-14), and 12-19 (6-13), respectively), which may have been caused by dilution, considering also the slightly lower salinities as a whole. Alternatively a local differential immobilization by sulphates-carbonates could be possible ("short term" geochemical modelling considering kinetic parameters is in progress, while "long term" modelling has been partially completed; Perkins et al., 2004). The WEY 5 well (corresponding to the vertical EnCana well 1-11 (6-14)) seems to be pertinent to slightly deeper strata.

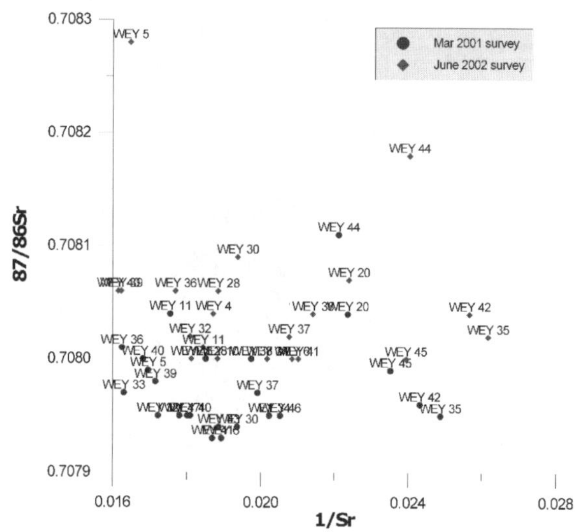

Figure 3. Binary diagrams of (a) $^{87}Sr/^{86}Sr$ versus 1/Sr; (b) $^{87}Sr/^{86}Sr$ versus Na/Cl of the Weyburn oil-brines (M1 and M5 surveys).

Albitization of K-feldspar has been used (Chaudhuri and Clauer, 1993) to explain high-slope points in the graph, in other words direct replacement of K-feldspar or replacement of K-feldspar by some other minerals (calcite, anhydrite, etc…) which in turn are replaced by albite. This effect of albitization would result in an elevation of the Sr isotopic ratio in the waters without any significant change in the Sr content of the water. In the Weyburn case the "end member" component (as defined by Chaudhuri and Clauer, 1993) common to all the analysed waters may be identified as one whose $^{87}Sr/^{86}Sr$ ratio is closer to the Mannville component used for water flooding, progressively changed during the CO_2 injection and consequent rock dissolution. This is also suggested by the $^{87}Sr/^{86}Sr$ versus Na/Cl binary diagram (Fig 3 b), where the Na/Cl ratio is higher as a whole for the M1-2001 waters than for the M5-2002 waters and vice versa for the $^{87}Sr/^{86}Sr$ ratios. As a whole, the Weyburn data do not display a trend of higher $^{87}Sr/^{86}Sr$ ratios if the Na/Cl molar ratio approaches 1:1, as found in the literature (i.e., solutes are increasingly dominated by halite dissolution, as could also occur for the Weyburn oil brines; Burke et al., 2004). Moreover the past water-flooding modified the original $^{87}Sr/^{86}Sr$ and Na/Cl binary relationship, now affected by the host rock dissolution linked to the new industrial CO_2 injection.

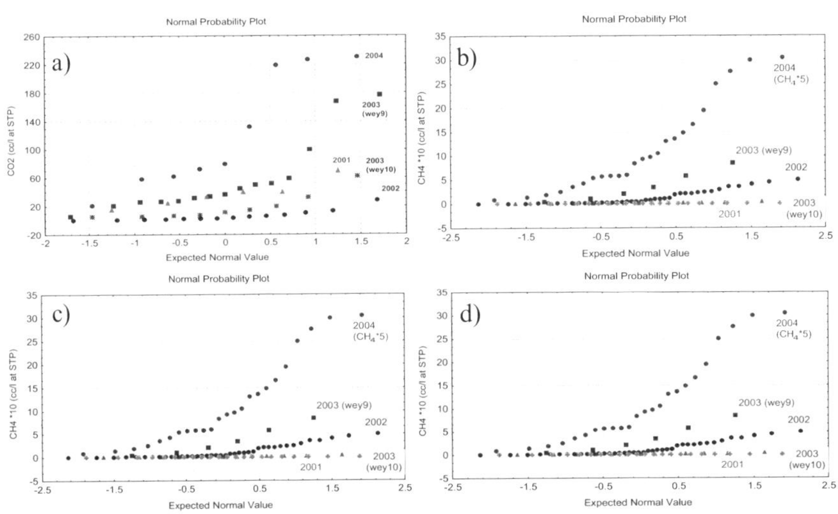

Figure 4. Normal probability plots (after Sinclair; 1991) for the Weyburn "gross composition" of dissolved gases (atm. P.); (a) dissolved CO2; (b) dissolved CH4; (c) dissolved H2S; (d) dissolved H2. All data are reported as cc/L (STP), except H2S which is reported as composition % in the head-space (not as total dissolved concentration).

Also the available "gross composition" of the dissolved gases (see the limitations in the methods paragraph detailed in Quattrocchi et al., 2003 and in Riding & Rochelle, 2005) are helpful in jointly interpreting the Sr isotopic data. Available dissolved gas data strengthens the second hypothesis, i.e. enhanced Mississippian rock dissolution. In oil brines CO_2 solubility is dependent on T, P, fugacity coefficient, salinity (eq. NaCl), and detailed chemical composition. Theoretically, from chemical equilibria calculations (Czernichowski-Lauriol et al., 2001; Perkins et al., 2004), more than 1 mole of CO_2 can dissolve into 1 kg of water for typical Weyburn reservoir brines (50°C, 14 MPa, salinity range from 35 to 110 g/l).

The steep salinity gradient in the Weyburn area will greatly affect CO_2 solubility, migration and reactivity as demonstrated by the regional 2D modelling of natural fluids in the Mississippian aquifer (BRGM deliverables within the EC Weyburn Project, Riding & Rochelle, 2005). Normal probability plots (Sinclair, 1991) of the 2001-2004 dissolved gases in the Weyburn oil-waters (Fig. 4) exhibit, as mentioned, the "gross composition" evolution, before the "degassing correction". We found a general increase of the main dissolved gases (CO_2, CH_4, H_2S) for the different surveys (2001-2004), parallel to the increase of the Sr isotopic ratio increases. In particular progressive increases of dissolved CO_2, CH_4 and H_2S were observed, while H_2 initially increased then it decreased after 2002. Dissolved He, after an abrupt early increase, decreased after 2003; the initial increasing stage was very probably due to the release of accumulated crustal He from the rock-matrix during CO_2-driven dissolution (Torgersen & Clarke, 1987).

Our dissolved CO_2 data are essentially consistent with the Tian et al. (2004) calculations, performed following the Material Balance Equation method (MBE). These authors stated that approximately 86% of injected CO_2 can be dissolved into reservoir oil, 7% into water and 7% will occur as free gas. This occurs mainly in the early stages of CO_2 injection (Gunter et al., 2000). Moreover with the increase of CO_2 injection time, the percentage of CO_2 dissolved in oil decreased by a small degree and the percentage of CO_2 dissolved in reservoir water and as free CO_2 increased slightly. This is because progressive oil recovery will result in a decrease in the average oil saturation and increase in the average water saturation. As a result, for the authors, the injected CO_2 had more chance to contact (dissolve in) the water phase than the oil phase, even though under reservoir conditions the solubility of CO_2 in oil is larger than that in water (b factor).

The increase of both dissolved gases and selected trace metals (Quattrocchi et al., 2003, reported in Riding & Rochelle, 2005) during the Phase 1 A is consistent with the second Sr isotope scenario outlined above, which highlights host-rock dissolution as the main process during the B0-M1-M9 (i.e. injection) period. During the first year of CO_2 injection preliminary trace element data in the Weyburn oil waters showed an initial increase of Al, Be, Ba, Fe, Cr and a decrease of Ni, Cu, Zn. This different behaviour among trace metals depends mainly on redox/acidic geochemical barriers (Perel'man, 1986) developing within the reservoir and mainly around the injection wells during CO_2 injection.

Our available trace metals data for the Weyburn oil-field could be interpreted within the framework of the Basinal Brine Theory (Kharaka et al., 1987), such that high dissolved metal concentrations in the presence of aqueous sulphide species require acidic solutions with pH values lower than those generally obtained in basinal brines. Very acidic solutions in some sectors of the reservoir are assured by CO_2 injection. From the mixture it will lead to non-equilibrium in the reaction: H_2S (gas) > H_2S (aqu) and will result in the transfer of gas to the water phase (Kharaka et al., 1987) in agreement to our dissolved H_2S data. This process will continue until essentially all the metals are precipitated as sulphides, as suggested by early PHREEQC chemical equilibria calculations (in progress and outlined in Quattrocchi et al., 2003). We suggest, in the frame of future Weyburn field monitoring, to merge more strictly the periodically measured parameters pertaining to different phases, i.e., dissolved H_2S together with the H_2S "free gas phase" (U of C data), as well as with the trace metal contents, to also study the sulphide precipitation/dissolution equilibria. Both chemical dissolution of the Mississippian Midale-(Marly)-Vuggy oil host-rocks and CO_2 dissolution as part of "solubility trapping" (Gunter et al., 1993; 1997, 2000) are clearly outlined by merging the Sr isotopic ratio evolution with other isotopic and chemical data collected for Weyburn oil waters (Hutcheon et al., 2003; Perkins et al., 2004; Shevelier et al., 2004).

4. CONCLUSIONS

From the Sr isotopic data of the Weyburn oil brines collected during the CO_2 EOR early injection period spanning from March 2001 to September, 2003 we found an evolution of the water-rock interaction processes. The progressively more positive (i.e., higher) trend in the Sr isotope ratios from 2001 to 2003 is suggested to be due to the dissolution of the Mississippian host rock. Carbonate minerals within the Midale beds precipitated from waters during the Mississippian, and therefore would have Sr ratios consistent with what would be considered a "baseline" isotopic composition. The leaching of strontium from the dissolution of these minerals would drive the Sr composition of the produced waters to heavier values, thereby masking the Mannville aquifer contribution. This hypothesis is strengthened by the $\delta^{13}C$ data, by the Sr isotopes of the reservoir rocks as well as by the evolution of other chemical data (dissolved gases, trace metals, major element chemistry).

In summary the Sr isotopic ratio in produced fluids corroborates the hypothesis of chemical and isotopic input from dissolved carbonates, strengthening the foreseen prevalent solubility trapping of CO_2 during the 2001-2004 Weyburn early CO_2 injection phase.

Results of this work suggest the usefulness of the Sr geo-chronological curves for the study of evolving CO_2 geological storage studies. The comparison between newly produced oil brine data and the data of Bruckschen et al., 1995, relative to the Dinatian Sr isotope record, show the potential of the Sr isotope geo-chronological curve, together with all the literature data.

ACKNOWLEDGEMENTS

We are grateful to the European Community which funded the Weyburn Project (Contract N. ENK5-2000-00304). A particular thank you is extended to Ken Ferguson and Dan Fradette working at the Weyburn EnCana oil field.

REFERENCES

Bachu, S., and Hitchon, B., 1996, Regional-scale flow of formation waters in the Williston Basin, *AAPG Bulletin* **80**(2): pp. 248-264.

Barnaby, R., Oetting, G.C. and Gao, G., 2004, Strontium isotopic signatures of oil-field waters: applications for reservoir characterization, *AAPG Bullettin* **88 (12):** 1677-1704.

Bruckschen, P., Bruhn, F., Veizer, J. and Buhl, D., 1995, $^{87}Sr/^{86}Sr$ isotopic evolution of Lower Carboniferous seawater: Dinatian of Western Europe, *Sedimentary Geology* **100:** 63-81.

Burke, W.H., Denison, R.E., Hetherington, E.A., Koepnick, R.B., Nelson, H.F. and Otto, J.B., 1982, Variation of seawater $^{87}Sr/^{86}Sr$ through Phanerozoic time, *Geology,* **16:** 516-519.

Capasso, G. and Inguaggiato, S., 1998, A simple method for the determination of dissolved gases in natural water. An application to thermal water from Vulcano Island, *Applied Geochem.* **13***:* 631-642.

Chapoy, A., Mohammadi, A.H., Tohidi, B., Richon, D., 2004, Gas solubility measurement and modeling for the Nitrogen + Water System from 274.18 to 363.02 K, *J., Chem. Eng. Data* **49**: 1110-1115.

Chaudhuri, S. and Clauer, N., 1993, Strontium isotopic compositions and potassium and rubidium contents of formation waters in sedimentary basins: clues to the origin of the solutes, *Geochemica Cosmoch. Acta* **57:** 429-437.

Connolly, C.A., Walter, L.M., Baadsgard, H. and Longstaffe, F.J., 1990, Origin and evolution of formation waters, Alberta Basin, Western Canada sedimentary basin: II. Isotopic systematics and water mixing, *Applied Grochemistry.* **5:** 397-413.

Czernichowski-Lauriol, I., Le Nidre, Y.M., Azaroual, M., Quattrocchi, F., Pearce, J.M., Sprinter, N., 2001, The Weyburn CO_2 Monitoring Project. Baseline hydrogeology, hydrochemistry and mineralogy, *Confidential Report BRGM/RP-51414-FR.*

Denison, R.E., Koepnick, R.B., Burke, W.H., Hetherington E.A. and Fletcher A., 1994, Construction of the Mississippian, Pennsylvanian and Permian seawater $^{87}Sr/^{86}Sr$ curve, *Chemical Geology* **112:** 145-157.

Gunter, W.G., Perkins, E.H. and Hutcheon, I., 2000, Aquifer disposal of acid gases: modeling of water-rock reactions for trapping acid wastes, *Applied Geochem.* **15**: 1086-1096.

Fisher, R.S. and Kreitler, C.W., 1987, Geochemistry and hydrodynamics of deep-basin, Palo Duro Basin, Texas, USA, *Applied Geochemistry* **2:** 459-476.

Franklyn, M.T., McNutt, R.H., Kamineni, D., Gascoyne, M. and Frape, S.K., 1991, Groundwater $^{87}Sr/^{86}Sr$ values in the Eye-Dashwa Lakes pluton, Canada: evidence for plagioclase-water reaction, *Geochimica Cosmoch. Acta* **86**: 111-122.

Hutcheon, I., Shevalier, M., Durocher K., Mayer B., Gunter, W. and Perkins, E., 2003, Carbon isotope evidence for CO_2 dissolution and fluid-rock interaction at the Weyburn CO_2 injection enhanced oil recovery site, *Proc. 2th Annual Conference on carbon Sequestration, May, 5-8, Alexandria, Virginia.*

Kharaka, Y.K., Maest, A.S., Carothers, W.W., Law, L.M., Lamothe, P.J. and Fries, T.L., 1987, Geochemistry of metal-rich brines from central Mississippi Salt Dome Basin, U.S.A., *Applied Geochemistry* **2**: 543-561.

Jones, C.E., Jenkyns, H.C. and Hesselblo, P., 1994, Strontium isotopes in Early Jurassic seawater, *Geoch. Cosmoch. Acta* **58:** 1285-1301.

Jones, D.G., Beaubien, S.E., Baubron, C., Cinti D., Davis, J.R., Emery, C., Fascetti, A., Lombardi, S., Michel, K., Morgantini., N., Penner, L., Quattrocchi, F. and Strutt, M.H., 2005. Continued soil gas monitoring at the Weyburn Unit in 2004. *British Geological Survey Commissioned Report CR/05/033, 68 pp.*

Le Nidre, Y.M. and Gaus, I. 2004, Characterization of the lower watrous aquitard as a major seal for CO_2 geological storage, (Weyburn Unit, Canada), In: *Proc. 7th Int. Conf. on Greenhouse Gas Control Technologies. Vol. 1 Peer-Reviewd Papers and Plenary Presentations, BC, September 5-9, 2004.*

McNutt, R.H., Frape, S.K. and Dollar, P. 1987, A strontium, oxygen and hydrogen isotopic composition of brines, Michigan and Appalachian basins, *Applied Geochemistry* **2:** 495-505.

McNutt, R.H., Frape S.K., Fritz, P., Jones, M.G. and McDonald, I.M., 1990, The $^{87}Sr/^{86}Sr$ values of Canadian shield brines and fracture minerals with applications to groundwater mixing, fracture history and geochronology, *Geochimica Cosmoc. Acta* **54:** 205-215.

Nakano, T., Kajiwara, Y. and Farrell, C.W., 1989, Strontium isotope constrain on the genesis of crude oils, oil-field brines, and Kuroko ore deposits from the Green Tuff region of northeastern Japan, *Geochimica Cosmoch. Acta* **53:** 2683-2688.

Nitao, J.J., 1996, The NUFT Code Modeling Non isothermal, Multiphase, Multicomponent Flow and transport in Porous Media, *EOS American Geophysical Union* **74 (3):** 3.

Notsu, K., Wakita H. and Nakamura Y., 1988, Strontium isotopic composition of oil-field and gas-field waters, Japan, *Applied Geochemistry* **3:** 173-176.

Parkhurst, L.D. and Appelo, C.A.J., 1999, User's guide to PHREEQC, (version 2), U.S.G.S. *Water-Resources Investigations report, 99-4259.*

Perel'man, A.I., 1986, Geochemical Barriers: theory and practical applications, *Applied Geochem.* **1**: 669-680.

Perkins, E., Czernichowski-Lauriol, I., Azoural, M. and Durst, P., 2004, Long term predisctions of CO_2 storage by mineral and solubility trapping in the Weyburn Midale Reservoir, In: *Proc. 7th Int. Conference on Greenhouse Gas Control Technologies. Poster.*

Popp, B.N., Podosek F.A., Brannon J.C., Anderson T.F. and Pier, J. 1986, $^{87}Sr/^{86}Sr$ ratios in Permo-Carboniferous sea water from the analysis of well-preserved brachiopod shells, *Geochimica Cosmoc. Acta* **50:** 1321-1328.

Quattrocchi, F., Barbieri, M., Bencini, R., Cinti, D., Galli, G., Pizzino, L. and Voltattorni, N., 2003, Water chemistry, isotopic chemistry ($^{87}Sr/^{86}Sr$) and dissolved gases variations at the Weyburn oil-field (Canada). Possible water-rock processes during the CO_2 injection, *Proced. Int. Workshop "Gas WRGI induced by reservoirs exploitation, CO_2 sequestration and other geological storages". November, 18-20, 2003, IFP, Paris, France.* p. 94.

Quattrocchi, F., Bencini, R., Cinti, D., Galli, G., Pizzino, L., Voltattorni, N. and Navarra, A., 2004, CO_2 geological sequestration in Italy: state of art and potential sites, In: *Proc. XI Inter. Conf. "Water Rock Interaction" "WRI-XI", Saratoga Spring, New York, USA, June 2004.* pg. 573-576.

Riding J. and Rochelle C. (2005), The IEA Weyburn CO_2 monitoring and storage project. Final Report of the European Research Team. *British Geological Survey Research Report, RR/05/03*, 54 pp (ISBM 085272 507 8). BGS, Keyworth, Nottingham.

Romer, R.L., Wawrzenitz, N. and Oberhansli, R., 2003, Anomalous unradiogenic $^{87}Sr/^{86}Sr$ ratio in ultrahigh-pressure crustal carbonates – evidence for fluid infiltration during deep subduction ? *Terra Nova* **15 (5):** 330-336.

Russell, C.W., Cowart J.B. and Russell G.S., 1988, Strontium isotopes in brines and associated rocks from Cretaceous strata in the Mississippi Salt Dome Basin (Southeastern Mississippi, USA), *Chemical Geology* **74:** 153-171.

Shevalier, M., Durocher, K., Perez, R., Hutcheon, I., Mayer, B. and Perkins, E. 2004 Geochemical monitoring of gas-watrer-rock interaction at the IEA Weyburn CO_2 Monitoring and Storage Project *Proc. 7th Int. Conf. on Greenhouse Gas Control Technologies (GHGT-7).* In progress on http://www.ghgt7.ca/

Sinclair, A.J., 1991, A fundamental approach to threshold estimation in exploration geochemistry: probability plots revisited, *J. of Geochemical Exploration* **41**: 10-20.

Smalley, P.C., Raheim, A., Dickson, J.A.D. and Emery D., 1988, $^{87}Sr/^{86}Sr$ in waters from Lincolnshire Limeston aquifer, England, and the potential of natural strontium isotopes as a tracer fro a secondary recovery seawater injection process in oilfields, *Applied Geochemistry* **3:** 591-600.

Starinsky, A., Bielski, M., Lazar, B., Steinitz, G. and Raab, M., 1983, Strontium isotope evidence on the history of oil field brines, Mediterranean coastal plain, Israel, *Geochimica Cosmochim. Acta,* **47:** 687-695.

Stueber, A.M., Pushkar, P. and Hetherington, E.A., 1987, A strontium isotopic study of formation waters from the Illinois Basin, USA, *Applied Geochemistry,* **2:** 477-494.

Sunwall, M.T. and Pushkar P., 1979, The isotopic composition of strontium in brines from petroleum fields of Southerm Ohio, *Chemical Geology,* **24:** 189-197.

Tian, S., Zhao, G. and Asghari, K., 2004, CO_2 storage during EOR process in the Weyburn oil pool: modeling and monitoring results, In: *Proc. 7th Int. Conf. on Greenhouse Gas Control Technologies. Vol. 1 Peer-Reviewd Papers and Plenary Presentations, BC, September 5-9, 2004.* http://www.ghgt7.ca/

Torgersen, T., and Clarke, W.B., 1987, Helium accumulation in groundwater, III. Limits on He transfer across the mantle-crust boundary beneath Australia and the magnitude of mantle degassing, *EPSL* **84**: 345-355.

Wilson, M. and Monea, M. (Eds.), 2004, IEA-GHG Weyburn CO_2 Monitoring and Storage Summary Report 2000-2004. In: *Proc. 7th International Conference on Greenhouse Gas Control Technologies. Vol. III, Petroleum Technology Research Center (PTRC Eds.)*, 243 pp.
Veizel, J. and Compston, W., 2004, $^{87}Sr/^{86}Sr$ composition of seawater during the Phanerozoic, *Geochimica Cosmoc. Acta* **38**: 1461-1484.

APPENDIX

Table 2. INGV geochemical surveying in the frame of the Weyburn IEA-EC Project, from August 2000 to September 2004. At the end of the M 8 survey the injected CO_2 amount since the beginning of the Phase A1 was around 1.7 Ml m^3 CO_2.

Survey	Dissolved gases	Chemical analyses	$^{87}Sr/^{86}Sr$ ratio
Baseline (B 0), Aug. 2000	no	no	partially
Monitor 1 (M 1), March 2001	yes (few)	yes	yes
Monitor 2 (M 2), July 2001	no	no	no
Monitor 3 (M 3), Sept. 2001	no	no	no
Monitor 4 (M 4), March 2002	no	no	no
Monitor 5 (M 5), June 2002	yes	yes	yes
Monitor 6 (M 6), Sept. 2002	no	no	no
Monitor 7 (M 7), April 2003	no	no	yes
Monitor 8 (M 8), June 2003	yes	no	no
Monitor 9 (M 9), Sept. 2003	yes	partially	partially
Monitor 10 (M 10), March 2004	yes	partially	partially
Monitor 11 (M 11), Sept. 2004	yes	partially	partially

OPTIMIZATION OF CO_2 INJECTION FOR SEQUESTRATION / ENHANCED OIL RECOVERY AND CURRENT STATUS IN CANADA

T. Babadagli
University of Alberta, Civil and Environmental Eng., School of Petroleum and Mining, 3-112 Markin/CNRL - NREF, Edmonton, Alberta, T6G 2W2, Canada

Abstract: In today's industrialized world, the generation (and emission) of greenhouse gases (GHG) is likely to continue. The reduction of greenhouse gas emissions through public outreach programs is one approach to mitigate this problem however, in practice, it has not yet been received well by the public due to economic costs. Therefore, permanent storage of these gases in underground reservoirs is believed to be one of the most suitable solutions for the mitigation of greenhouse gases. Sequestration of GHG is not cheap, however, and thus the injection of greenhouse gases into oil or gas reservoirs to enhance production may offset some of these associated costs. The use of CO_2 for purely EOR purposes versus injection of CO_2 primarily for sequestration are technically two different problems. Proper design practices and technology need to be developed and applied in order to inject CO_2 into oil reservoirs not only for the purpose of tertiary oil recovery but also for permanent sequestration. In conventional CO_2 injection projects the main purpose is to increase the amount of oil produced per amount of CO_2 injected. In contrast optimization of CO_2 injection for sequestration purposes requires "maximum oil production with the highest amount of CO_2 storage". Breakthrough time is a critical parameter in this exercise as recycling CO_2 is undesirable due to economic and environmental constraints. This paper summarizes on-going research into the conditions that will maximize oil recovery while maximizing the underground sequestration of CO_2. Results obtained from numerical modeling of the injection process are discussed. Current efforts on CO_2 injection in Canada are also presented.

Key words: CO_2 sequestration, enhanced oil recovery, optimisation, miscible/immiscible CO_2 injection, flue gas injection, numerical modeling

S. Lombardi et al. (eds.),
Advances in the Geological Storage of Carbon Dioxide, 261–270.

1. INTRODUCTION

One of the most suitable means for geological CO_2 sequestration is in enhanced oil recovery (EOR) operations, which lower the cost of CO_2 disposal by recovering incremental oil (Fanchi, 2001; Jessen et al., 2001; Brush et al., 2002). According to the U.S. DOE, 2,000 scf of CO_2 are stored per barrel of currently produced oil, with targets set for an eventual 10,000 scf per barrel of oil (http://www.netl.doe.gov/coalpower/sequestration).

CO_2 injection for EOR is a proven technology and many applications are in progress all around world. Note, however, that the sequestration of CO_2, while enhancing oil production in partly depleted reservoirs, is technically a different problem. Detailed analyses of the process are needed as early breakthrough of CO_2 is not desired.

This paper focuses on the determination of optimal conditions that maximize both oil recovery and underground CO2 sequestration. This was investigated by using a numerical model and considering different injection schemes, such as miscible and immiscible CO_2 and flue gas. Evaluations were done for breakthrough times.

2. STATUS IN CANADA

Depleted or nearly depleted oil reservoirs are good candidates for CO_2 storage as they offset some of the costs of sequestration by producing additional oil recovery. This is of particular importance in Alberta, Canada, as many oil reservoirs here are nearing depletion and the needed infrastructure is already in place. A recent study has shown that 4,470 oil reservoirs out of 8,637 in Alberta have passed the screening criteria for CO_2 injection (Shaw and Bachu, 2002). Preliminary calculations show that up to $558x19^6$ m^3 of additional oil could be recovered while sequestering 1,118 Mt of CO_2. Injecting CO_2 into oil reservoirs, however, should be done in a controlled way to improve oil recovery (Rojas et al., 1991, Babadagli, 2003-a and b).

Existing, proven oil reserves in Canada (~ 6.5 billion barrels) are diminishing. On the other hand, Alberta owns the largest oil sands reserves in the world and they are expected to provide approximately half of Canada's crude oil production by 2005. For both cases, enhanced oil recovery (EOR) applications are inevitable, but still more effort is needed for the development of technology which can recover the great amount of oil left in the depleted pools and oil sands.

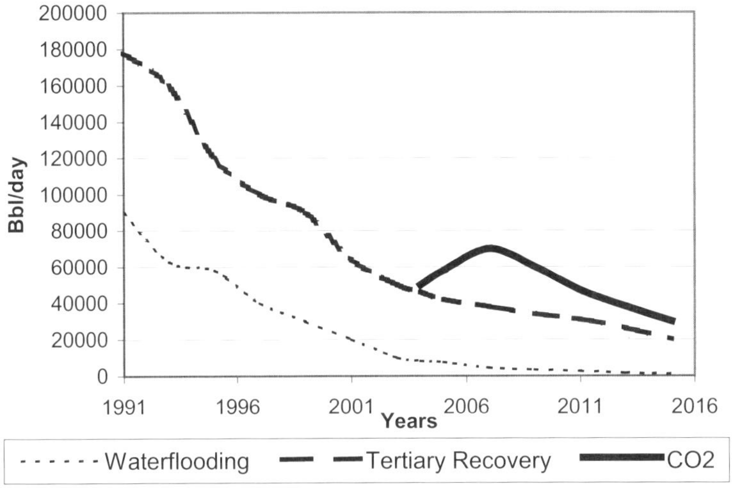

Figure 1. Enhanced oil production in Canada as of 2001. The values beyond 2001 are forecast. Source: www.ptac.org, CAPP, ADOE, and AEUB.

The current status and prospect of Canadian secondary-tertiary oil recovery are shown in Figure 1 in comparison with potential CO_2 applications.

More than half of the existing oil pools in Alberta have shown a potential for CO_2 recovery (Shaw and Bachu, 2002), however a well known problem is CO_2 capture and transportation from different sources. Having oil and gas reservoirs near dependable CO_2 sources makes sequestration applicable in this region. The total amount of anthropogenically-produced CO_2 in Canada was 277 Mt/year in 1997. The prospect in 2010 is 333 Mt/year and the great percentage of this comes from power plants, as shown in Figure 2. Applications in the eastern part of Canada are also under consideration (Shafeen et al., 2004).

Emission of impure CO_2 other than flue gas also exists in Alberta. Acid gas has been produced in Zama (80% CO_2 – 20 % H_2S) and disposal of this gas has been a concern (Davison et al., 1999). It has also been noted that acid gas is an effective enhanced oil recovery agent as it reduces the minimum miscibility pressure of CO_2 as opposed to N_2. There is no doubt that storing H_2S with CO_2 in underground reservoirs will double the benefit for the environment (Trivedi et al., 2005).

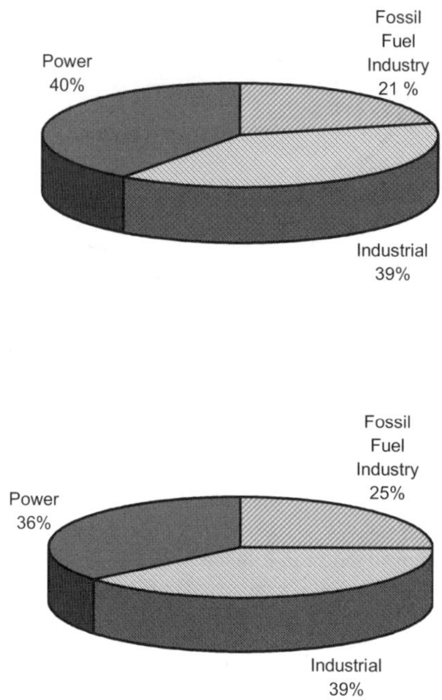

Figure 2. CO_2 emissions in Canada (source: www.nrcan.gc.ca/es/etb/cetc/combustion/co2network).

Early breakthrough is a serious problem in naturally fractured reservoirs since the fractures serve as a high permeability pathway which results in the faster movement of injected fluids (Beliveau et al., 1993 and Babadagli, 2003-a and b). Beliveau et al. (1993) observed that water injected into a fractured carbonate reservoir (Midale, Canada) yielded very early breakthrough (order of hours) whereas CO_2 breakthrough took a month. This indicates the viability of EOR (due to diffusive matrix-fracture interaction) during the sequestration of CO_2. Oil recovery and sequestration potential should be identified in such weakly-water-wet fractured systems, and the Weyburn field in Canada is a good example of this type of reservoir.

3. OBJECTIVES AND METHODOLOGY

The objective of this study is to test numerous different injection schemes and to determine the conditions which yield the highest oil recovery while storing the maximum amount of CO_2. To achieve this, the compositional option of a commercial reservoir simulator (CMG) was used. Runs were performed for different injection schemes and comparisons were made for the amount of oil produced against the amount of CO_2 stored at breakthrough time.

4. MODELING CO_2 SEQUESTRATION DURING ENHANCED OIL RECOVERY

4.1 Numerical Model

A hypothetical reservoir was considered by adapting CO_2 injection data from the Wasson field (Hsu et al., 1997). The model consists of 7x7x3 grids. The injector was located at the centre and four producers were in the four corner grids representing a typical 5-spot injection pattern.

Twelve components (C1-C29, CO_2, NO_2) were considered. Other reservoir, rock and fluid properties used as simulator variables are given in Table 1. Relative permeabilities for oil, gas and water phases were obtained from Hsu et al. (1997).

Table 1. Reservoir, rock and fluid properties used as simulator variables.

Properties	
Depth	5200 ft.
Average porosity	12%
Permeability	5 md
Net thickness	460 ft
Reservoir temperature	110 °F
Oil gravity	33 API
Min. miscibility press.	1300 psi

Reported minimum miscibility pressure (MMP) values for Wasson crude oil range from 1250 to 1400. For pressures less than 0.9 times the MMP the displacement is treated as immiscible, while for pressures equal to or greater than the MMP the displacement is assumed miscible. Note that the runs were performed after a long water-flooding period.

4.2 Results

Evaluations were performed using cumulative oil recovery vs. cumulative CO_2 storage plots and cumulative oil recovery-cumulative CO_2 storage vs. time plots. Examples of these graphs for pressures less than the

MMP are shown in Figures 3 and 4. Increasing pressure yielded higher oil recovery and CO_2 storage (Fig. 3). Obviously the breakthrough time of CO_2 is critical and it is expected that the earlier breakthrough time will be obtained with increasing pressure. Note that this plot is not sufficient to decide whether an optimum pressure (or conditions) exists.

Another performance indicator was considered to be the plot of cumulative oil vs. CO_2 stored to explain the efficiency of the process (Fig. 4). For any given oil recovery, higher storage was obtained with increasing pressure. For any given amount of CO_2 storage, cumulative oil produced increased with decreasing pressure. Thus any decision should be made on what the final expectation or economic limitations are. If the purpose of injection is CO_2 sequestration, one should pay attention to the latter. If oil recovery is the primary concern, however, lower pressure would be more suitable for the application. In other words, the weighing factor is a critical issue in this decision. Kovscek and Cakici (2004) proposed an objective function ($f = wN_p + (1-w)V_{CO_2}$) and analysed the CO_2 injection cases where the storage (V_{CO2}) and the recovery (N_p) are equally important ($w = 0.5$).

The above process was repeated for other injection scenarios and the cumulative oil recoveries were compared to the CO_2 storage at breakthrough time. The results are shown in Figures 5 through 8. The immiscible case, which occurs under the MMP (1500 psi), was considered first. Both oil recovery and CO_2 storage showed a similar trend (Fig. 5). The lowest recovery and storage occurs at a pressure of 1300 psi, but both parameters increase with increasing pressure when the pressure is above the MMP. The continuation of this trend can be found in Figure 6, where recovery and storage continue to increase with increasing pressure. Obviously the highest pressure yielded the best solution at breakthrough. One can conclude that miscible injection is desirable for storage and recovery if the required pressure is practically applicable.

Another case considered was the injection of flue gas rather than pure CO_2. Flue gas injection is a proven enhanced oil recovery technique (Iijima, 1998 and Dong and Huang, 2002). It is practical and economical to store CO_2 in this form as the purification of CO_2 generated in power plants is costly. The results for the flue gas injection case are illustrated in Figure 7, where the trend is similar to the miscible case. Note that maximum storage with maximum recovery was obtained for the highest pressure (2500 psi). The amount of storage below this pressure is significantly lower even though the recovery did not change significantly. Thus injection pressure would be selected based on the primary purpose, i.e., storage or additional oil recovery.

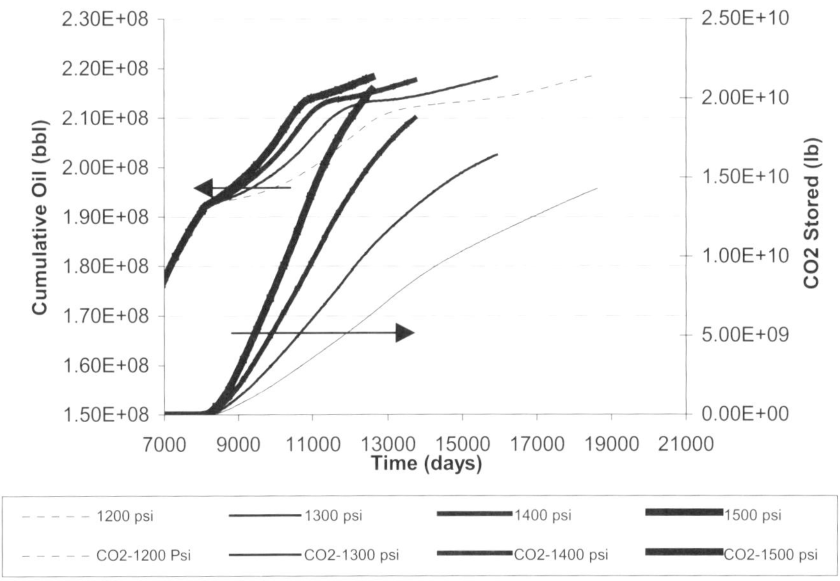

Figure 3. Cumulative oil and CO_2 stored against time for different injection pressures.

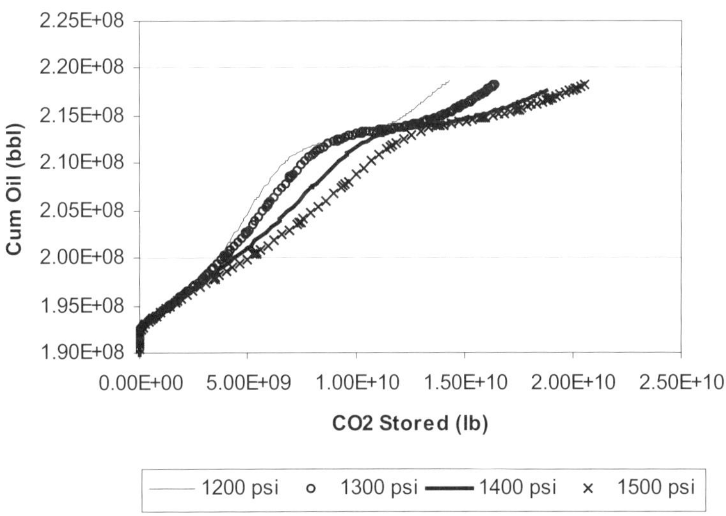

Figure 4. Cumulative oil against CO_2 stored at different injection pressures.

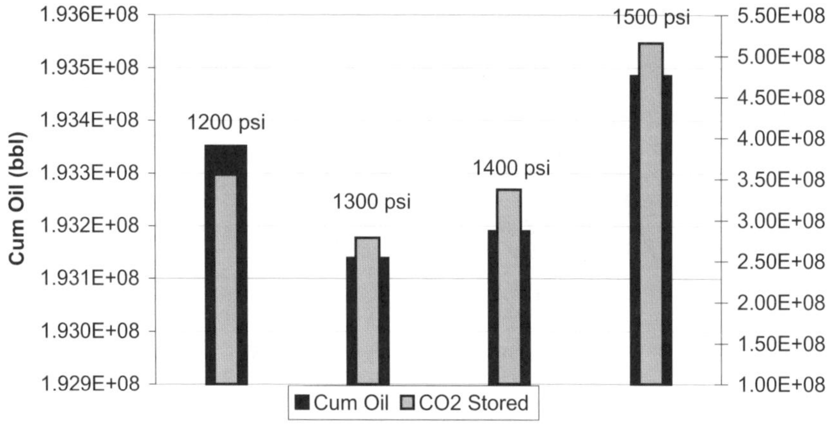

Figure 5. Comparison of cumulative oil recovery and CO_2 storage at breakthrough time for the immiscible case.

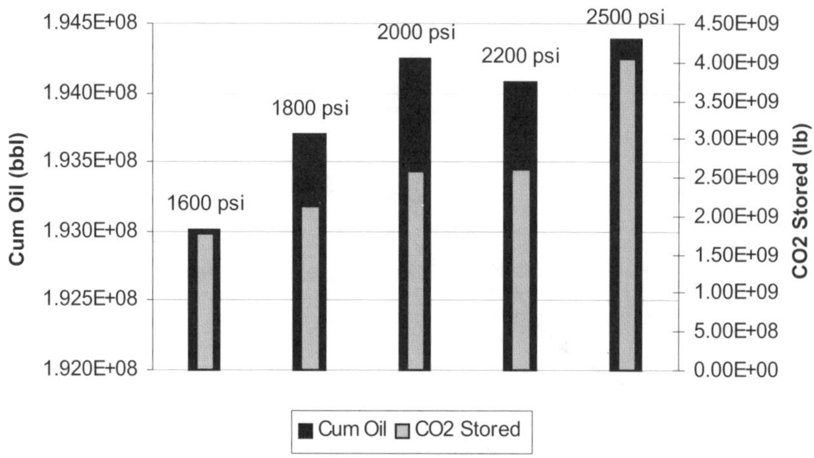

Figure 6. Comparison of cumulative oil recovery and CO_2 storage at breakthrough time for the miscible case.

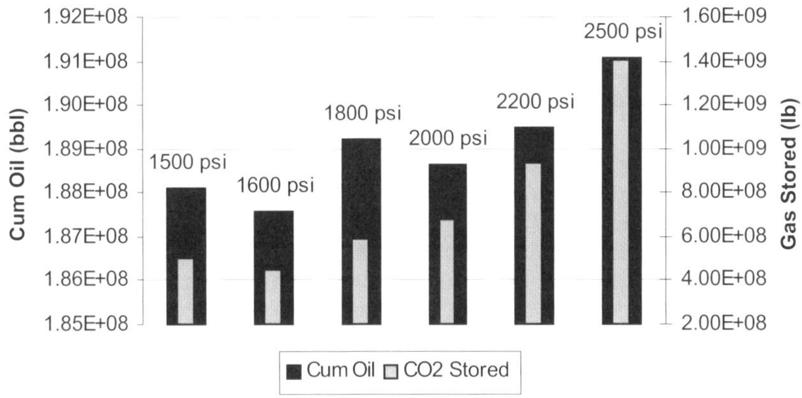

Figure 7. Comparison of cumulative oil recovery and CO_2 storage at breakthrough time for flue – gas injection.

5. CONCLUDING REMARKS

This paper reports the preliminary results of a numerical study on the optimization of CO_2 storage during enhanced oil recovery. It was observed that the immiscible and miscible cases exhibit different recovery/storage behavior. Although the miscible case showed better recovery performance and storage capacity with pressure increases, it is difficult to reach MMP due to technical difficulties and/or regulatory restrictions in many circumstances. One should select the injection pressure depending on the priorities, i.e., more additional oil or more or CO_2 storage.

Identification of the causes of CO_2 entrapment is critical but it varies reservoir by reservoir. Answering what really causes CO_2 to get trapped in the reservoir needs special studies for any particular reservoir, with focus being given to the effects of pore types, fractures and chemical interactions with the reservoir rock. The need for investigations into the long-term behavior of injected and trapped gas are also obvious (Ennis-King and Paterson, 2003).

ACKNOWLEDGEMENTS

The author would like to thank the Computer Modeling Group (CMG) for providing the simulator used in this research. Mr. Japan Trivedi is gratefully acknowledged for his help during simulation runs.

REFERENCES

Babadagli, T., 2003a, Evaluation of EOR methods for heavy-oil recovery in naturally fractured reservoirs, *J. of Petroleum Sci. and Eng.*, **37** (1-2): 25-37.

Babadagli, T., 2003b, Viability of EOR during CO_2 sequestration," PTAC, CO_2 from Industrial Sources to Commercial Enhanced Oil & Gas Rec. Workshop and Forum, Calgary, AB, Canada, Oct. 1-2.

Beliveau, D., Payne, D.A. and Mundry, M., 1993, Waterflood and CO_2 flood of the fractured Midale field, *JPT*, Sept. 1993, 881-887.

Brush, R. M. et al., 2000, Immiscible CO_2 flooding for increased oil recovery and reduced emission, SPE 59328, SPE Imp. Oil Rec. Sym., Tulsa, OK, 3-5 April.

Davison, R.I., Mayder, A., Hladiuk, D.W. and Jarrell, J, 1999, Zama acid gas disposal/miscible flood implementation and results, *JCPT*, vol. 38, no 2 (Feb.).

Dong, M. and Huang, S., 2002, Flue gas injection for heavy oil recovery, *JCPT*, **41** (9): 44-50.

Ennis-King, J. and Paterson, L., 2003, Role of convective mixing in the long-term storage of carbon dioxide in deep saline formations, SPE 84344, SPE Annual Tech. Conf. and Exh., Denver, CO, 5-8 Oct.

Fanchi J. R., 2001, Feasibility of monitoring CO_2 sequestration in a mature oil field using time-lapse seismic analysis, SPE 66569, SPE/EPA/DOE Exp. and Prod. Env. Conf., San Antonio, TX, 26-28 Feb.

Hsu, C.-F, Morell, J.J. and Falls, A.H., 1997, Field-scale CO_2-flood simulations and their impact on the performance of the Wasson Denver unit, *SPE RE*, Feb., 4-11.

Iijima, M., 1998, A feasible new flue gas CO_2 recovery technology for enhanced oil recovery, SPE 39686, SPE/DOE Imp. Oil Rec. Symp., Tulsa, OK, 19-22 April.

Jessen, K., Kovscek, A.R. and Orr, F., 2005, Increasing CO_2 storage in oil recovery, *Energy Conversion and Management*, **46**: 293-311.

Kovscek, A.R. and Cakici, M.D., 2004, Geological storage of carbon dioxide and enhanced oil recovery: II Cooptimization of storage and recovery, *Energy Conversion and Management*, to appear.

Rojas, G.A., Zhu, T., Dyer, S.B., Thomas, S. and Farouq Ali, S.M., 1991, Scaled model studies of CO_2 floods, *SPE Res. Eval. & Eng.*, May, 169-178.

Shafeen, A., Croiset, E., Douglas, P.L. and Chatzis, I, 2004, CO_2 sequestration in Ontario, Canada: Part I: Storage evaluation of potential reservoirs, *Energy Conversion and Management*, **45**: 2645-2659.

Shaw, J. and Bachu, S., 2002, Screening, evaluation, and ranking of oil reservoirs suitable for CO_2 flood EOR and carbon dioxide sequestration, *JCPT*, **41**(9):51-61.

Trivedi, J., Babadagli, T., Lavoie, R., and Nimchuk, D., 2005, Acid Gas Sequestration During Tertiary Oil Recovery: Optimal Injection Strategies and Importance of Operational Parameters," 56th Canadian Int. Annual Tech. Meet., Calgary, 7-9 June.

THE USE OF CO_2 AND COMBUSTION GASES FOR ENHANCED OIL RECOVERY IN RUSSIA

V.A. Kuvshinov
Institute of Petroleum Chemistry SB RAS, Tomsk, Russia

Abstract: The use of carbon dioxide and combustion gases in EOR technologies is of interest from the point of view of CO_2 geological sequestration. During the period of 1980-1990 large-scale pilot tests were carried out in Russia to utilize carbon dioxide and combustion gases, formed at different petrochemical production plants, to enhance oil recovery in different hydrocarbon fields. The analysis of field data indicates that the greater part of CO_2 injected into oil reservoirs was recovered from production wells during the 1-1.5 years of the experiment. The total amount of CO_2 which was irreversibly stored in the reservoir was not measured. To reveal mechanisms of CO_2 geological sequestration it is recommended to study in detail carbon dioxide interactions with oil, formation water and with the reservoir rocks under the conditions of miscible and immiscible oil displacement with carbon dioxide.

Key words: EOR in Russia; sequestration; carbon dioxide

1. INTRODUCTION

The use of carbon dioxide and combustion gases in EOR technologies is of great interest from the point of view of CO_2 geological sequestration, as there is the potential for both decreased disposal costs while increasing resource recovery.

During the period from 1980-1990 large-scale pilot tests were performed in Russia to inject carbon dioxide and combustion gases, formed at different petrochemical plants, into petroleum hydrocarbon reservoirs in order to enhance oil recovery (Table 1). This paper summaries some of the results obtained during this project from one site.

S. Lombardi et al. (eds.),
Advances in the Geological Storage of Carbon Dioxide, 271–275.

Table 1. The results of CO_2 injection into oil pools in Russia.

	Radaevskoye	Kozlovskoye	Sergeevskoye	Elabuzhskoye
Start of CO_2 injection	August 1984	May 1986	September 1986	July 1987
Injection of CO_2 for 01.07.89, thou. t.	787.2	110.1	73.8	58.3
Oil after production, thou. t.	218	12.6	17.7	no data
Specific oil after production, t./t. CO_2	0.277	0.114	0.24	no data

2. RESULTS

Most often the technical approach consisted of continuous injection of CO_2 as carbonated water and alternate injections of CO_2 and water slugs. This technology proved to be effective in fields containing heavy oils, highly mineralised waters, and in steeply dipping strata having lower reservoir permeabilities. The most successfully carbonated-water EOR project was in the Aleksandrovskay area of the Tuymazinskoye oil field, where the final oil recovery increased by 12%.

An inexpensive source of CO_2 and a heterogeneously structured oil deposit are considered as necessary conditions for continuous CO_2 injection. During pilot tests two problems complicated the continuous injection of carbon dioxide, one organizational and the other technical. The organizational problem lied in establishing a reliable system of CO_2 collection from the plants, where it is a waste product, and transportation to the oil fields. The technical problem, which was not completely solved, consisted of strong carbon dioxide corrosion of the equipment and communications in the CO_2 collection and transportation system.

The results obtained during the injection of CO_2 into a Devonian reservoir in the Elabuzhskoye oil field in 1987 have been analysed and are presented below as an example. The CO_2 used for this project was formed as a waste by-product during ethylene oxide production at the Nizhnekamskneftekhim petrochemical plant, located 5 km away from the oil field. The Elabuzhskoye oil field has been developed by perimeter water-flooding since 1972. The test area is show schematically in Figure 1, where average inter-well spacing is 500 m. From the beginning of development of the test area a total of 89.2% of the initial recoverable reserves have been recovered; the current oil recovery factor equals 0.437 (finite - 0.490) and the water cut exceeds 61%.

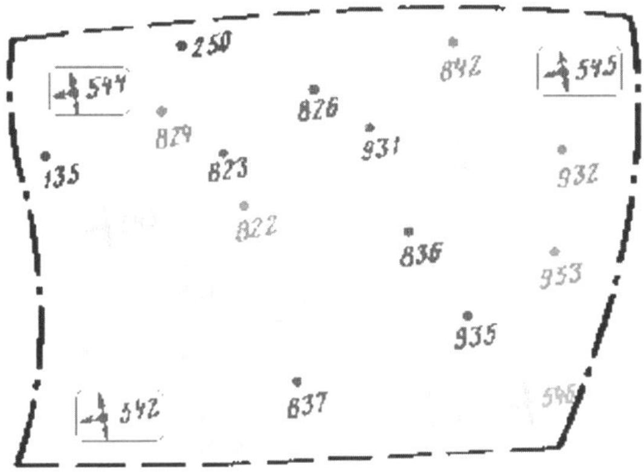

Figure 1. Scheme of well spacing in the test area.

CO_2 was injected into four injection wells (numbers 542, 544, 545 and 546) located along the test area perimeter. Monthly dynamics of the CO_2 injection are presented in Figure 2.

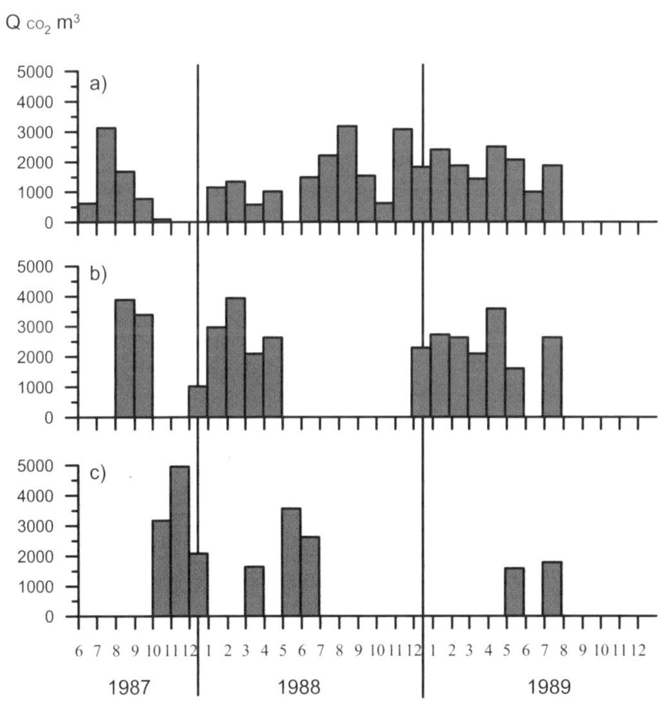

Figure 2. Dynamics of CO_2 injection into injection wells 542 (a), 545 (b) and 544 (c).

Production wells in the test area were divided into 4 groups based on their response to the CO_2 injection (Fig.3):

1. appearance of CO_2 in a well without any change in water cut (wells 842, 932 and 824)
2. appearance of CO_2 in a well and decrease in water cut (wells 250 and 135);
3. absence of CO_2 in a well, decrease in water cut (wells 826, 836, 837, 931, 935, 949 and 823) and increase in water cut (wells 822 and 933);
4. no changes were observed in well operations (wells 820, 821, 828, 848, 938, 949, 951).

The analysis of the field data indicates that the greater part of CO_2 injected into the oil pool was recovered from production wells during the 1-1.5 years of the test. The total amount of CO_2 which was irreversibly trapped within the pool was not estimated.

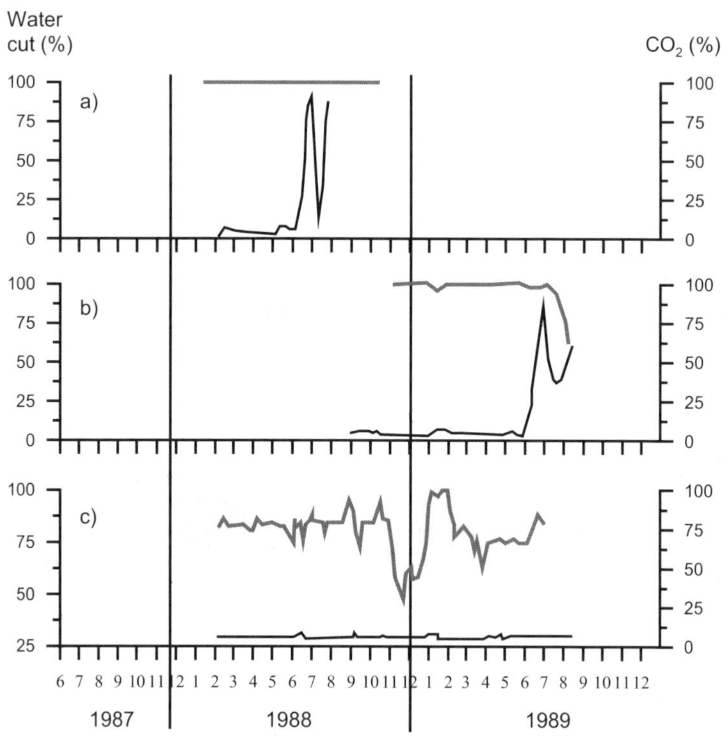

Figure 3. Dynamics of water cut and CO_2 content in wells 932 (a), 250 (b) and 836 (c).

3. CONCLUSIONS

To better understand the mechanisms of CO_2 geological sequestration it is necessary to study in detail the interactions of carbon dioxide with oil, formation water and with the reservoir lithologies under the conditions of miscible and immiscible oil displacement with carbon dioxide. It would be promising to study the physical-chemical mechanisms of long-term geological retention of CO_2 in the form of insoluble organo-carbonate complexes in the oil reservoir at a later stage of its development. It is possible to form carbonate and organo-carbonate compounds in the oil reservoir via CO_2 interaction with the mineralised formation water and any additionally injected chemical reagents. Such studies could result in the development of EOR technology which combines enhanced oil recovery and long-term geological retention of CO_2 due to the creation of flow-deflecting screens via insoluble carbonate and organo-carbonate complexes in the oil reservoir.

REFERENCES

Surguchev, M.L., Gorbunov, A.T., Zabrodin, D.P., Ziskin, E.A., and Malyutina, G.S., 1991, Methods for residual oil recovery, *M.: Nedra*, 347 p.

CONTROLS OF CO_2 FILTRATION IN HETEROGENEOUS RESERVOIRS WITH FOAM-EMULSION SYSTEMS

A.G. Telin
The Institute of Mechanics of Ufa Research Center of Russia450078, Ufa, Revolutionary Str., 96/2 (E-mail: telin@ngt.ru)

Abstract: Pumping CO_2 into oil reservoirs is one of the most promising methods to increase oil recovery. A significant drawback of the method is channelled flow of CO_2 through the most permeable intervals in heterogeneous reservoirs. To control flow we used foam-emulsion systems generated both before pumping into the reservoir model and directly in the reservoir. This allows us to double the driving medium from high- to low-permeability reservoirs and presents a likely mechanism of flow redirection. Studying the regulating effect of foam-emulsion systems using a transparent micro-model of the reservoir and the Hele-Shaw cell revealed a previously unknown effect of dynamic blocking of porous mediums and fractured structures during transformation of emulsions filtering through it.

Key words: channelled flow during CO_2 flooding; heterogeneous reservoir; porous structure; foam-emulsion system; driving medium; redirection of flow; interchanged injection; dynamic blocking

1. INTRODUCTION

Pumping carbon dioxide into oil reservoirs is one of the most promising and quickly developing methods to increase oil recovery. A large-scale introduction of the method in the USA showed its effectiveness for oilfields during different stages of development and in reservoirs having a wide range of geology, structure and oil properties.

Data published in the international press and results of trial CO_2 injection into Russian oilfields shows that the main technological factor that limits the

S. Lombardi et al. (eds.),
Advances in the Geological Storage of Carbon Dioxide, 277–287.

development of the method is the quick inrush or channelling of CO_2 into producing wells along strata or structures with high permeability. The main instrument to control the process of oil extraction is sequential injection of CO_2 and water slugs. Recently, surfactant-water solutions have been used to push CO_2 to increase the effectiveness of the technology.

2. RESULTS AND DISCUSSION

Trial pumping of CO_2 in the Radayevskoye and Sergeyevskoye oilfields has shown the technical effectiveness of this method. Injection of CO_2 has also been conducted in the Kozlovskoye (PO Kuybyshevneft) and Yelabuzhskoye (PO Tatneft) oilfields, while preliminary studies are being performed on the Olkhovskoye oilfield (PO Permneft). Technical indices are listed in table 1.

Table 1. Technical indices of carbon dioxide injection.

Parameters	Oilfield				
	Radayevs-koye	**Kozlovs -koye**	**Sergeevs-koye**	**Yelabuzh-skoye**	**Olkhovs-koye**
Predicted oil output increase, %	12.80	10.40	10.40	8.00	12.40
Additional output, million tons	6.90	3.21	0.89	0.40	3.10
Yearly injection of CO_2, thous. tons	430.0	400.0	165.0	140.0	400.0
Actual injection by 01/07/89, thous. tons	787.2	110.1	73.8	58.3	-
Injection started	August 1984	May 1986	September 1986	July 1987	-

The largest quantity of CO_2 was injected into the reservoir within the Radayevskoye oilfield. Results of a four year trial indicate the effectiveness of the method, with the injection of 787.2 thousand tons of CO_2 during this period. Despite the fact that this was 2.6 times less than the planned amount for this time period an additional 218 thousand tons of oil were produced due to CO_2 injection by 01/07/89, which amounts to 0.28 tons per 1 ton of injected reactant. As a result the obtained recovery rate was higher than the predicted rate (0.21 ton/ton).

In this paper results are presented on water-saturated model experiments of non-communicated heterogeneous reservoirs having water permeability ratios equal to 15.7. Experiments were conducted at a pressure of 10.5 MPa, temperatures equal to 26.5 and 40°C, and an average fluid flow velocity of 201 m/year.

The experimental setup (Figure 1) used to study the distribution of CO_2 in heterogeneous reservoirs consists of standard equipment and devices used to investigate processes of fluid and gas filtration under reservoir conditions. Detailed descriptions of the equipment, as well as the preparation of the porous mediums can be found in OST (1986).

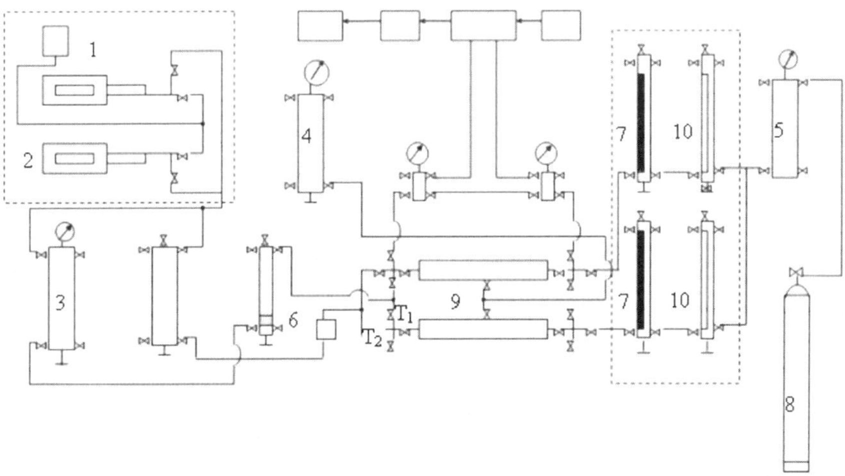

Figure 1. The experimental setup used to investigate the distribution of CO_2 in heterogeneous reservoirs: 1, 2 –constant flow rate sensors; 3,4,5 – columns for distilled water, reservoir water and compressed air or nitrogen, respectively; 6 – filters; 7, 10 – measuring tanks to determine the yield rate of the reservoir and volume of the delivered CO_2 visually; 8 – balloon with air or nitrogen; 9 – core-holder.

The plant, excluding the constant flow rate sensors (SCF), is thermostatted. Each experiment consisted of three stages. The first stage was to deliver water (using SCF-1 through assembly T_1) with a constant flow rate (measured using measuring tank 7) up to the establishment of a constant yield rate. The second stage was to study the influence of one of the investigated parameters. Water was delivered into reservoir models as done during the first stage, while CO_2 was delivered using SCF-2 from tank 10 through assembly T_2. The third stage was to wash the porous medium separately using water to remove the CO_2.

Table 2. Characteristics of reservoir models.

Model number	Length, cm		Permeability, mcm^2		Permeability ratios, K_0
	HPR	LPR	K_B	K_H	
1	58	54.8	0.144	0.018	7.8
2	52	54.8	0.158	0.026	6.0
3	54.8	54.8	0.133	0.039	3.41
4	101.8	54.8	5.001	0.039	128.2

All the presented dependencies were produced via no fewer than 5 experiments. A total of 4 porous medium models were used in the experiments (Table 2). A conditional value was introduced and used to simplify the calculation and plotting of the graphs; this consists of the CO_2 (R_{CO2}) and water (R_B) distribution parameters, which are equal to the ratio of yield rates for high- (Q_B) and low-permeability (Q_H) reservoirs.

The obtained results (Figure 2) show that using the non-ionic surfactants sufficiently increases the reservoir coverage of both carbon dioxide and water, via alternating injection of the two reactants. The supply of liquid CO_2 into the low-permeability reservoir increased from 16.67% and 17.01% (after injection of mineralised (140 g/l) and distilled water, respectively) to 36-38% when using the solution of Neonol AF_9-12 in concentrations of 0.25-2.00%. The injection of gaseous carbon dioxide (10.0 MPa, 40°C) also increases into low-permeability reservoirs from 15.4% (water) to 36.5-37.7% (0.25-1.00% solution of AF_9-12). Coverage of the reservoir in the case of alternating injection of CO_2 slugs and surfactant water solutions increases due to the formation of water-carbon dioxide emulsions and foams at the water/CO_2 interfaces within the porous medium. Because of the different mobilities of carbon dioxide and water, the volume of emulsion formed in pores will be maximal in intervals with higher permeabilities. Accumulation of specific volumes of emulsion (foam) in these intervals will cause partial (temporary) blocking of the pores, which allows for the redistribution of flow from high- to low-permeability intervals.

The influence of gas mixture composition on its reservoir distribution was studied in two models (1 and 2). The characteristics of the porous mediums (represented by natural sandstones) are listed in Table 3. Experiments were conducted at 26.5°C and 10.5 MPa.

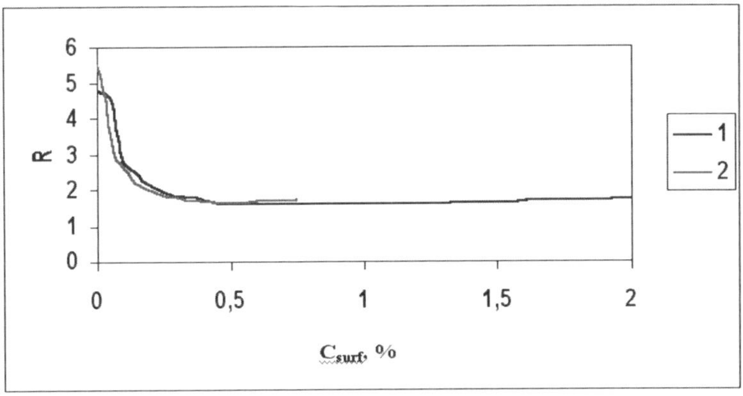

Figure 2. Dependence of CO_2 and water distribution parameters on the concentration of Neonol AF9-12, where R is a dimensionless distribution parameter, while QB and QH are the quantity of fluid (gas) entering the low- and high-permeability porous mediums, respectively.

Table 3. Characteristics of reservoir models.

№ of the well	Porous medium*		Ratio K_B/ K_H
	Length, cm	Permeability, mcm^2	
1	57.9/57.1**	0.55/0.035	15.7
2	57.6/57.7	0.48/0.096	5.0

* Diameter of porous mediums – 50 mm;
** The numerator – high permeability (K_B), the denominator – low permeability porous medium (K_H).

The results of complex filtration testing to optimise the technical parameters during alternating carbon dioxide – water injection were used as the basis for the current work. These include a total gas-mixture volume of 12% ΣV_n, a gas-mixture / water volume ratio of 3 and the number of gas mixture divisions equal to 20.

As a result of eight experiments (Figure 3) (four for both reservoir models) it can be stated that the distribution parameter increases linearly with the increase of petroleum gas content, i.e. the intake of gas mixture into the low-permeable reservoir decreases linearly. Obtained data are in good agreement with data described in previous works which studied the viscous and rheological properties of water-gas systems and experiments to determine the stability of foam-emulsion systems.

The next group of experiments involved regulating gas mixture filtration using the foam-emulsion mixtures. Parameters of the alternating injection in these experiments were kept constant. The CO_2 content in the gas mixture was 50%, while a mixture of Neonol AF_9-12 and KOBS was used for the surfactant (the mass ratio of each component in water was 0.125%).

It was found that the alternating injection of the gas mixture and water solution of the indicated composition into model №2 caused the distribution parameter R to decrease from 3.30 to 2.91, which corresponds to an increase of reactant ingress into the low-permeable reservoir from 23.26 to 25.58% (Figure 3). If one compares the obtained redistribution of gas mixture filtration in the heterogeneous reservoirs with the distribution of the clear carbon dioxide in model №1, it can be seen that regulation of the gas mixture filtration has limited effectiveness. For clear CO_2 the distribution parameter R decreased from 5.50 to 2.06 with the injection of the AF_9-12 / KOBS solution in water instead of simple water slugs.

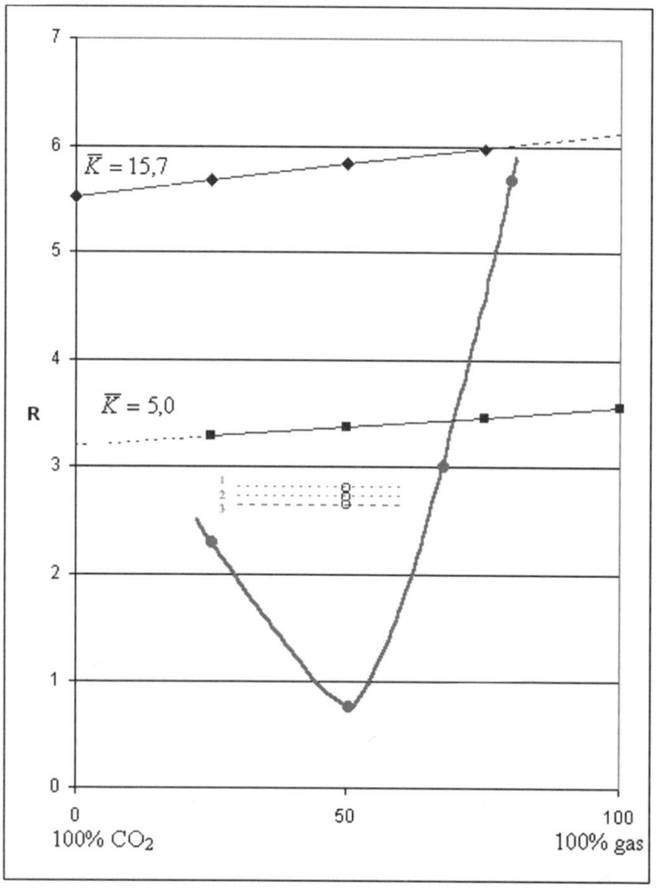

Figure 3. Distribution of CO_2 and gas mixtures during alternating water and surfactant water solutions injection and combined injection, respectively.

The increase of AF_9-12 concentration to 0.25% did not lead to a sufficient improvement of the gas mixture filtration alternating with the surfactant solution. The distribution parameter in this case was 2.88, which corresponds to 25.77% ingress of reactant into the low-permeable reservoir.

The increased division of the gas mixture into 40 portions also did not bring about sufficient increases in the coverage of the reservoir model. The distribution parameter in this case decreased to only a value of 2.82 (Q_H=26.18%).

Thus earlier results that allowed the increase of reservoir coverage by CO_2 with the alternating injection of surfactant water solution slugs was not repeated for the mixture of CO_2 and petroleum gas. As gases have a low solubility in water foam-emulsion systems form on the interface of the slugs of the filtering phases. Evidently in this case the decrease of the foam-emulsion system stability, consisting of gas mixture and water compared to the dispersion of CO_2 and water, has a decisive impact on the regulation of the filtration process.

The following experiments on the technology of water-gas mixture combined-injection were conducted to increase the contact surface of gas mixture and surfactant water solution. The content of CO_2 in the gas mixture was in all cases 50%, the gas mixture slug volume was 12% ΣV_{pore} while the detergent was a 0.25% Neonol AF_9-12 water solution. The investigated parameter in these experiments was the ratio of water solution to gas mixture volumes.

The increase of the gas mixture ratio from 25 to 50% at the combined injection with 0.25% surfactant solution decreased the distribution parameter from 2.26 to 0.89 and increased the intake of the reactant into the low-permeable reservoir from 30.67% to 52.91%, respectively. This result clearly supports the previous assumption regarding the primary influence of foam-emulsion system stability on the surfactant solution regulating properties. In this case the combined injection of the gas mixture and the surfactant water solution results in direct foam filtration, and the pore cells work as the centers of foam formation.

An additional increase in the foam's aeration extent (ratios of water and gas mixture volumes 1:2.3; 1:3) causes the sudden decrease of reactant intake into the low-permeable reservoir – 25.32 and 8.05% respectively. The dependence of the distribution parameter on the water and gas mixture volumes ratio has a strongly pronounced maximum. Its presence can be explained by the increase of the foam-emulsion system's stability which causes the decrease of R from 2.26 to 0.89 with the increase of the gas mixture ratio during injected dispersion. Appearance of a free gas phase during the continued increase of the gas ratio in the injected mixture to 75% causes the temporary blocking of small pores and the growth of the distribution parameter. The distribution parameter minimum, R=0.89, was obtained with the combined injection of the gas mixture slug and the 0.25% Neonol AF_9-12 solution into reservoir model №2.

The technical parameters have been optimised based on the results of the conducted filtration experiments, and the main principles needed to increase coverage of reservoirs with heterogeneous permeability using foam-emulsion systems have been shown.

The next experiments were devoted to studying the mechanism of emulsion filtration in fractured and porous structures (reservoir micromodel experiments were conducted under the direction of Dr. A.T. Akhmetov).

The reactant Neftenol NZ 40M (designed by the company "Chimeko-GANG") was used to stabilize the inverted water-oil emulsions. The experiments conducted on physical models of the reservoir elements (i.e. fractured and porous structures) were conducted in order to understand what happens with the emulsion during flow through the reservoir.

A Hele-Shaw cell was used to study processes occurring in a fracture. It consisted of two parallel optical glass plates with a clearance of 17 or 35 microns and a size of 4 x 2 cm, restricted by a foil. The processes taking place in the porous structure were studied using a micro-model.

The porous structure micro-model is a two-dimensional transparent porous system, representing the pore structure in the plane of the polished section of the oil-bearing core sample. The micro-model working section is the same as in the Hele-Shaw cell (i.e. 4 x 2 cm) and has channel depths equal to 15 microns.

To measure the liquid flow rate an HM200 scale was used (interval 0.1 mg) and the data were automatically transferred to a computer. The fluid flowing out of the model displaced distilled water from a tube and was deposited into a receptacle on the scale, thereby allowing the volume flow through the model to be calculated. To avoid capillary effects the end of the outlet tube was put into liquid, while evaporization errors were minimized by covering the water with a thin layer of light oil.

Injection of the highly-concentrated oil emulsion, stabilized by Neftenol, into the Hele-Shaw cell at pressure drops of 150–200 kPa caused exuding of some aqueous phase volume from the emulsion. Decreasing the pressure drop to 50–100 kPa resulted in a stable emulsion. In all cases, however, cell blockage occurred over time, though the pressure drop remained constant.

The emulsion volume which flowed through the model over time was calculated via the computer according to the scale data. The flow structure changed over time, and it was most precisely exhibited at the greater gap of the cell, equal to 35 microns. The curve describing the blocking of the cell is shown in Figure 4, as are various video images of the flow structure at different emulsion discharge rates. The flow structure changes are shown at a greater scale in Figure 5. Complete blockage, discovered using the scale data and verified with careful microscopic study, has shown the presence of small fluid flow at the inlet and outlet (Figure 6.).

The flow rate value is four orders of magnitude lower than the initial flow rate and does not increase with time, although the emulsion structure in the cell and in the micro-model change considerably after a long period of time (a day and more) and resemble the composite emulsion, formed during flow of the non-stabilized emulsion. As emulsion micro-flow is always present, the observed "blocking" effect has been termed "*the dynamic blocking effect*".

The same dynamic blocking effect was found in the study of stabilized emulsion flow in the porous structure micro-model. In this case the consumption of the emulsion is much less, but the relationship between emulsion volume flow and time have the same nature and blocking as in the case of the Hele-Shaw cell, which happens within about one hour. Under dynamic blocking the emulsion has the property of "the fluid nipple", i.e. at the pressure build-up in the reverse direction the motion of the emulsion is restored (curve 4 in Figure 7). In the case of both the micro-model and the Hele-Shaw cell there is micro-flow at the inlet and outlet of the model.

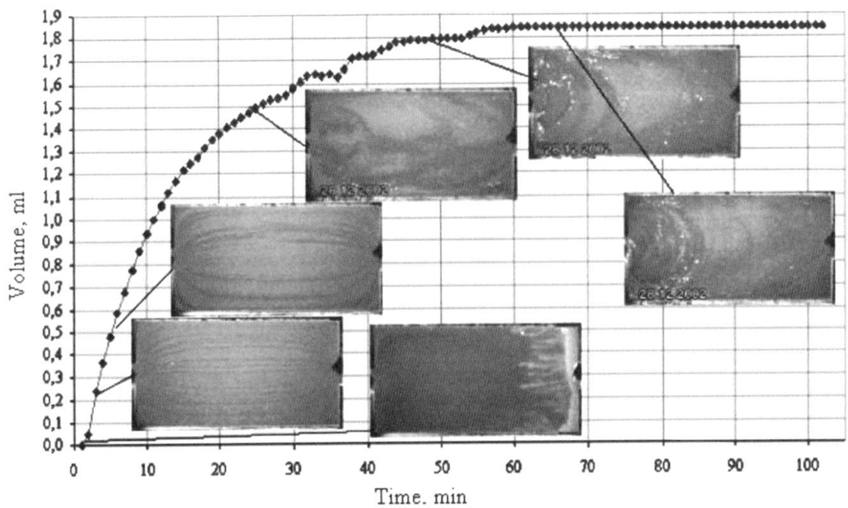

Figure 4. Stabilized emulsion volume passed through the fracture model versus time at a constant pressure drop. The video images of flow structure are shown at different moments of time.

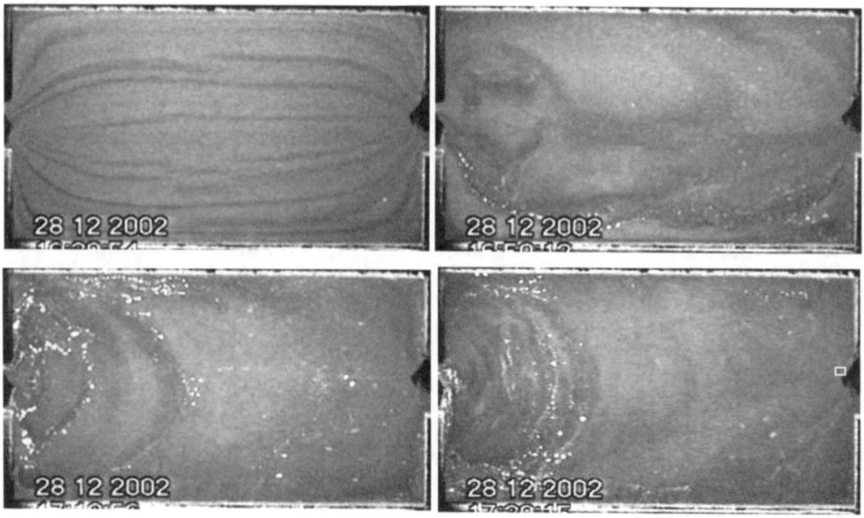

Figure 5. The modification of the flow structure due to the dynamic blocking of the Hele-Show cell (pressure drop is 100 kPa, on the right lower video image the rectangle at the outlet corresponds to the field observed under a microscope).

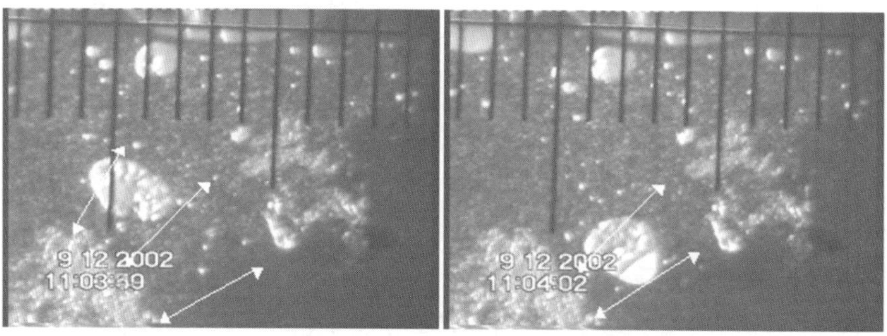

Figure 6. Micro video image of the single site, at the outlet, on which the motion is observed, the flow parameters are indicated by arrows. Screen width 700 microns, flow 200 microns, movement 80 microns per 12 seconds.

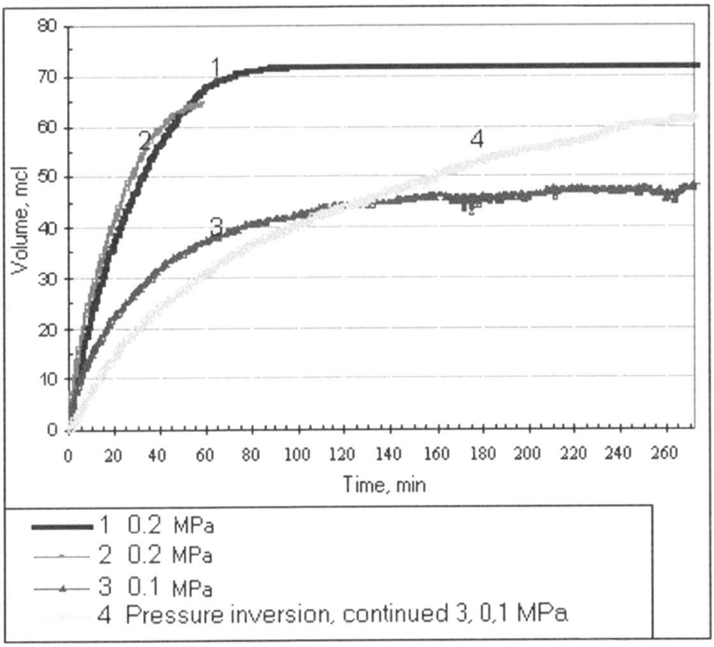

Figure 7. The volume of the stabilized emulsion, passed through the micro model at a constant pressure drop.

3. CONCLUSIONS

In conclusion it can be stated that one of the most interesting experimental result is the observed dynamic blocking effect, which practically ceases flow despite the considerable (up to 5 MPa/m) constant average pressure gradient. It is hoped that this effect can be successfully exploited in different versions of the technologies of the levelling of the well injection capacity profile and waterproofs in porous, fractured porous and fractured reservoir types.

REFERENCES

OST, 1986, Oil. The method to determine the index of oil displacement by water in laboratory conditions, *OST Report 39-195-86*.

STATE OF CO_2 CAPTURE AND SUBSURFACE STORAGE ACTIVITIES IN GERMANY

B.M. Krooss[1] and F. May[2]
[1]*RWTH-Aachen University, Institute of Geology and Geochemistry of Petroleum and Coal, Lochnerstr. 4-20, D-52056 Aachen, Germany;* [2]*Bundesanstalt für Geowissenschaften und Rohstoffe, Stilleweg 2, D-30655 Hannover, Germany*

Abstract: The Federal Republic of Germany (FRG) is the greatest carbon emitter in Europe and is responsible for ~ 3.4% of the world's total fossil fuel-based carbon emissions. In the Kyoto Protocol and national climate protection programs Germany has committed to substantial reductions of CO_2 emissions. Because of the German government's priority on sustainable energies and reduction of fossil energy consumption, R&D activities on CO_2 capture and subsurface storage have so far played only a minor role. These activities are, however, increasing and several projects have been launched recently.

Key words: CO_2-emissions, CO_2 capture, subsurface storage, Germany, EU Projects, COORETEC, GEOTECHNOLOGIEN

1. INTRODUCTION

Between 1990 and 2003 worldwide annual emissions of CO_2 increased from 22,700 Megatons (Mt) to 25,700 Mt. The share of the OECD countries in these emissions increased during this period from 3,700 Mt to 4,500 Mt.

The Federal Republic of Germany (FRG) is the greatest carbon emitter in Europe with a 24.2 % share of the total European Community (EC) emissions in 2001 (993.5 Mt CO_2) which corresponds to ~3.4% of the world's total fossil fuel-based carbon emissions. Germany joined the Kyoto Protocol to the United Nations Framework Convention on Climate Change (Kyoto Protocol) on 4th March 2002 as a member of the EC, which at that time comprised 15 countries. The EC committed itself to reduce its annual

S. Lombardi et al. (eds.),
Advances in the Geological Storage of Carbon Dioxide, 289–301.

average greenhouse gas emissions by the time period 2008 – 2012 by eight percent with respect to the base 1990 level (4204 Mt). Reduction targets were allocated to the individual member countries in the "Burden Sharing Agreement" of June 1998 (Figure 1). According to this Agreement, Germany has to reduce emissions by 21% with respect to the 1990 level (1216.2 Mt CO_2 equivalent emissions). In the national FRG climate protection program of 18th October 2000 the target was set to a reduction of CO_2 emissions by 25% with respect to 1990 by 2005. By 2001 annual CO_2 in emissions Germany had decreased by 18.3 % to 993.5 Mt putting the country ahead of the reduction target and in line with the United Kingdom, Sweden and Luxembourg.

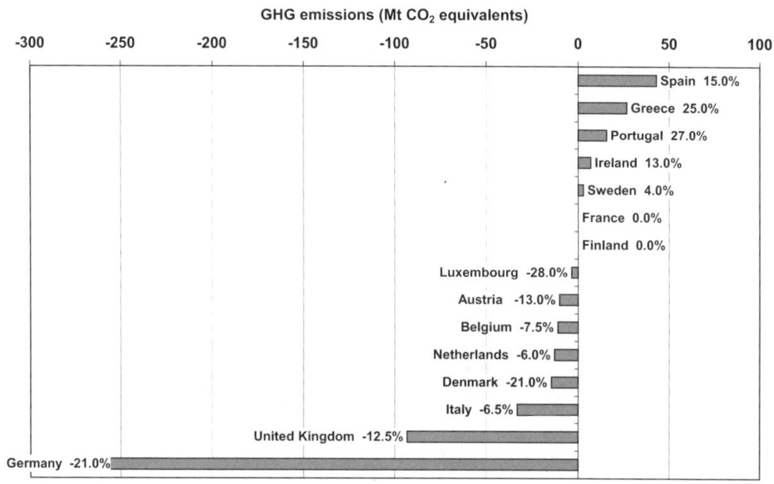

Figure 1. Greenhouse gas emission targets of EU Member States for 2008-2012 under the EU Burden-Sharing Agreement.

The decision of the German government to phase out the use of nuclear power by 2020 will have serious implications on the energy supply and requires alternative solutions. For a transition period fossil fuels will remain an indispensable source of energy.

To further reduce greenhouse gas emissions, the climate policy of the German government focuses on sustainable energies and the reduction of fossil energy consumption. R&D activities on CO_2 sequestration and subsurface storage have played a comparatively minor role. Activities have, however, increased recently both on the capture and storage fields, in view of the start of trading of CO_2 Emission certificates in 2005.

2. GERMAN R&D ACTIVITIES ON CO_2 CAPTURE AND STORAGE

German research institutes have successfully participated in various international research projects at the European Union (EU) level. These comprise Joule II (1993-1995; RWE-AG), GESTCO (1999-2003; BGR), NASCENT (2000-2004; BGR, RWTH-Aachen), RECOPOL (2001-2005; DBI-GUT, RWTH-Aachen), CASTOR (2004-2006; BGR, University of Stuttgart), and CO_2GeoNet (BGR). The EU project CO2SINK, which started in 2004, was the first major EC project of this kind initiated by a German institution (GFZ-Potsdam). A worldwide overview of former and ongoing projects on CO_2 capture and underground storage can be found at the IEG GHG website (http://www.co2captureandstorage.info/).

More recently three major national R&D programs have been launched to promote CO_2 capture from coal-fired power plants (COORETEC), to support research on greenhouse gas emission reduction and adaptation to climate change (Forschung für den Klimaschutz und Schutz vor Klimawirkungen), and to explore methods and facilities for underground storage of CO_2. The latter activity is embedded in the GEOTECHNOLOGIEN Program of the Federal Ministry of Education and Research (BMBF) and The German Research Foundation (DFG).

2.1 German involvement in international projects

2.1.1 Joule II

The Joule II Project "Underground Disposal of Carbon Dioxide" was the first European research project examining issues associated with underground CO_2 disposal. Its goal was to assess quality and quantity of CO_2 available from fossil fuel power plants, to examine pipeline transport of CO_2 and to examine various options for underground disposal. Furthermore it addressed safety and stability issues, investigated the geochemistry associated with underground disposal and carried out techno-economic modeling.

The Joule II consortium comprised the Geological Surveys of the UK (BGS), France (BRGM) and The Netherlands (TNO-NITG), the IKU Petroleum Research Institute of Norway and the University of Sunderland (UK). The German energy industry was represented in this project by the company RWE AG.

2.1.2 GESTCO

The GESTCO (GEological STorage of CO_2) project started in 2000. It was initiated to assess the European potential for geological storage of CO_2 from fossil fuel combustion. Participants were the Geological Surveys of eight European countries (Belgium, Denmark, France, Germany, Greece, Norway, Netherlands, UK, and the research institute ECOFYS).

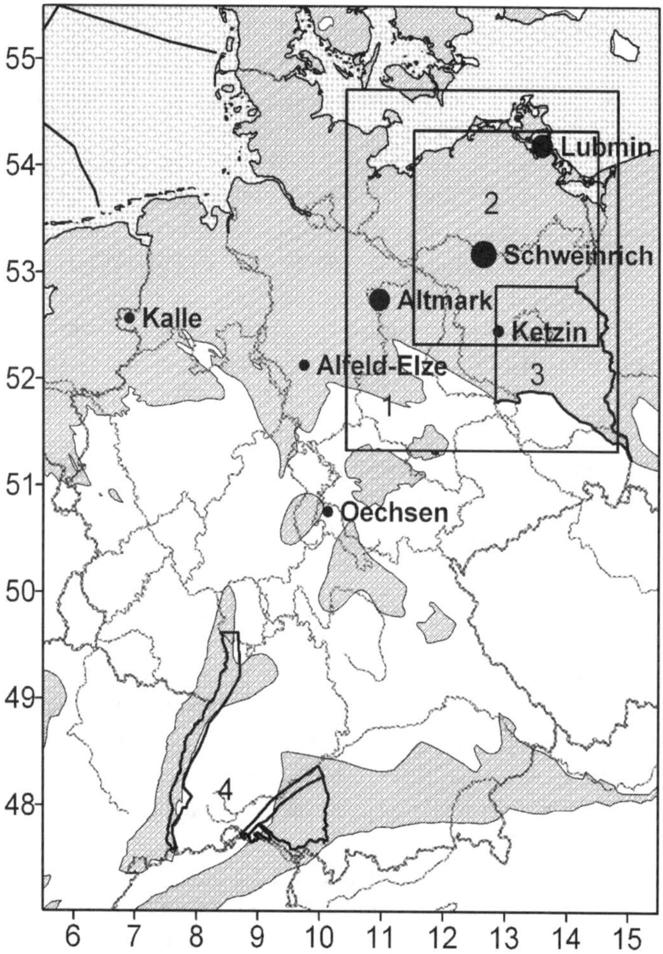

Figure 2. Sedimentary basins in Germany > 1 km depth, hosting natural gas fields and aquifer structures that could provide storage capacity. Case study sites and regional storage capacity investigations (1-4) (BGR, GESTCO).

The German Federal Institute for Geosciences and Natural Resources (BGR) calculated CO_2-emissions for large industrial point sources and mapped the extent of potential storage areas in aquifers and deep unminable coal seams. Storage capacities were estimated for the various underground storage options: aquifers, deep unminable coal fields, oil/gas fields and abandoned coal and salt mines in Germany (cf. Figure 2). Finally, a descriptive geologic overview was compiled of potential CO_2 storage formations in Northern Germany (Permian and Mesozoic) and an overview of the regional distribution of seals.

The data compiled were incorporated into a common CO_2 Storage Geographic Information System (GIS) for Europe, established within the GESTCO project and covering the territory of the participating countries and their off-shore areas as well. This GIS comprises the first systematic compilation of CO_2 point sources and potential storage facilities in Germany.

Two case studies, the abandoned natural gas field Alfeld-Elze and the Buntsandstein aquifer close to the planned power plant Lubmin have been selected for numerical simulations of CO_2 injection into the reservoirs and long-term migration, thermodynamic modeling of potential reactions between formation water, CO_2, and reservoir rocks, and for cost calculations with a decision support system developed by TNO- NITG.

BGR and the British Geological Survey (BGS) jointly prepared a report on CO_2 storage safety and conflicts of use.

2.1.3 NASCENT

Two German institutions, BGR and RWTH-Aachen University, were project partners in the EU project NASCENT (Natural Analogues for the Storage of CO_2 in the Geological Environment) from 2001 to 2004. Various locations with natural occurrences and different intensities of CO_2 emission were selected in France, Germany, Greece, Hungary and Italy and studied in great detail to analyse the conditions, effects and processes related to long-term underground storage of CO_2.

In the Vorderrhön area in central Germany natural CO_2 occurs below and locally within Permian Zechstein salts and was produced commercially until the 1980s. Here BGR conducted detailed geological studies, analyses of fluid obtained from wells and sampled in salt mines in the overburden of the CO_2 reservoir Oechsen (Fig. 2). Soil gas surveys were performed to identify potential leakage pathways of CO_2 to the surface. In addition, BGR operated automatic soil gas monitoring systems and provided expertise for surveys at the other NASCENT sites.

Within the NASCENT project, RWTH-Aachen performed fundamental experimental laboratory work to assess the sealing efficiency of cap rock sequences overlying potential CO_2 storage targets. The experiments comprised permeability tests, assessment of the capillary gas-sealing efficiency of water-saturated rocks and diffusion experiments with CO_2 in water-saturated rocks. Selected results from this project work are shown in Figure 3. The experimental data were subsequently used to develop models for the prediction of leakage rates from CO_2 reservoirs.

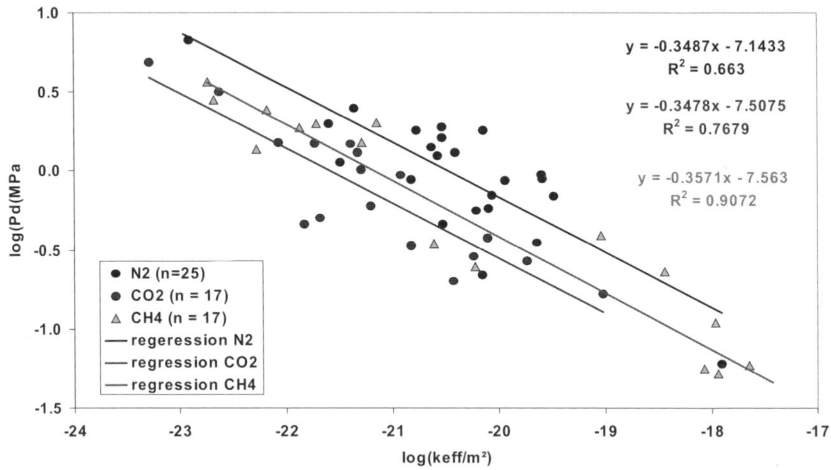

Figure 3. Capillary gas sealing efficiency (breakthrough pressure) vs. gas permeability of cap rocks (RWTH-Aachen, NASCENT).

2.1.4 RECOPOL

RWTH-Aachen University and DBI-GUT, a company specialized in underground storage of natural gas, were the two German partners in the EU RECOPOL project (Reduction of CO_2 emission by means of CO_2 storage in coal seams in the Silesian Coal Basin of Poland) that started in November 2001 and will finish in 2005. RECOPOL was initiated and co-ordinated by the Netherlands Institute of Applied Geoscience -National Geological Survey (TNO-NITG) in co-operation with the Polish Central Mining Institute (GIG) to investigate the feasibility of CO_2 underground storage in unminable coal seams in combination with the production of coal bed methane (CBM). The project comprised extensive laboratory studies on high-pressure sorption and transport processes of methane and CO_2 in coals, performed mainly at the Technical Universities of Delft (The Netherlands) and RWTH-Aachen, in close co-operation with the Institut Français du Pétrole (IFP). The main achievement of RECOPOL was the first European field test on CO_2 storage in deep coal seams in the Silesian Coal Basin of Poland near the city of Katowice.

RWTH-Aachen contributed extensive experimental data sets on single gas and mixed gas sorption capacity, selectivity and kinetics which were incorporated into reservoir simulation software in close collaboration with the Australian partner CSIRO Petroleum.

DBI-GUT was in charge of planning, preparation and supervision of the engineering activities at the RECOPOL CO_2 injection site.

Figure 4 shows the scheme of the RECOPOL field test with CBM production from an existing well (MS-4) and CO_2 injection into the newly drilled MS-3 well. Liquid CO_2 is delivered by trucks and stored on-site in two 30 t storage tanks.

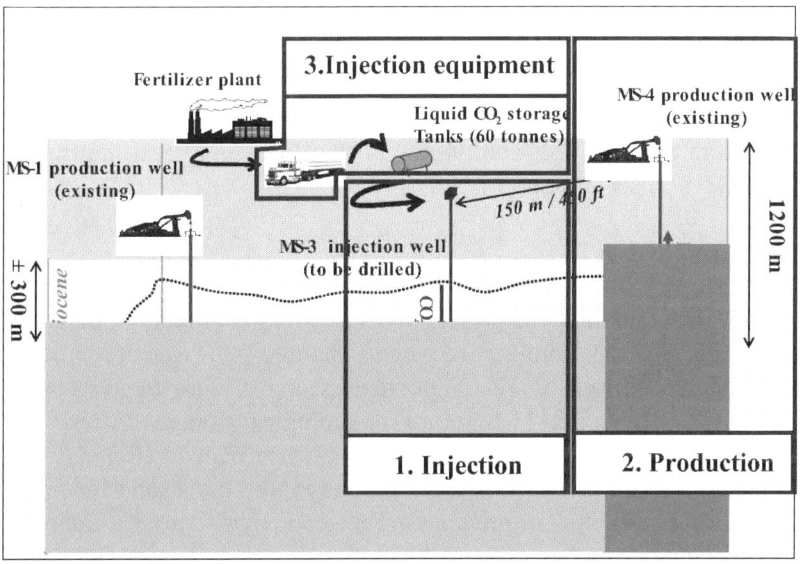

Figure 4. The RECOPOL field test.

2.1.5 CASTOR

The CASTOR (CO_2 from Capture to Storage) project started in 2004 and aims at the development of new technologies for the separation of CO_2 from flue gases and its geological storage. The development of tools and methods to quantify and minimize the uncertainties and risks linked to the storage of

CO_2 is another aim of this project. Thirty parties from industry, research organisations and universities participate in this integrated project. Three off-shore and one on-shore hydrocarbon reservoirs have been selected as case study sites. On the CO_2 storage side, BGR is the only German project partner but has subcontracted experimental work on the characterisation of seal efficiency to RWTH-Aachen University. At the Atzbach-Schwanenstadt natural gas field in Austria, BGR conducts soil gas investigations, including flux measurements, monitoring and isotopic and geochemical analyses of gases. Furthermore, BGR coordinates the studies on preventive and corrective action technologies for leaking wells and faults. RWTH-Aachen investigates CO_2 transport processes in core samples from the cap rock sequences above the gas reservoirs. In the context of the demonstration of subsurface storage of CO_2 in a nearly depleted gas reservoir and enhanced gas recovery, RWTH will perform further experiments on cap rock sequences of an offshore hydrocarbon reservoir in the Netherlands (K12-B).

On the capture side, the University of Stuttgart provides a mini power plant for the development of efficient methods for CO_2 extraction from flue gas. German companies involved in the project are Siemens AG and RWE Power, participating in the evaluation, optimisation and integration of post-combustion capture processes in power plants and the process validation of a pilot capture plant. BASF contributes to the developing and improving of solvents for CO_2 separation.

2.1.6 CO2SINK

CO2SINK is the first EU project on CO_2 emission reduction initiated and co-ordinated by a German research institute. It was proposed by GeoForschungsZentrum (GFZ, National Research Centre for Geosciences) Potsdam as an *in-situ* R&D laboratory for geological storage of CO_2.

The target reservoir is located near the town of Ketzin (Fig. 2) west of Berlin and Potsdam at a depth of ~700m and represents a sandstone aquifer of several tens of metres of thickness. The scope of the project comprises a baseline geologic survey with a detailed risk assessment to ensure safe injection and storage of CO_2. Detection and monitoring of the injected CO_2 is another focus in this project.

A total of 14 organisations from 8 European countries, including universities, research institutes and industrial companies are presently participating in this project. Its aim is to integrate CO_2 injection into a framework of CO_2 capture facilities including a biomass power plant.

2.1.7 CO2STORE

CO2STORE is a European integrated project with the aim of transferring the experience gained in the scientific projects SACS I+II from Statoil`s offshore CO_2 injection to other potential storage sites with different geological conditions. One of the four additional cases proposed is the lignite fired power plant Schwarze Pumpe in East Germany. For this power plant BGR had to find potential storage sites capable of taking in 400 Mt of CO_2. The only gas field with a storage capacity of this magnitude is located in the Altmark (Fig. 2). Numerical modelling to simulate CO_2 injection and enhanced gas recovery have been performed for this reservoir. One of the identified aquifer structures, Schweinrich (Fig. 2), has been selected for further characterisation of the reservoir structure and caprock properties. Risk assessments are carried out together with BRGM (the simulation of geochemical reactions), NITG-TNO (long-term CO_2 migration simulation) and Vattenfall/NITG-TNO.

2.1.8 CO2GeoNet

The EU has initiated a Network of Excellence for CO_2 capture and storage, intending to facilitate a better co-ordination of research activities in this field. The network comprises 12 institutes from 8 European Countries was formed in 2004. BGR is the sole German participant. After initial activities to create inventories of relevant research infrastructure and to identify gaps in knowledge and future research needs, longer term joint research begins in 2005.

2.2 National research programmes and initiatives

Germany has recently initiated national research activities for the reduction of CO_2 emissions from fossil fuel utilization. One of these, the COORETEC initiative, is aiming at improved power-plant technologies while the GEOTECHNOLOGIEN program investigates underground storage options for CO_2. Other individual research and development activities of universities, industry and research institutes exist. Their description is, however, beyond the scope of this paper.

2.2.1 COORETEC

The COORETEC initiative was launched in 2002 by the Federal Ministry of Economics and Labour (BMWA) to promote CO_2-reduction-technologies for fossil fuel-fired power plants. Four working groups (WG) were established to address technical, scientific, socio-economic and legal aspects of CO_2 capture and storage:

- WG1: Boundary conditions (legal framework, economic conditions)
- WG2: CO_2-capture, utilisation and storage
- WG3: Short and mid term power plant technologies
- WG4: Future technologies

In 2004 the report of these four COORETEC working groups was published under the title: "Research and Development Concept for Zero-Emission Fossil-Fuelled Power Plants". This report can be downloaded from the COORETEC website (http://www.cooretec.de).

As shown in Figure 6, the short-term priority target of the COORETEC project is to increase the efficiency of steam (DKW) and gas & steam (GuD)electric power stations and thus decrease the CO_2 emissions per unit of electric energy produced from fossil fuel combustion. Thus, R&D activities of COORETEC projects focus on:

- Measures to improve efficiency of combined cycles
- Measures to improve efficiency of conventional steam power plants
- Integrated Gasification Gas Combined Cycle (IGCC) power plant development with and without CO_2 capture
- Novel technology developments with CO_2 capture, especially oxyfuel concepts

With the OXYCOAL-AC project, funded in the context of COORETEC, the Institute of Heat and Mass Transfer (www.wuek.rwth-aachen.de) of RWTH-Aachen is conducting front-end R&D work in this field.

CO_2 capture and storage technologies are also in the scope of the COORETEC program, but with lower priority and on an intermediate and long-term R&D horizon. The compilation of a national storage inventory is considered a prerequisite for near-zero-emission power generation.

2.2.2 GEOTECHNOLOGIEN

In November 2003 a call for project applications was launched by the GEOTECHNOLOGIEN Programme with the research focus: "Exploration, Usage and Protection of the Subsurface". GEOTECHNOLOGIEN is a geoscientific research and development programme funded jointly by the Federal Ministry of Education and Research (BMBF) and the Deutsche Forschungsgemeinschaft (DFG — German Research Council).

The call solicited project applications related to "Utilisation of the Subsurface for Global Climate Protection Purposes" with a specific focus on the "appraisal of procedures for subsurface storage of the greenhouse gas CO_2 and the development of corresponding exploration and monitoring technologies".

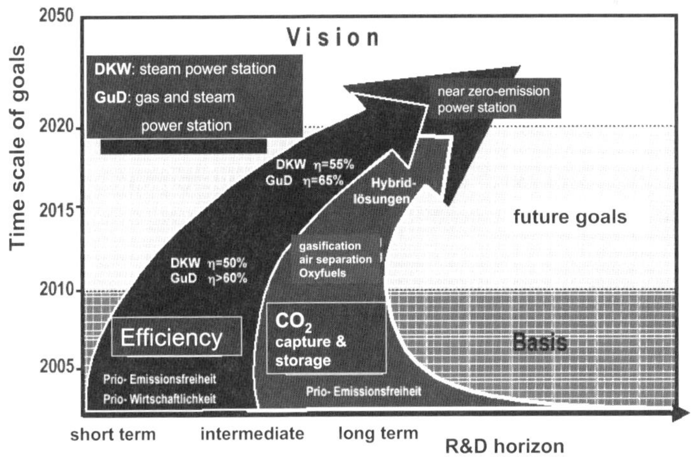

Figure 5. The COORETEC program.

The scope of R&D activities envisaged by this programme comprised the evaluation of appropriate rock formations, such as deep saline aquifers, depleted oil and natural gas reservoirs and deep, unminable coal seams for underground storage of CO_2 with special consideration of long-term integrity and safety. Specifically, applications were requested related to the investigation of reactive transport of CO_2, the reaction kinetics of this gas with reservoir rocks and seals and potential effects on their petrological and mechanical properties. Furthermore, this programme aimed at promoting research into new injection procedures and geophysical and geochemical monitoring techniques ensuring a safe long-term deposition and precise survey of CO_2 migration in the subsurface during and after injection. Project applications were also requested for the development and application of numerical models for the prediction of CO_2 transport and the assessment of potential risks of groundwater contamination.

Among the applications filed in response to this call, approximately 10 have been selected for funding over a period of three years. Details on the selected project and the total funding will be released by April 2005.

2.2.3 Research for Climate Protection and Protection from Effects of Climate Changes

This programme, launched by the Federal Ministry of Education and Research in 2004, has two focal points: (i) Reduction of CO_2 emissions, especially in industrial applications (power generation is a focus of the COORETEC Programme) and (ii) Adaptation to climate change and extreme weather events. Projects probably will not commence before 2006.

3. CONCLUSION

Although the climate policy of the Federal Republic of Germany is aiming at an increased use of renewable energy and reduction of energy consumption, fossil fuels will remain an important energy source in the foreseeable future, even more so because the production of nuclear energy within Germany will be phased out over the next 20 years. In this context one strategy, namely the reduction of CO_2 emissions from fossil fuel power plants is reflected in the COORETEC programme. Here the first priority is the development of modern, highly efficient power plant technology. New procedures for CO_2 separation and zero-emission fossil power plants are also considered within COORETEC but will invariably reduce power generation efficiency and increase energy costs. Provided that this approach continues to find public acceptance, the issue of long-term geological storage of large quantities of CO_2 arises as an immediate consequence of CO_2 separation. Experience gathered by several German research groups in European projects is now integrated into the national GEOTECHNOLOGIEN R&D programme which takes into account the specific geological, infrastructural, socio-economic and legal situation in Germany. Involvement of German research groups in previous (GESTCO, NASCENT, RECOPOL) and ongoing EU projects (CASTOR, CO2SINK, CO2STORE, CO2GeoNet) will ensure the continuity of exchange of expertise on the international level. Additionally, according to the second call in the EU 6th Framework Programme, joint research projects on CCS with the participation of German research institutes involved in current EU projects are in preparation.

ACRONYMS AND PROJECT LINKS

- CASTOR: CO2 from capture to storage
- COORETEC: CO_2 Reduction Technologies (http://www.cooretec.de)
- CO2SINK: In-situ R&D laboratory for geological storage of CO_2 (http://www.co2sink.org/)
- GEOTECHNOLOGIEN: (http://www.geotechnologien.de/)
- GESTCO: GEological STorage of CO_2 (http://www.nitg.tno.nl/projects/eurogeosurveys/projects/GestcoWeb/)

- JOULE II: Underground Disposal of Carbon Dioxide (http://www.dti.gov.uk/energy/environment/eerp/reports/ps_001.pdf)
- NASCENT: Natural Analogues for the Storage of CO_2 in the Geological Environment (http://www.bgs.ac.uk/nascent/home.html)
- RECOPOL: Reduction of CO_2 emission by means of CO_2 storage in coal seams in the Silesian Coal Basin of Poland (http://recopol.nitg.tno.nl/)

INSTITUTION NAMES AND CONTACT INFORMATION

- BGR: Bundesanstalt für Geowissenschaften und Rohstoffe (Federal Institute for Geosciences and Natural Resources); (http://www.bgr.de/)
- BMBF Bundesministerium für Bildung und Forschung (Federal Ministry of Education and Research); (http://www.bmbf.de/)
- BMWA Bundesministerium für Wirtschaft und Arbeit (Federal Ministry of Economics and Labour); (http://www.bmwa.bund.de/)
- CSIRO Petroleum: Commonwealth Scientific and Industrial Research Organisation – Petroleum Resources (http://www.dpr.csiro.au/)
- DBI-GUT: DBI Gas- und Umwelttechnik GmbH (http://www.dbi-gut.de)
- GFZ-Potsdam: GeoForschungsZentrum Potsdam (National Research Centre for Geosciences); (http://www.gfz-potsdam.de/)
- GIG: Główny Instytut Górnictwa (Central Mining Institute), Katowice, Poland (http://www.gig.katowice.pl/)
- RWTH-Aachen University (http://www.rwth-aachen.de/)
- TNO-NITG: Netherlands Institute of Applied Geoscience - National Geological Survey (TNO-NITG) (http://www.nitg.tno.nl/eng/)
- Universität Stuttgart (http://www.uni-stuttgart.de/)

GEOPHYSICAL MONITORING OF THE CO_2 PLUME AT SLEIPNER, NORTH SEA

An outline review

A. Chadwick[1], R. Arts[2], O. Eiken[3], P. Williamson[1] and G. Williams[1]

[1]British Geological Survey, Kingsley Dunham Centre, Keyworth, Nottingham, NG12 5GG, United Kingdom; [2]Netherlands Institute of Applied Geoscience TNO - National Geological Survey, Kriekenpitplein 18, PO Box 80015, 3508 TA Utrecht, The Netherlands; [3]Statoil Research Centre, Rotvoll, N-7005 Trondheim, Norway.

Abstract: CO_2 produced at the Sleipner gas field is being injected into the Utsira Sand, a major saline aquifer some 1000m beneath the North Sea. The injection plume is being monitored by geophysical methods. 3D seismic data were acquired in 1994, prior to injection, and again in 1999, 2001 and 2002; seabed gravimetric data were also acquired in 2002. The CO_2 plume is imaged on the seismic data as a number of bright sub-horizontal reflections, growing with time, underlain by a prominent velocity pushdown. Quantitative modelling is based on plume reflectivity largely comprising tuned responses from thin layers of CO_2 trapped beneath thin intra-reservoir mudstones, with layer thicknesses being mapped according to an amplitude-thickness tuning relationship. Between the layers a lesser component of much lower-saturation, dispersed CO_2 is required to match the observed velocity pushdown. However, reservoir temperatures are subject to significant uncertainty, and inverse models of CO_2 distribution, based on lower and higher temperature scenarios, can produce both the observed plume reflectivity and the velocity pushdown. Higher temperature models however require that the dispersed component of CO_2 has a somewhat patchy rather than uniform saturation. Analysis of the datasets suggests that accumulations of CO_2 as small as 500 tonnes may be detectable under favourable conditions, providing a basis for setting leakage criteria. To date, there is in fact no evidence of migration from the primary storage reservoir.

Key words: CO_2 sequestration, Sleipner, geophysical methods, 4D seismic, gravity, velocity pushdown, quantitative modeling, accumulation detection

S. Lombardi et al. (eds.),
Advances in the Geological Storage of Carbon Dioxide, 303–314.

1. INTRODUCTION

The carbon dioxide injection at the Sleipner field in the North Sea (Baklid et al. 1996), operated by Statoil and the Sleipner partners, is the world's first industrial scale CO_2 injection project designed specifically as a greenhouse gas mitigation measure. CO_2 separated from natural gas is being injected into the Utsira Sand (Fig. 1), a major saline aquifer of late Cenozoic age (Chadwick et al., 2004a; Zweigel et al., 2004). The injection point is at a depth of about 1012 m bsl, some 200 m below the reservoir top. Injection started in 1996 and a total of more than 7 million tonnes of CO_2 are presently in situ at the time of writing.

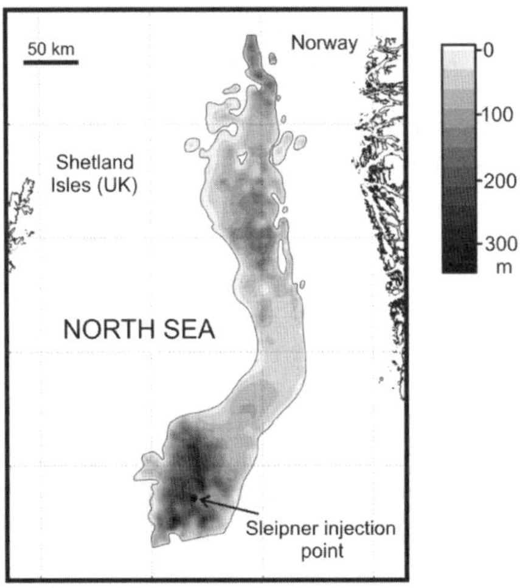

Figure 1. Location map showing Sleipner and the extent/thickness of the Utsira Sand aquifer.

Since 1998 the injection operation has been linked to a number of research projects, notably SACS, SACS2 and CO2STORE. These projects, funded by the EU, industry and national governments, aim to show that underground storage is a safe and verifiable technology. Specifically they have carried out scientific research into the geological aspects of the Sleipner injection operation by monitoring and modelling the injected CO_2 plume.

Key aims of the monitoring programme at Sleipner are outlined below:

1. To show that the CO_2 is being confined safely within the primary storage reservoir.
2. To image the distribution and migration of CO_2 throughout the reservoir and, should it occur, into adjacent strata.

3. To provide early warning of any potentially hazardous migration towards the seabed.

Baseline 3D seismic data were acquired in 1994, prior to injection, with repeat surveys in 1999 (2.35 million tonnes of CO_2 in the reservoir), 2001 (4.26 Mt) and 2002 (4.97Mt). In addition, to complement the information available from the seismic datasets, a seabed gravimetric survey was acquired in 2002.

2. TIME-LAPSE SEISMIC DATASETS

This paper provides a brief outline of current interpretive work on the seismic datasets. Fuller details are given in Arts et al. (2004a, 2004b) and Chadwick et al. (2004b, 2005).

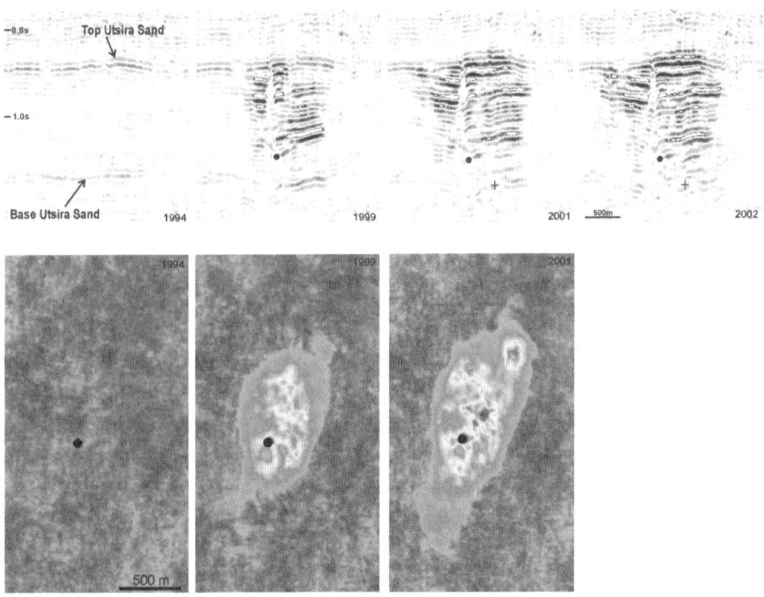

Figure 2. Time-lapse seismic images of the CO_2 plume: a) N-S inline through the 1994 dataset prior to injection and through the 1999, 2001 and 2002 datasets. b) Maps of integrated absolute reflection amplitudes from the plume showing its elliptical form in plan view and growth from 1999 to 2001. Black disc denotes injection point.

2.1 Time-lapse images

The CO_2 plume is imaged as a number of bright sub-horizontal reflections within the reservoir, growing with time (Figure 2a). The reflections are interpreted as arising from thin (< 8 m thick) layers of CO_2 trapped beneath thin intra-reservoir mudstones and the reservoir caprock. The plume is roughly 200 m high and elliptical in plan, with a major axis increasing from about 1500 m in 1999 to about 2000 m in 2001 (Figure 2b). The plume is underlain by a prominent velocity pushdown (Figure 3) caused by the seismic waves travelling much more slowly through CO_2-saturated rock than through the virgin aquifer.

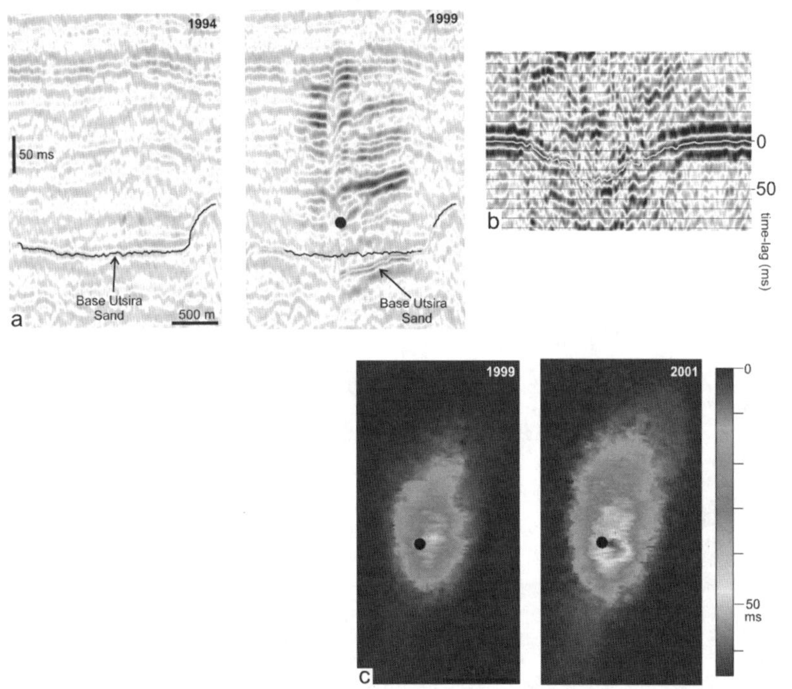

Figure 3. Velocity pushdown. a) Inline through the reservoir in 1994 and 1999 showing pushdown of the Base Utsira Sand beneath the plume. b) Cross-correlogram of a reflection window beneath the central part of the 2001 plume. Pick follows the correlation peak and defines pushdown. c) Pushdown maps in 1999 and 2001. Black disc denotes injection point.

2.2 Seismic modeling

Seismic modelling aimed at verifying the in situ injected mass of CO_2 has utilised both inverse and forward modelling techniques. Forward modelling, via history-matched reservoir simulations of the CO_2 plume produces a reasonable match to the observed data (Figure 4), though the detailed geometry of the plume layering remains uncertain.

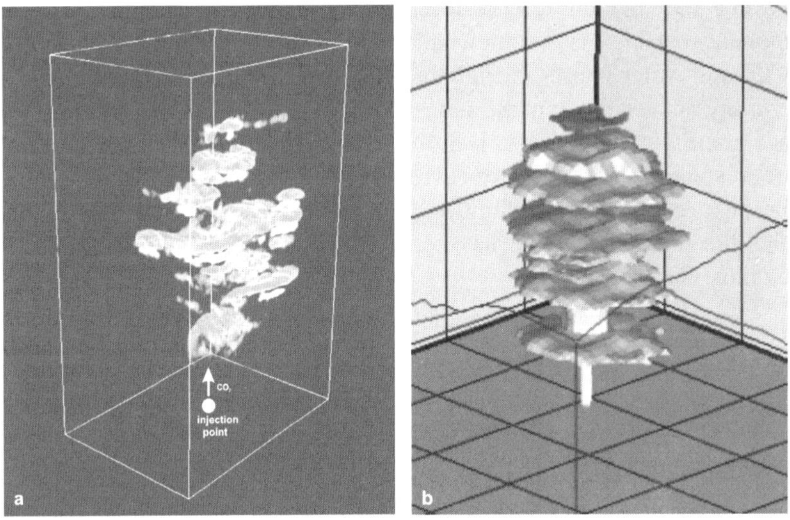

Figure 4. The 2001 plume: a) observed seismic data (opacity display of the 2001 minus 1994 difference cube) b) reservoir simulation.

Inverse modelling aims to quantify amounts of CO_2 from layer reflectivity and velocity pushdown. Because fluid pressures are believed to have changed very little during injection, modelling is based solely on fluid saturation changes. The observed plume reflectivity most likely comprises tuned responses from thin layers of CO_2 whose thickness varies directly with reflection amplitude. Inverse modelling takes as a starting point, thin, high-saturation layers of CO_2, mapped according to an amplitude-thickness tuning relationship. This is supported by structural analysis of the topmost CO_2 layer, whose thickness, estimated directly from the top reservoir topography, varies directly with reflection amplitude. In addition, in order for the modelled CO_2 distributions to produce the observed velocity pushdown, a minor, intra-layer component of much lower saturation CO_2 is required.

A measured formation temperature of 36°C is available for the Utsira reservoir, but it is poorly-constrained. Regional temperature patterns suggest that the reservoir may be up to 10°C warmer. At the higher temperatures, CO_2 would have significantly different physical properties. In particular its density would be significantly lower, giving a correspondingly larger in situ volume. Inverse models of CO_2 distribution in the 1999 plume have been generated, based on both the measured and a possible higher temperature scenario (Figure 5). The distribution of CO_2 in both models is consistent with the known injected mass (allowing for parameter uncertainty) and can replicate the observed plume reflectivity and the velocity pushdown. However, the higher temperature model requires that the dispersed component of CO_2 has a somewhat patchy, rather than uniform, mixing of the CO_2 and water phases (Sengupta & Mavko 2003). This highlights a key uncertainty in verification estimates; the velocity behaviour of the CO_2 – water – rock system, which is heavily dependent on the (poorly constrained) nature of small-scale mixing processes between the fluid phases.

Because of these uncertainties, a modelling solution which uniquely verifies the injected volume has not yet been obtained. Work on reducing uncertainty is ongoing.

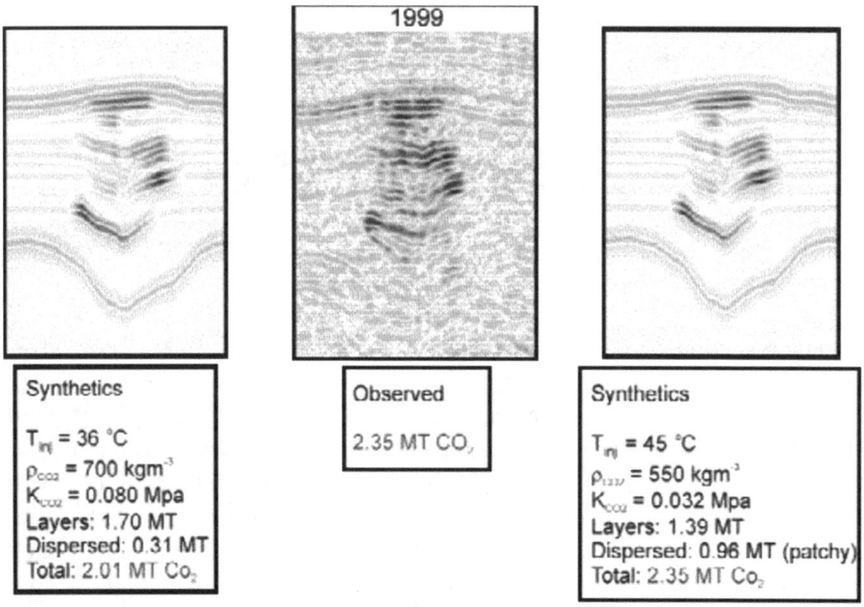

Figure 5. Inverse modelling of the 1999 plume. Observed data compared with synthetic seismograms based on inverse models for two plume scenarios: Injection point at 36°C with fine-scale mixing throughout; Injection Point at 45° C with patchy mixing in the intra-layer dispersed component of CO_2.

2.3 Detecting migration from the storage reservoir

The seismic data indicate that no detectable leakage of CO_2 into the caprock has occurred so far. The potential detection capability of the Sleipner data can be illustrated by examining the 1999 plume (Figure 6). The topmost part of the plume is marked by two small CO_2 accumulations trapped directly beneath the caprock seal.

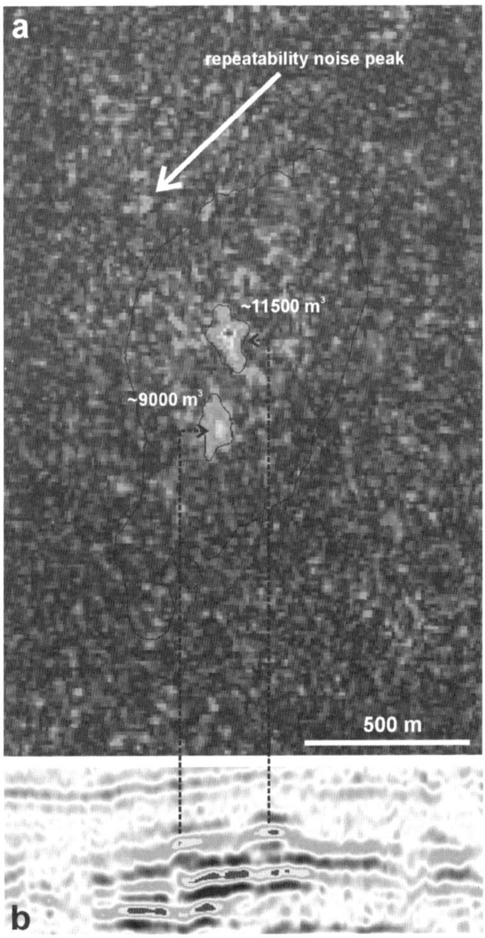

Figure 6. Detection limits for small amounts of CO_2 a) Map of the 1999-94 difference data showing integrated reflection amplitude in a 20 ms window centred on the top Utsira Sand. Note high amplitudes (paler greys) corresponding to the two small CO_2 accumulations. Note also scattered amplitudes due to repeatability noise. b) Part of seismic line showing the topmost part of the plume and the two topmost CO2 accumulations.

From the reflection amplitudes, the volumes of the two accumulations can be estimated at 9000 and 11500 m^3, respectively. Other seismic features on the timeslice can be attributed to repeatability noise, arising from slight intrinsic mismatches between the 1999 and 1994 (baseline) surveys. It is clear that the level of repeatability noise plays a key role in determining the detectability threshold. Thus for a patch of CO_2 to be identified on the data it must be possible to discriminate between it and the largest noise peaks. Preliminary analysis suggests that accumulations larger than about 4000 m^3 should fulfill this criterion. This corresponds to about 2000 tonnes of CO_2 at the top of the reservoir (where CO_2 has a density of about 500 kgm^{-3}), but less than 600 tonnes at 500 m depth where the density is considerably lower. Seismic detection depends crucially on the nature of the CO_2 accumulation. Small thick accumulations in porous strata would tend be readily detectable. Conversely, distributed leakage fluxes through low permeability strata may be difficult to detect with conventional seismic techniques. Similarly, leakage along a fault within low permeability rocks would be difficult to detect. Fluxes of CO_2 such as these may well be associated with changes in fluid pressure, in which case shear-wave seismic data is likely to prove useful as a detection tool.

3. TIME-LAPSE SEABED GRAVIMETRY

Measurements of the gravitational acceleration due to mass distributions within the earth may be used to detect variations in subsurface rock or fluid density. Although of much lower spatial resolution than the seismic method, gravimetry offers some important complementary adjuncts to time-lapse seismic monitoring. Firstly, it can provide independent verification of the change in subsurface mass during injection via Gauss's Theorem. This potentially important capability may enable estimates to be made of the amount of CO_2 going into dissolution, a significant source of uncertainty in efforts to quantify free CO_2 in the reservoir (dissolved CO_2 is effectively invisible on seismic data). Secondly, deployed periodically, gravimetry could be used as an 'early warning system' to detect the accumulation of CO_2 in shallow overburden traps where it is likely to be in the low density gaseous phase with a correspondingly strong gravity signature.

The possibility of monitoring injected CO_2 with repeated gravity measurements is strongly dependent on CO_2 density and subsurface distribution. A feasibility study of time-lapse gravimetry at Sleipner (Williamson et al., 2001) modelled plume scenarios with CO_2 densities ranging from over 700 kgm^{-3} (corresponding to the lower reservoir temperature scenario) to less than 350 kgm^{-3} (corresponding to possible higher reservoir temperatures). The modelling indicated that future changes in the CO_2 plume could theoretically be detectable by seabed gravimetry. For example it was shown that the addition of 2 million tonnes to the plume would produce a change in peak gravity signal of between -8 and -33 mGal, corresponding to CO_2 densities of 700 kgm^{-3} and 350 kgm^{-3}, respectively (Figure 7). Longer-term predictions suggest that the peak gravity signature of the plume will gradually decrease as it thins by lateral migration at the reservoir top. On the other hand, if CO_2 leaked to shallower levels where it would have a still lower density, gravity changes could well exceed -100 mGal.

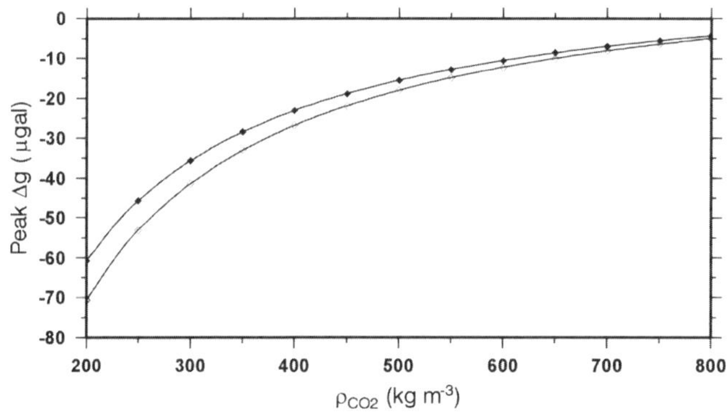

Figure 7. Peak gravity anomaly as a function of CO_2 density, predicted for the Sleipner CO_2 plume in 1999 (2.3 MT in situ). Gravity changes computed for the sea surface (solid symbols) and the seabed (open symbols).

A seabed gravity survey was acquired at Sleipner in 2002 (Eiken et al. 2003), with approximately 5 million tonnes of CO_2 in the plume. The survey was based around pre-positioned concrete benchmarks on the seafloor that served as reference locations for the (repeated) gravity measurements. Relative gravity and water pressure measurements were taken at each benchmark by a customised gravimetry and pressure measurement module mounted on a Remotely Operated Vehicle (Figure 8).

Thirty concrete benchmark survey stations were deployed in two perpendicular lines, spanning an area some 7 km east-west and 3 km north-south and overlapping the subsurface footprint of the CO_2 plume. Each survey station was visited at least three times to better constrain instrument drift and other errors. Single station repeatability was estimated to be 4 mGal. For time-lapse measurements an additional uncertainty of 1 – 2 mGal is associated with the reference null level. The final detection threshold for Sleipner therefore is estimated at about 5 mGal.

A repeat gravity survey is planned for the summer of 2005, with a projected 8 million tonnes of CO_2 in the plume. The additional 3 million tonnes of CO_2 are expected to produce a gravity change of between about -10 and -43 mGal depending on density. Such a change should theoretically be detectable. In the event that acceptably accurate measurements are obtained, it will be possible to derive the average density of CO_2 in the plume. This will help to constrain plume temperatures, which will in turn reduce uncertainty in the seismic analysis.

Figure 8. The seabed gravimetry operation at Sleipner showing the seabed gravimetry / pressure instrumentation and the remotely operated vehicle being lowered into the sea.

4. CONCLUSIONS

Time-lapse seismic monitoring has proved notably successful in imaging the growing CO_2 plume at Sleipner. Quantitative analysis of the1999 dataset has shown that the observed seismic signature is consistent with known injected amounts, but a complete verification has not been possible due to a number of uncertainties. These are related both to reservoir properties and conditions and also to the seismic properties of the CO_2 - water - rock system. Gravimetry has so far been restricted to an initial survey, but it is hoped that future repeat datasets will provide additional complementary information that can be used to further reduce seismic uncertainty.

The Utsira Sand is a relatively shallow, thick reservoir with notably high porosity and permeability. In this respect it is very suitable for both seismic and gravimetric monitoring, with injected CO_2 giving rise to particularly pronounced geophysical signatures. Other storage scenarios are likely to prove more challenging from a monitoring standpoint. In addition to

Sleipner, industrial-scale CO_2 injection projects are ongoing at Weyburn in Canada (Wilson and Monea, 2004), and at In Salah in Algeria, with another planned for the Snohvit field in the Barents Sea. These will further test the efficacy of geophysical monitoring methods in a range of storage situations. Reservoir depths range from ~ 1500m to nearly 3000 m, with widely different reservoir types, including carbonates. Major research projects are linked to all of these projects and important new insights into monitoring capabilities are anticipated in the coming years.

ACKNOWLEDGEMENTS

We thank the CO2STORE consortium for permission to publish this work. SACS/SACS2 and CO2STORE have been funded by the EU Thermie Programme, by industry partners Statoil, BP, ExxonMobil, Norsk Hydro, TotalFinaElf, Vattenfall, Schlumberger and by national governments. R&D partners are BGR (Bundesanstalt fur Geowissenschaften und Rohstoffe), BGS (British Geological Survey), BRGM (Bureau de Recherches Geologiques et Minieres), GEUS (Geological Survey of Denmark), IFP (Institute Francais du Petrole), TNO-NITG (Netherlands Institute of Applied Geoscience – National Geological Survey) and SINTEF Petroleum Research. Andy Chadwick publishes with permission of the Executive Director, British Geological Survey (NERC).

REFERENCES

Arts, R., Eiken, O., Chadwick, R.A., Zweigel, P., Van Der Meer, L., and Kirby, G.A., 2004, Seismic monitoring at the Sleipner underground CO_2 storage site (North Sea). in: *Geological Storage for CO_2 emissions reduction*, S. Baines, J. Gale and R.J. Worden, ed., Special Publication of the Geological Society, London, 233, pp. 181 - 191.

Arts, R., Eiken, O., Chadwick, R.A., Zweigel, P., Van Der Meer, L., and Zinszner, B., 2004b, Monitoring of CO_2 injected at Sleipner using time-lapse seismic data. *Energy*, **29:**1383-1393.

Baklid, A., Korbøl, R., and Owren, G., 1996, Sleipner Vest CO_2 disposal, CO_2 injection into a shallow underground aquifer. *SPE Annual Technical Conference and Exhibition*, Denver, Colorado, USA, SPE paper 36600, 1-9

Chadwick, R.A., Zweigel, P., Gregersen, U., Kirby, G.A., Johannessen, P.N., and Holloway, S., 2004a, Characterisation of a CO_2 storage site: The Utsira Sand, Sleipner, northern North Sea. *Energy*, **29:**1371-1381.

Chadwick, R.A., Holloway, S., Brook, M., and Kirby, G.A., 2004b, The case for underground CO_2 sequestration in northern Europe, in: *Geological Storage for CO_2 emissions reduction*, S. Baines, J. Gale and R.J. Worden, ed., Special Publication of the Geological Society, London, 233, pp. 17 - 28.

Chadwick, R.A., Arts, R., and Eiken, O., 2005, 4D seismic quantification of a growing CO_2 plume at Sleipner, North Sea, in: *Petroleum Geology: North West Europe and Global Perspectives - Proceedings of the 6th Petroleum Geology Conference.* A.G. Dore and B. Vining, eds, Published by the Geological Society, London, pp. 1385 – 1399.

Eiken, O., Stenvold, T., Zumberge, M., and Nooner, S., 2003, *Gravity monitoring at the Sleipner CO_2 injection site: Report on 2002 baseline survey.* Statoil Technical Report, 22pp.

Sengupta, M., and Mavko, G., 2003, Impact of flow simulation parameters on saturation scales and seismic velocity. *Geophysics*, **68**(4):1267-1280.

Williamson, J.P., Chadwick, R.A., Rowley, W.J., and Eiken, O., 2001, *Saline Aquifer Storage: A Demonstration Project at the Sleipner Field. Work Area 5 (Geophysics) – Gravity modelling of the CO_2 bubble.* BGS Commissioned report CR/01/063. 25pp.

Wilson, M. and Monea, M., 2004, (eds.) I*EA GHG Weyburn CO_2 Monitoring & Storage project Summary Report 2000 – 2004.* Petroleum Technology research Centre, Regina. 273pp.

Zweigel, P., Arts, R., Lothe, A. E., and Lindeberg, E., 2005, Reservoir geology of the Utsira Formation at the first industrial-scale underground CO_2 storage site (Sleipner area, North Sea) in: *Geological Storage for CO_2 emissions reduction*, S. Baines, J. Gale and R.J. Worden, ed., Special Publication of the Geological Society, London, 233, pp. 165-180.

4-D SEISMICS, GAS-HYDRATE DETECTION AND OVERPRESSURE PREDICTION AS A COMBINED METHODOLOGY FOR APPLICATION TO CO_2 SEQUESTRATION

Combined seismic methods for CO_2 monitoring

S. Persoglia, J.M. Carcione, G. Rossi And D. Gei
Istituto Nazionale di Oceanografia e di Geofisica – OGS-, Trieste, Italy

Abstract: Seismic surveys have proven to be useful for monitoring injected CO_2 in the subsurface. In this work, we show how rock physics, poro-elastic modeling and 3D seismic tomography can be combined to detect the subtle changes in seismic properties related to changes in pore-fill. 3D seismic tomography yields the P- and S-wave velocity cubes, which are converted to petro-physical properties by using rock-physics models of partial saturation under varying temperature and pressure conditions, and seismic numerical modeling. The methodology is illustrated with field examples of time-lapse analysis and gas-hydrate detection.

Key words: Seismic properties, tomography, poro-elasticity.

1. INTRODUCTION

The underground storage of CO_2 involves technologies developed and widely used by the oil and gas industries for exploration, exploitation and monitoring of hydrocarbon reservoirs. In CO_2 sequestration, the type of monitoring, its duration and frequency are key elements to verify the storage efficiency and distribution of the injected carbon dioxide. Seismic surveys have proven to be capable of monitoring CO_2 in the subsurface (e.g. Arts et al., 2004; Ghaderi and Landrø, 2004). To guarantee a similar success in less favourable geological contexts, it is therefore important to apply the most advanced methodologies to the specific case of CO_2 monitoring. In

S. Lombardi et al. (eds.),
Advances in the Geological Storage of Carbon Dioxide, 315–323.

particular, the combination of 3D seismic tomography and poro-elastic seismic modeling are powerful tools to tackle the injection problem.

In this work, we show field examples of time-lapse analysis, free-gas and gas-hydrate evaluation, and the detection of overpressure conditions in an effort to extend these techniques to CO_2 detection and monitoring.

2. SEISMIC TOMOGRAPHY

Time-lapse analysis, required to detect small changes in the reservoir due to fluid movements, often has the problem that the seismic data has been collected with different acquisition configurations and technologies, and thus special processing aimed at rendering the different data sets equivalent is required (e.g. Magesan et al., 2005). Another problem that may mask the velocity variations in the reservoir are the seasonal changes of the overburden (the sea-water layer in marine surveys or the water table on-land). The use of 3D time-lapse tomography has proven to be a flexible and powerful tool in solving both problems, enabling the removal of the related small velocity variations and giving a set of coupled models to highlight the reservoir changes (Vesnaver et al., 2003).

The tomographic inversion consists of minimizing the differences between the observed and calculated travel times, thus allowing to obtain the wave velocities. The travel time of seismic waves recorded by a geophone is obtained by integrating the slownesses along the ray path. The integral is replaced by a summation over the voxels, since a discrete blocky model is usually adopted. The travel time difference can be expressed as

$$\Delta t = t^{OBS} - t^{CALC} = \sum_j \Delta s_j \Delta u_j \ . \qquad 1$$

In practice, the tomographic inversion follows the scheme of Figure 1. The rays are first traced on an initial model, using the same acquisition geometry of the real experiment, and the obtained travel times are subtracted from those of the real experiment. The residuals are minimized to upgrade the velocity and geometrical features of the reflecting/refracting interfaces, until the model does not change compared to the previous one. The separate inversion of the velocity and interface depths and geometry makes the procedure less sensitive to cross talk between velocity and depth errors (Bishop et al., 1985; Bickel, 1990; Delprat-Jannaud and Lailly, 1992; Docherty, 1992; Lines, 1993; Tieman, 1994). The non-uniqueness of the solution is reduced by finding the optimal grid for the velocity field via adaptive irregular or staggered grids (e.g. Vesnaver and Böhm, 2000).

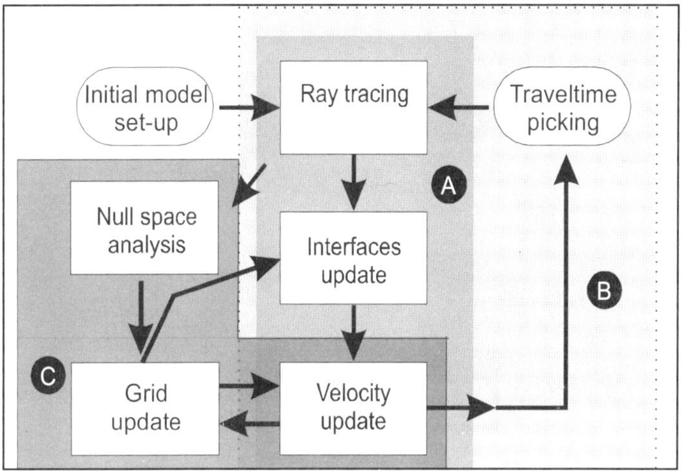

Figure 1. Scheme of the three main loops of the tomographic inversion. A) Update of the velocity field and interface depth; B) Picking and comparison with travel times from the tomographic model; C) Grid upgrading, based on velocity field and reliability.

The final result of the seismic tomography is therefore a velocity model in depth, where the geologic complexities are preserved. This output can be an optimal input for pre-stack depth migration to obtain an accurate imaging of the various geological features.

In the case of time-lapse analysis we can impose the additional constraint that the geometry and velocity of most of the model does not change over time, while leaving free the reservoir and the shallowest layer (which may be affected by seasonal variations). This procedure constitutes the basis of "Time-lapse tomography", successfully applied to different vintages acquired offshore Norway (Vesnaver et al., 2003). The data was acquired in 1989, with a single vessel towing two streamers and a central air gun, and in 1992, using one vessel with three streamers and a sleeve gun as source while a second vessel towed two additional streamers. In 1989 the hydrophone spacing was double that of 1992. These acquisition differences made a direct comparison of the seismic profiles difficult in the time domain, however the use of 3D tomography made it possible to overcome these difficulties (Vesnaver et al., 2003).

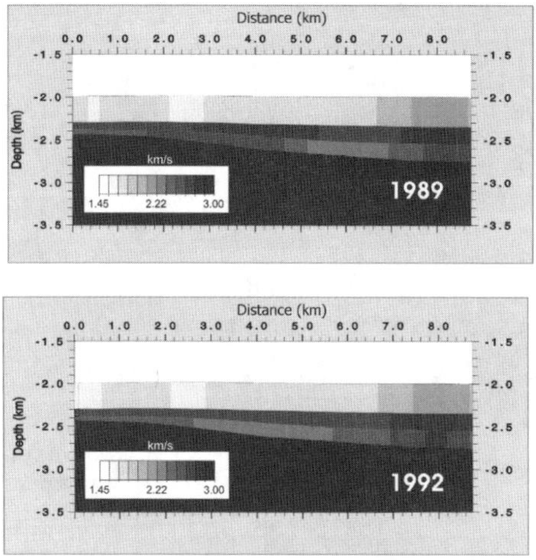

Figure 2. Vertical sections of the 1989 and 1992 3D tomographic models, obtained by a coupled time-lapse inversion.

After independently estimating the velocity models of the two vintages, both the velocities and geometries of the layers were assumed constant in time. They were averaged and used as initial models to improve the velocity fields within the overburden and the reservoir, resulting in two coupled models (see Figure 2). These models have been used as input for a pre-stack depth migration, obtaining the images shown in Figure 3. These sections and the velocity models allow us to verify the changes which occurred within the reservoir during production time.

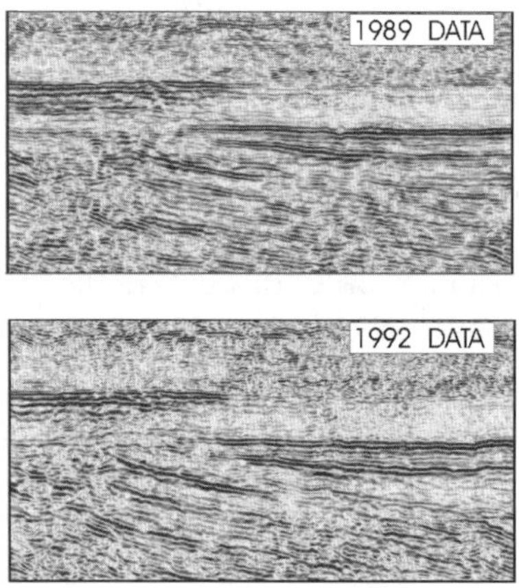

Figure 3. Pre-stack depth-migrated images of the reservoir for the 1989 and 1992 vintages, using coupled velocity models.

3. ROCK PHYSICS

A detailed 3D P and S wave velocity field versus depth is essential to evaluate fluid content, type and saturation, as well as the geopressure conditions. In fact, the seismic response of the subsoil depends on the elastic properties of the subsoil components. In particular, the velocity of seismic waves depends directly upon lithology, saturation and the in-situ conditions.

Generally, rocks in the subsoil are saturated with brine. When other fluids are partially saturating the pore space the elastic properties of the medium may change, resulting in a velocity anomaly relative to the normal trend of a brine saturated rock.

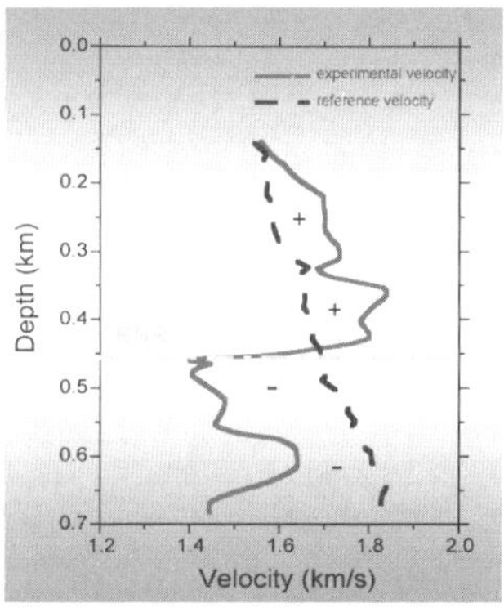

Figure 4. Theoretical velocity of a sedimentary sequence fully brine saturated versus depth (dashed line) and actual velocity (continuous line). Positive velocity anomalies (+) can be related to the presence of gas-hydrate in the pore space and negative anomalies (-) to free gas (Tinivella, 1999).

In Figure 4 the dashed line represents the theoretical velocity versus depth in a brine-saturated sedimentary sequence, while the continuous curve is the actual velocity trend inferred with either tomographic inversion or well-log measurements. Positive velocity anomalies (+) can be related to the presence of gas-hydrate in the pore space and negative anomalies (-) to free gas. Gas-hydrate is an ice-like solid compound where gas molecules, mainly methane, are occluded in a lattice of host water molecules. It forms at low temperature and high pressure conditions and occupies the porous space of the rock, causing a stiffening of the medium and consequently an increasing of the seismic velocity. Layers rich in gas-hydrates may act as cap rocks for trapped free gas underneath. The strong acoustic impedance contrast between high-velocity hydrate-rich and low-velocity free-gas saturated layers causes a strong reflection on seismic sections. Such a signal is called bottom simulating reflector (BSR) and is often used to infer the presence of gas-hydrate within sediments. Above the BSR the pores of the rock are filled with gas hydrate and water, whilst below the BSR the rocks are saturated with free gas and water.

Rock physics is used to establish relationships between rock properties and the observed seismic response. Rock-physics theories can be used to compute theoretical seismic velocities in sediments saturated with fluids or containing gas-hydrates. Rock compositions and fluid saturations can be estimated by comparing the theoretical and observed velocities. There are several theories predicting the elastic properties of composite media. Among them the Biot theory is one of the most commonly used and tested by comparison with laboratory measurements.

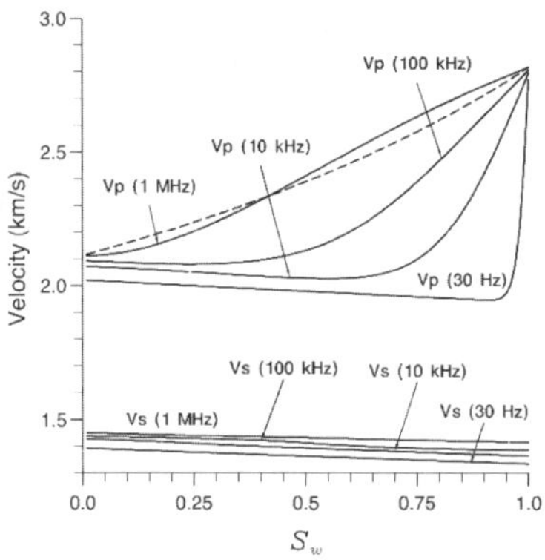

Figure 5. P- and S-wave velocities versus water saturation and different frequencies, using Biot's theory and White's model. The other saturating fluid is free gas and the gas-hydrate fraction is 0.3. Also shown is the P-wave velocity obtained using Hill's equation (dashed line) (Gei and Carcione, 2003).

Figure 5 shows the computed P and S-wave velocities as a function of water saturation (S_w) and frequency of the seismic signal. The matrix is a sandstone and the pore-saturating fluid is a mixture of water and methane whose proportions range from pure gas (S_w=0) to pure water (S_w=1); 30% of the pore space is occupied by gas hydrate. The S-wave velocity is slightly affected by the fluids. The velocity decrease as water saturation increases is due to variations in the density of the composite fluid.

Figure 6 shows the P-wave tomographic velocities of two seismic lines acquired in an area located to the north of the Knipovich Ridge (western Svalbard margin). The dashed line represents the BSR. These velocities have been used to compute the free-gas (methane) saturation and gas-hydrate concentration using a poro-elastic model based on a Biot-type approach (the interaction of the rock frame, gas hydrate and fluid is modeled from first physical principles). By fitting the tomographic velocity fields to theoretical velocities, average hydrate concentrations of 7% and maximum free-gas saturation of 0.4% (uniform saturation) and 9% (patchy saturation), have been obtained. These results are shown in Figure 7.

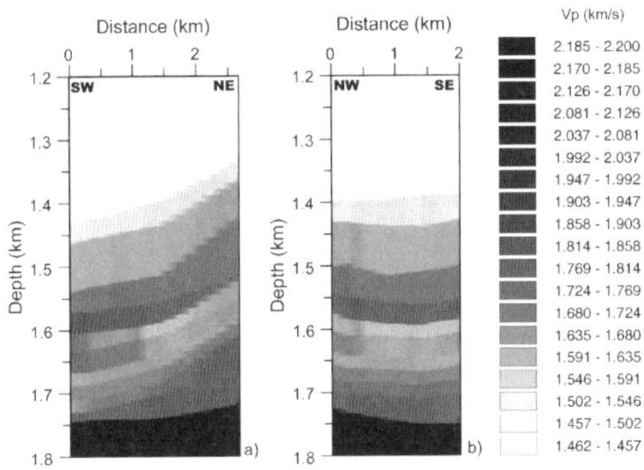

Figure 6. P-wave tomographic velocities of two seismic lines acquired in an area located at the northern side of the Knipovich Ridge (western Svalbard margin). The dashed line represents the BSR (Carcione et al., 2005).

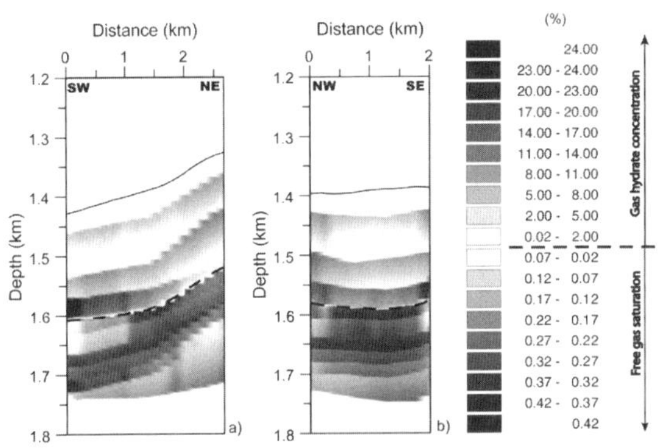

Figure 7. Sections of gas-hydrate concentration (blue color) and free-gas saturation (red color) corresponding to the velocity sections shown in Figure 6. The solid lines indicate the sea bottom and the dashed lines represent the BSR. (Carcione et al., 2005).

Another example of the usefulness and capability of the joint application of tomographic inversion and modeling is the detection of overpressure (Carcione and Tinivella, 2001). A rock is said to be overpressured when its pore pressure (in situ fluid pressure in pores) is significantly greater than the hydrostatic pressure (weight of the overlying pore fluids-mainly brine).

Acoustic and transport properties of rocks generally depend on effective pressure, which is a linear combination of pore and confining pressures. Using the Biot theory of poroelasticity (Biot, 1962; Carcione, 1998) it is possible to calculate the seismic velocities for different combinations of the pore and confining pressures. In particular, Carcione and Tinivella (2001) analysed a site offshore Norway, calibrating a model of the various formations on the basis of well data, tomographic inversion of seismic lines, AVO analysis, and laboratory measurements. The modeling indicated a substantial decrease in velocities with increasing pore-pressure, mainly because of the opening of compliant cracks. The decrease in P-wave velocity is higher for dry rocks than for brine-saturated rocks, while the S-wave velocity is generally higher for the dry case than for the fully brine-saturated case. The presence of gas does not have a major effect on velocities.

4. CONCLUSIONS

Two advanced seismic tools have been presented, namely 3D seismic tomography and poro-elastic modeling. The analysis of the time-lapse North-sea data sets demonstrates the flexibility of tomographic inversion, due to changes in the acquisition geometry and overburden conditions. Analogously, in the case of CO_2 injection, time-lapse tomography can be used to evaluate the subtle changes due to the bubble expansion and possible leakage. The pre-stack depth-migrated sections in depth can be subtracted to evaluate the differences in terms of pressure and saturation. Furthermore, the detection of gas-hydrates and free gas offshore Svalbard has shown that poro-elastic modeling is essential to interpret the velocity changes.

REFERENCES

Arts, R., Chadwick, A., and Eiken, O., 2004, Recent time-lapse seismic data show no indication of leakage at the Sleipner CO2-injection site, *7th Conference on Greenhouse Gas control Technologies*, Vancouver, Canada, 5-9 Sept. 2004.

Bickel, S. H., 1990, Velocity-depth ambiguity of reflection traveltimes, *Geophysics,* **55**:266-276.

Biot, M. A., 1962, Mechanics of deformation and acoustic propagation in porous media. *Journal of Applied Physics* **33**:1482-1498.

Bishop, T., Bube, K., Cutler, R., Langan, R., Love, P., Resinik, J., Shuey, R., Spindler, D., and Wyld, H., 1985, Tomographic determination of velocity and depth in laterally varying media, *Geophysics* 50:903-923.

Böhm, G., Camerlenghi, A., Lodolo, E., and Vesnaver, A., 1995, Tomographic analysis and geological context of a bottom simulating reflector on the South Shetland Margin (Antarctic Peninsula), *Boll. Geof. Teor. Appl.*, **37**:3-23.

Carcione, J. M., 1998, Viscoelastic effective rheologies for modelling wave propagation in porous media, *Geophysical Prospecting* **46**:249-270.

Carcione, J. M., and Tinivella, U., 2001, The seismic response to overpressure: a modelling study based on laboratory, well and seismic data., *Geophysical Prospecting* 49:523-539.

Carcione, J. M., Gei, D., Rossi, G. and Madrussani, G., 2005, Estimation of gas hydrate concentraion and free-gas saturation at the Norvegian-Svalbard continental margin. *Geophysical Prospecting*, accepted.

Delprat-Jannaud, F., and Lailly, P., 1992, Ill-posed and well-posed formulations of the reflection traveltime tomography problem, *J. Geophys. Res.*, **97(**B13):19827-19844.

Docherty, P., 1992, Solving for the thickness and velocity of the weathering layer using 2-D refraction tomography, *Geophysics* **57**:1307-1318.

Ghaderi, A., and Landrø, M., 2004, Pre-stack estimation of time-lapse seismic velocity changes- an example from the Sleipner CO2-sequestration project, 7th Conference on Greenhouse Gas control Technologies, Vancouver, Canada, 5-9 Sept. 2004.

Gei, D., and Carcione, J. M., 2003, Acoustic properties of sediments saturated with gas hydrate, free gas and water, *Geophysical Prospecting*, **51**:141-157.

Lines, L., 1993, Ambiguity in analysis of velocity and depth, *Geophysics,* **58**:596-597.

Magesan, M., Depagne S., Nixon, K., Regel, B., Opich, J., Rogers, G. and Hudson, T., 2005, Seismic processing for time-lapse study: Genesis field, Gulf of Mexico, *The Leading Edge*, 364-373.

Tieman, H. J., 1994, Investigating the velocity-depth ambiguity of reflection traveltimes, *Geophysics,* 59:1763-1773.

Tinivella, U., 1999, A method for estimating gas hydrate and free gas concentrations in marine sediments, *Boll. Geof. Teor. Appl.,* **40**:19-30.

Vesnaver A. and Böhm G., 2000, Staggered or adapted grids for seismic tomography? *The Leading Edge* **19:**944-950.

Vesnaver, A. L., Accaino, F., Böhm, G., Madrussani, G., Pajchel, J., Rossi, G., and Dal Moro, G., 2003, Time-lapse tomography, *Geophysics* **68:**815-823.

PART V

THE WAY FORWARD

Our Future Dependancy on Fossil Fuels and Possible Approaches to Make the Geological Storage of Carbon Dioxide a Viable Tool to Mitigate Anthropogenic Greenhouse Gas Loading to the Atmosphere

THE ROLE OF FOSSIL FUELS IN THE 21ST CENTURY

Energy Scenario and Climate Aspects

I. Lakatos and J. Lakatos-Szabó

Research Institute of Applied Chemistry, University of Miskolc, 3515 Miskolc-Egyetemváros POB 2, Hungary

Abstract: According to a widely accepted forecast the global energy consumption, which is roughly 400 EJ today, will quadruple by the end of the century and the use of fossil fuels will probably increase until the middle of the century. Hence, the energy scenario definitely implies that the emission of greenhouse gases will also increase by a minimum 30%, leveling off at that value for the coming decades. Unfortunately, a simplified idea is that the use of fossil fuels is solely responsible for global warming, and hence climatic changes. Although the anthropogenic impact on climate is represented by only 15% of carbon dioxide emissions, this is the area where science and engineering can focus all efforts to influence its detrimental effects. The small anthropogenic effect on climate, however, clearly proves that the earth is extremely vulnerable to even marginal changes in the atmosphere, hydrosphere and lithosphere.

Key words: energy scenario, fossil fuels, emission of greenhouse gases, global warming, climate, emission trading, sequestration

1. INTRODUCTION

Using fossil fuels for energy produces enormous quantities of CO_2 which is transferred to the atmosphere. The fossil fuels, particularly coal, have been and still are the workhorse of energy production. Replacing coal and oil by natural gas can substantially reduce CO_2 emissions, however unfortunately it cannot eliminate it completely. Although the "clean energy" concept forms a very important strategy for all nations, we must always keep in mind that this effort has serious limitations. The forecasted energy scenario predicts that CO_2 emissions will increase by a minimum of 30%, leveling off at that

S. Lombardi et al. (eds.),
Advances in the Geological Storage of Carbon Dioxide, 327–344.

value for two or three decades in the first half of the century. The main question now is how this increased emission will influence the climatic state of the earth? The popular, simplified answer is that the use of fossil energy is alone responsible for the rise of average temperatures, and hence climatic changes. The real scientific fact is that natural factors contribute 96% of climatic changes, while the effect of anthropogenic factors are marginal. As far as the natural factors are concerned, water-induced processes are responsible for 65% of the changes and the influence of CO_2 in the atmosphere is estimated to be less than 30%. In contrast, man-made energy production represents only 15% of the total CO_2 emissions, however it is within this area where science and engineering can and must focus all efforts to decrease CO_2 emissions. For a better understanding of natural and anthropogenic factors this paper analyses various aspects and consequences of energy production based on fossil fuels, its possible effect on climate and the actions that science, society and individuals must take to address this problem.

2. ENERGY SCENARIOS: PAST, PRESENT AND FUTURE

"Energy demand is expected to grow in the 21st century. The energy demand will be met by a global energy mix that is undergoing a transition from the current dominance of fossil fuels to a more balanced distribution of energy sources. Motivation of energy diversification includes population growth, quality of life, clean energy, resources/reserves. An understanding of the energy options available to us in the 21st century requires an understanding of a range of scientific theories. The scientific revolution will be based on fundamental changes of old paradigms" - *John R. Fanchi* in "Energy in the 21st Century", Gulf Professional Publisher, Austin (in print). This quote not only strongly demonstrates the conceivable energy scenario, but also the well-based prediction of the future role of crude oil and natural gas production/consumption. The world energy demand was about 400 EJ (10^{18} J) at the end of the last century (OECD, 2002; US-DOE, 2004) and the share of hydrocarbons was roughly 60% (Fig. 1). A major part of the energy, however, came from utilization of crude oil, which in itself represented about 50%. That fact clearly demonstrates that the western world and the highly developed countries are oil and gas dependent, and it is known that this dependency is steeply increasing today. Thus, Europe and North America are net hydrocarbon importers, with imports representing more than 60% of their oil and gas consumption.

According to various UN reports (UN-ESC, 1996; 2003) the present global energy demand will quadruple assuming that the world population, in an optimistic prediction, will be 8 billion at the end of the century (Fig. 2). Since there is a close correlation between the quality of life index (Fig. 3) and energy consumption (UN-ESC, 1996; 1998; 2003), which will also likely increase from the present average value of 60 to 200 MJ/y per capita, the total global energy demand will probably be 1,600 EJ (10^{18} J) in 2100 (UN-ESC, 1998).

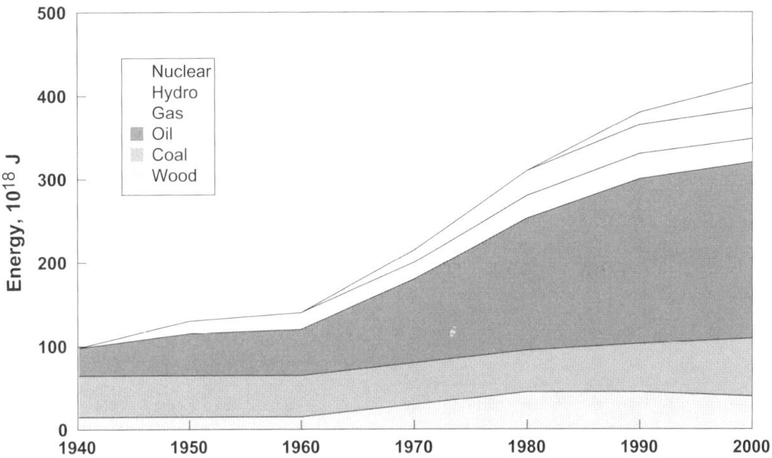

Figure 1. Absolute contribution of different sources to energy production (1940-2000).

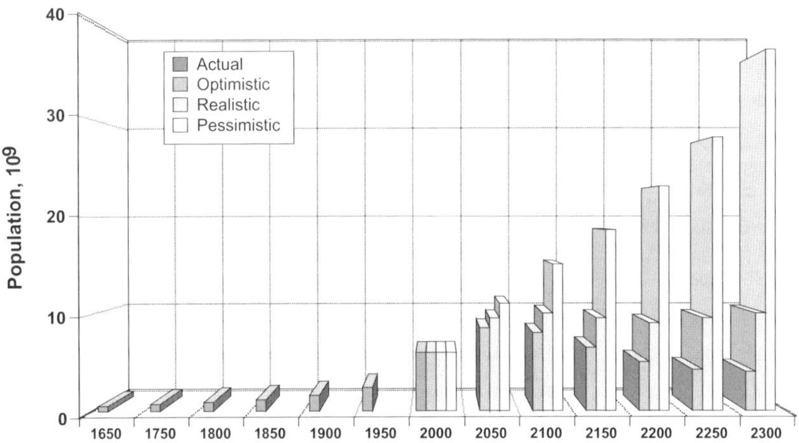

Figure 2. The actual and anticipated growth of world population between 1650 and 2300.

Converging data have been reported by various organizations (OECD, 2002; US-DOE, 2004; BP, 2005; WEC, 1995; Al-Fattah and Starzman, 2000; Skov, 2003; Arscott, 2003; Holditch, 2003; *SPE Review,* 2004; ExxonMobil, 2004; Shell, 2001) detailing the most probable diversification of energy sources in the future (Fig. 4). Although the relative importance of fossil fuels will apparently decrease, the stunning fact is that a total of 250-260 billion t of crude oil (on average 2.5 billion t per year through the whole century) must be produced to meet the world demand.

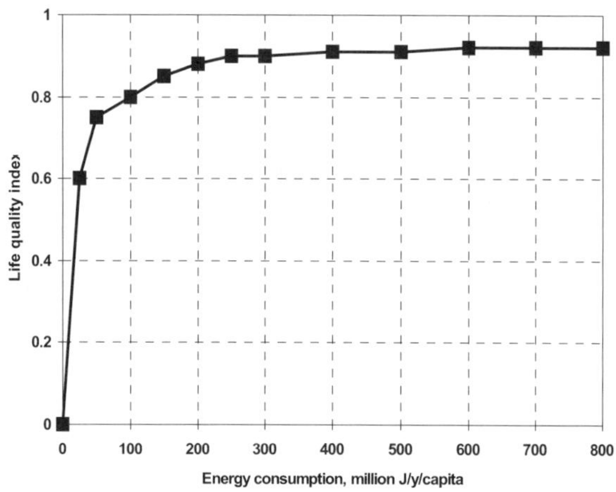

Figure 3. Correlation between the quality of life quality index and the energy consumption.

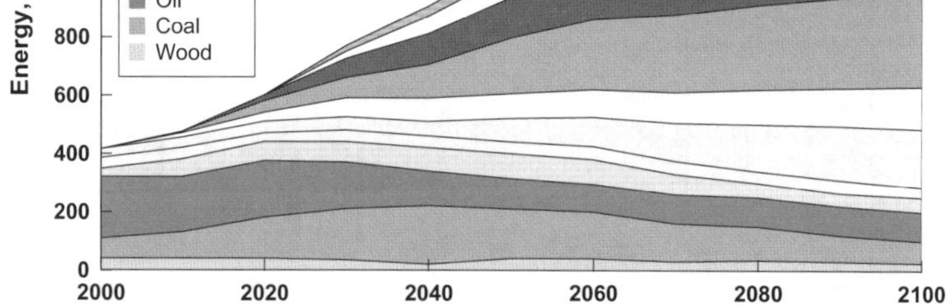

Figure 4. Absolute contribution of different sources to energy production (2000-2100).

This task is enormous if we consider that less than 100 billion t of crude oil has been produced since 1850 and that the average production rate was less than 0.7 billion t/y. As far as the reliable hydrocarbon supply is concerned the real challenge we are faced with is that the available oil reserves (proven and to be explored) amount to only 360-370 billion t. Consequently, the world demand can only be met by the supply if the recovery factor can be increased, on average, from 35% to 67%. Competent professionals and agencies say that the recovery factor can be increased up to 50% by improving existing reservoir engineering technology. Logically, the gap between 50% and 67% can only be surmounted by the extensive use of chemical IOR/EOR methods. The other option is that "dwindling supplies of oil and gas, obsolete power net-works and new environmental regulations threaten the Western world with a new energy crisis. Consequently, mankind is becoming again vulnerable to shortages in hydrocarbons, price shocks, supply interruptions, and in the worst case, political and military blackmail", *Emerson T.*: Newsweek, April 2002. Taking the hydrocarbon reserves and production rates into account, it is also obvious that these countries are dependent on sources from the Middle East which implies a global economic and political dependency which will remain for the coming decades.

3. CONTRIBUTION OF FOSSIL FUELS TO CO_2 EMISSIONS

Based on the energy scenario forecast to the end of the 21st century the probable and necessary demand of fossil fuels can be calculated (BP, 2005; WEC, 1995; Skov, 2003; ExxonMobil, 2004; Shell, 2001; US-DOE, 2005). The absolute and relative contribution of coal, oil and gas to energy production, per decade, are shown in Figs 5 and 6. The data in these diagrams suggest that the global role of fossil fuels, in contrast to various information in the world media, will increase until the middle of the century and then decrease monotonously until 2100.

The stacked columns in Fig. 5, however, indicate that utilization of these energy-bearing materials will represent at least 60-70% of the present amount. Analyzing the structure of the forecasted demand it is predicted that the relative contribution of coal and gas will reach a maximum (and oil a minimum) in the forties and fifties, while their share will stabilize through the last decades at 40-40% (coal and oil) and 20% (gas) (Fig. 6).

Technological improvements in the energy sector had already started well before the Kyoto accord. Significant efforts have been made to increase the efficiency of power stations and to decrease their air pollution. As far as the utilization of fossil fuels is concerned oil, and later natural gas, have replaced enormous amounts of coal. Thus, the extensive use of oil and gas, as a side effect, has also resulted in a substantial reduction in CO_2 emissions, since the average CO_2 emission depends fundamentally on the type of fuel used, be it coal (225 kg/MWh), oil (190 kg/MWh) or gas (128 kg/MWh).

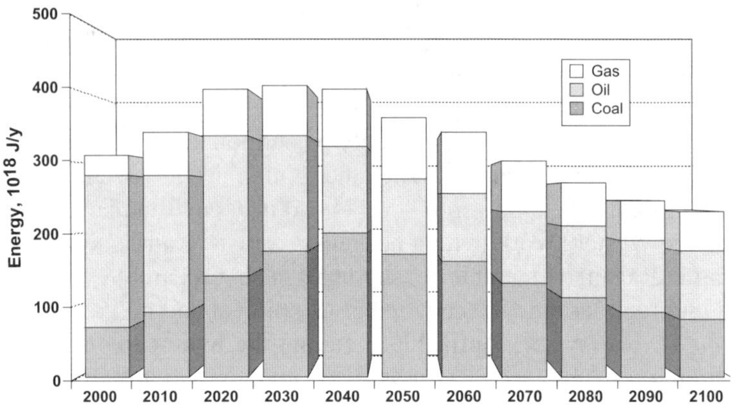

Figure 5. Absolute contribution of fossil fuels to energy production (2000-2100).

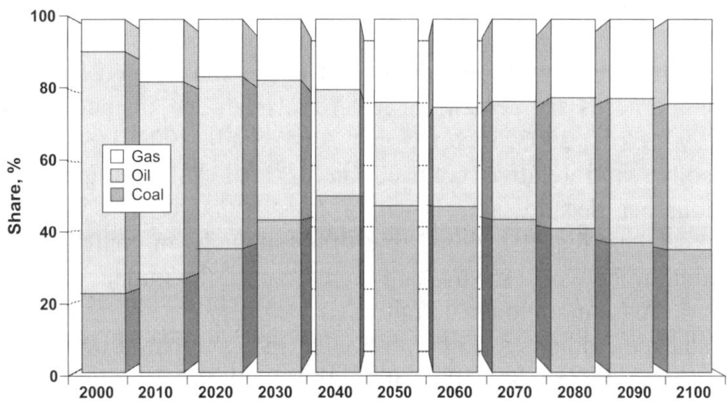

Figure 6. Relative contribution of fossil fuels to energy production (2000-2100).

Thus the hydrocarbons became attractive not only technologically, but also from an environmental point of view. Using these data and the anticipated use of fossil fuels in the future, predicted CO_2 emission rates can be calculated. The breakdown and summarized values of relative emissions are shown in Figs 7 and 8. As shown in Fig. 7 the CO_2 emissions from the use of oil and other liquid hydrocarbons will gradually decrease by about 50% until 2050, after which it will remain constant. A more extensive use of natural gas and coal, however, will unfavourably influence the CO_2 content of the atmosphere. A decreasing trend in CO_2 emissions can be expected only in the second half of the century (Fig. 8). This fact contradicts expectations of many environmental and "green" organizations that are looking forward to a substantial reduction of CO_2 emissions in the near future.

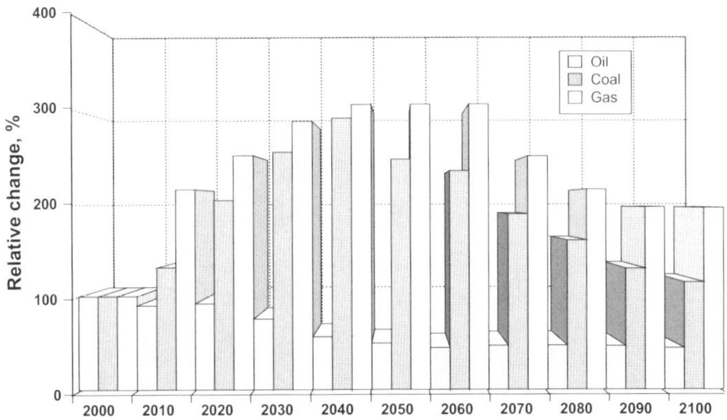

Figure 7. Relative CO_2 emissions of fossil fuels used for energy production (2000-2100).

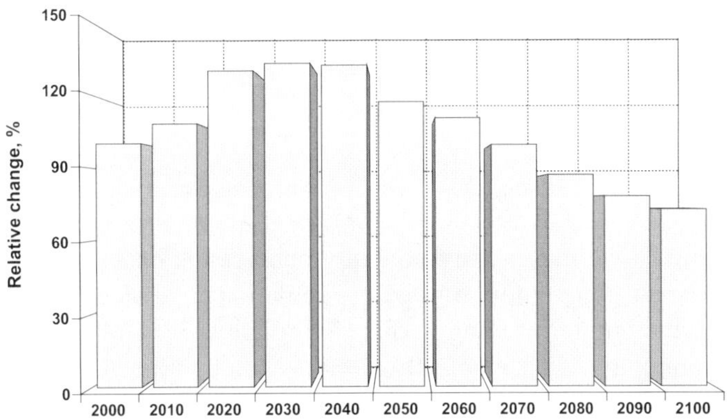

Figure 8. Summarized relative CO_2 emissions from fossil fuels used for energy production (2000-2100).

4. OVERVIEW OF GLOBAL CO_2 EMISSIONS

CO_2 emissions alone increased from 4 Gt/y to more than 10 Gt/y between 1970 and 2003 (ExxonMobil, 2004). The discrepancy in emissions of CO_2 and greenhouse gases can be analysed from different points of view. The Kyoto accord imposes limited or "capped" quotas on each country, taking special circumstances into account. As the natural, economic, industrial and even historical conditions of the nations are significantly different so are the absolute and relative CO_2 emissions. For instance, a surprising conclusion can be drawn from the data in Fig. 9, where specific CO_2 emissions are shown for several regions and countries. According to these data the Gulf countries are the biggest per-capita emitters, while the highly developed countries, like the USA, Germany, etc., rank lower on the list. Obviously, the sequence of countries can be explained by the fact that the energy demand rapidly increases in the OPEC countries and the energy is solely produced from hydrocarbons.

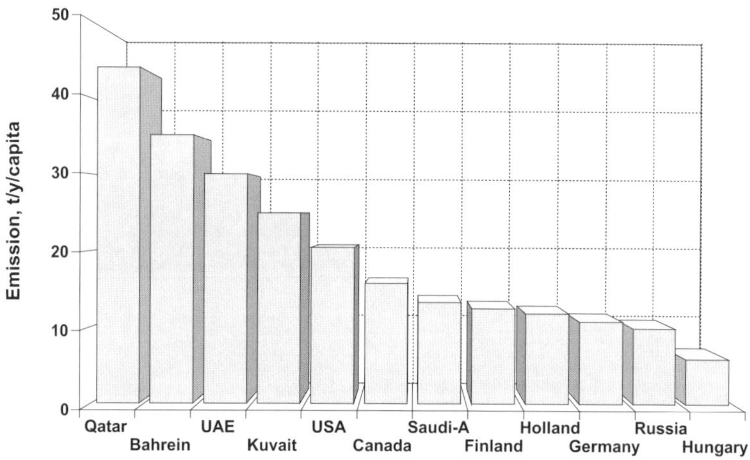

Figure 9. Absolute CO_2 emission per capita in several countries.

In the media we may more frequently observe the data presented in Fig. 10. Since the absolute output of CO_2 is population dependent, the USA is by far the biggest emitter, having a share of roughly 30% of the global CO_2 emissions annually (US-DOE, 2004; 2005). Although the government of the USA defends its decision not to sign the Kyoto accord claiming a similar CO_2 absorption (i.e. zero release), this explanation is not widely accepted yet. Consequently, environmental agencies are supporting efforts to find and use alternative energy sources, instead of fossil fuels. In 2002 a remarkable part of the global energy production already came from alternatives, like hydro, solar, wind and geothermal sources (Fig.11). It is also hopeful that the annual growth rate of renewable energies, particularly of non-polluting solar and wind, has been accelerating recently, while the use of fossil fuels is basically stagnant.

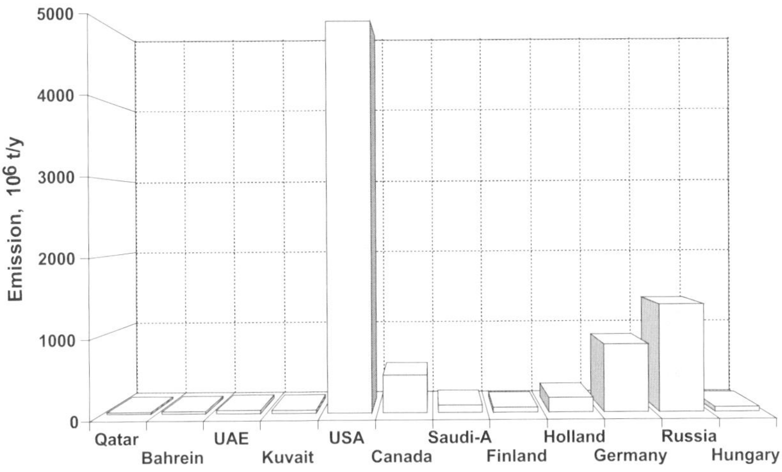

Figure 10. Absolute CO_2 emission in several countries.

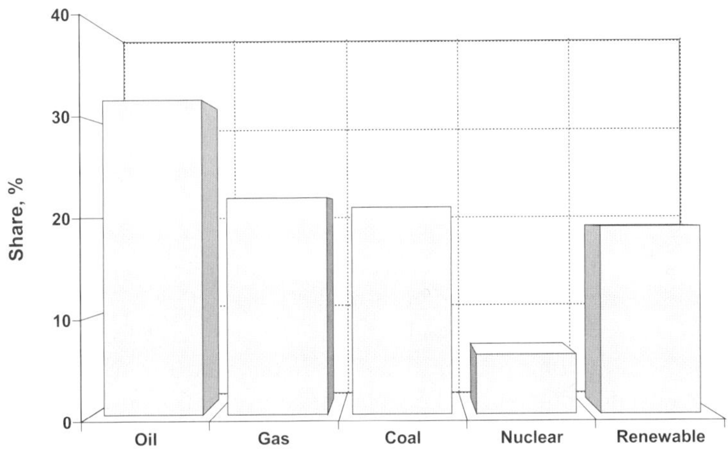

Figure 11. Contribution of different sources to global energy production in 2002.

Recently, a promising decision by the European countries has been to pledge to significantly increase the contribution of renewable energy sources to help meet national energy demand (US-DOE, 2004; Shell, 2001). But the situation, shown in Fig. 12, is not consistent. In some countries, such as Denmark, a radical increase in the use of renewable energy is planned, while in other countries the incremental change will be negligible. The options are partly hindered because the use of renewable energy is already high (Austria and Sweden), or because of unfavourable natural and climatic conditions. For instance, the morphological conditions, number of sunny days and wind conditions are definitely unfavourable for wide scale application of solar and wind renewables in Hungary and thus only geothermal energy offers a realistic possibility to increase the percentage of "clean energy". Unfortunately, extensive agricultural research programs focussed on wood and biofuels (being less harmful than coal and hydrocarbons) will also result in a measurable CO_2 emission.

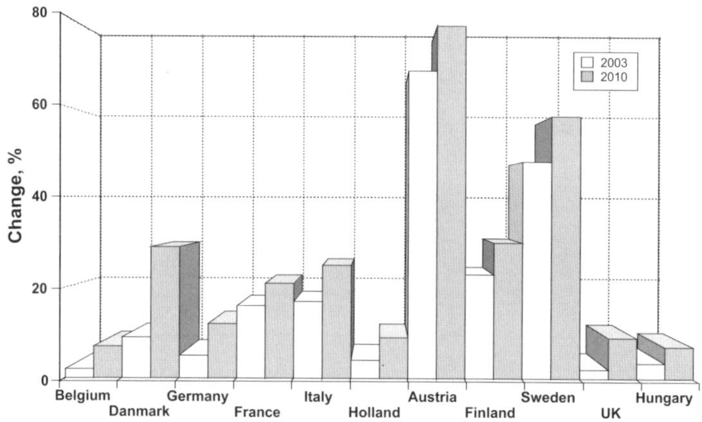

Figure 12. Decision of the European countries to increase the share of renewable energy.

At the present time the annual carbon emission rate is in the range of 5-7 10^9 t/y and great efforts have been made to predict global emission over the coming years. Diverse mathematical models have been developed which address not only technical but also social factors which influence emission rates to anticipate the equivalent carbon emissions. According to the most probable prognosis (Arscott, 2003), carbon emissions will develop as shown in Fig. 13. Although the simulation has a gradually increasing uncertainty with time, the results indicate that the summarized equivalent carbon emission rate can be as high as 30 10^9t/y, approximately six times higher than the present value, by the end of the century. This disturbing forecast puts added urgency on the need to use clean, non-polluting fuels. Parallel with that trend, there is also a need to implement emission trading and apply efficient and innovative CO_2 sequestration methods world-wide.

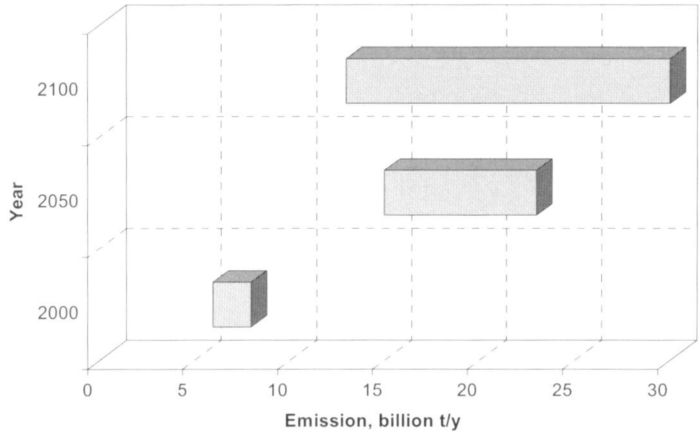

Figure 13. Anticipated global CO_2 emission in 2050 and 2100.

5. THE EFFECT OF CO_2 AND GREENHOUSE GAS EMISSIONS ON CLIMATE AND GLOBAL WARMING

The CO_2 content of the atmosphere has been essentially constant throughout the past centuries (Fig. 14), but measurable change can be detected from 1950 and today the characteristic concentration is about 380-400 ppm (Arscott, 2003; IPCC, 1995; Falkowski, 2002). According to a reliable forecast the CO_2 content in the atmosphere might be as high as 750 ppm by the end of the 21st century. Since CO_2 directly influences the average temperature of the earth its accumulation in the atmosphere results in global warming, as is well demonstrated by the fact that temperatures have been rising since 1970 (Fig. 15).

Responsible organizations (IPCC, 1995; Falkowski, 2002; Glick et al., 2004) have painted global warming in the darkest colour: oceans warm, glaciers melt, sea levels rise, permafrost thaws, lakes shrink, etc. All these unfavourable and accelerating processes may have serious and detrimental impact on flora, fauna and human life. Although the essence of the problem is not questioned, there is no consensus concerning the role of anthropogenic factors. Opponents of the anthropogenic factors usually refer to the distribution of CO_2 on earth. It is a fact that only 2% of the total amount of CO_2 is found in the atmosphere, while the majority (93%) is absorbed in the hydrosphere (particularly in the deep ocean water) (Ónodi, 2003). Hence, the anthropogenic influence may have a negligible effect on CO_2 equilibrium, or in other words, any positive or negative man-made interference in water, CO_2 and other gases might have only limited impact on natural processes and climate.

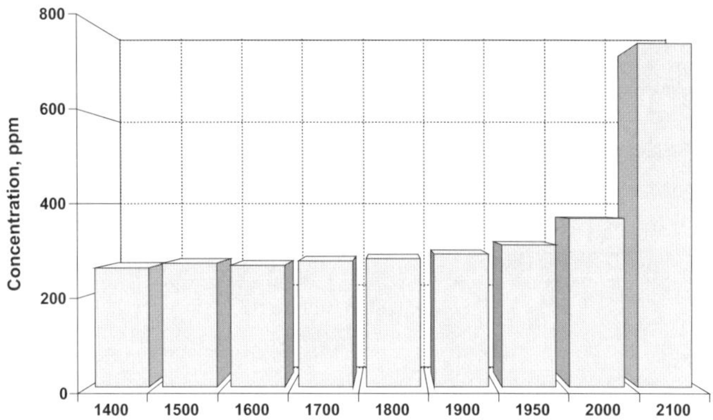

Figure 14. Concentration of carbon dioxide in the atmosphere.

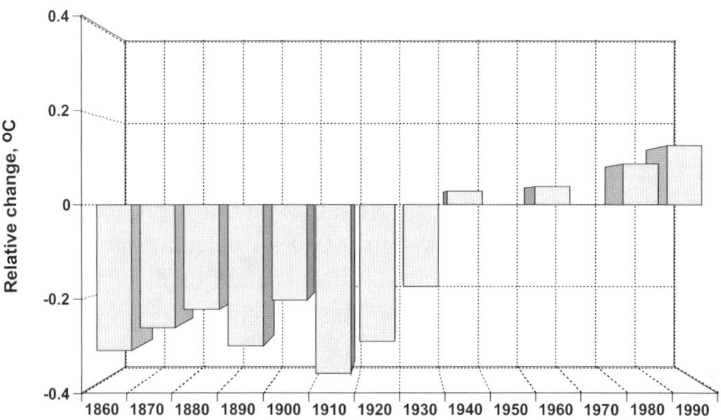

Figure 15. Deviation from average temperature on earth (basis: 1950-1970).

Some sceptic remarks can also be traced back to the cosmic and geological evolution of the earth (Petit, 1999; Gerhard et al., 2001). It is well know that the global climate is fundamentally influenced by long-term factors. In geoscience the following two factors are often mentioned:

1. Extraterrestrial (astronomical) phenomena, like decreasing solar radiation, irregular emission and distribution of cosmic dust and
2. Terrestrial phenomena, like continental drift, polar wandering and changes in the rotation axis of the planet.

These factors may result in a rhythmic fluctuation of glacial (cooling) and interglacial (warming) periods in each cosmic (190-200 million terrestrial) year, which means 15-20 global climate cycles (ice age or global warming) during the past 3 billion years. It is certain that the climate drastically changed in each cycle, but the crucial question today is how fast did these changes proceed in the past. Today it is increasingly apparent that the present rate of global warming is unusual and that the anthropogenic factor is probably decisive in the unfavourable processes.

The net CO_2 production always depends on the rates of emission and absorption. Some countries are net emitters, others absorbers. A detailed study carried out in Hungary (Fig. 16) has definitely shown that the CO_2 balance is negative and that the net CO_2 emission is estimated to be 40 TgC/y. Obviously, all countries must strive for zero CO_2 release or positive absorbance.

Despite different arguments it seems evident that anthropogenic CO_2 emissions are relatively small, but that their decrease may beneficially influence the global warming.

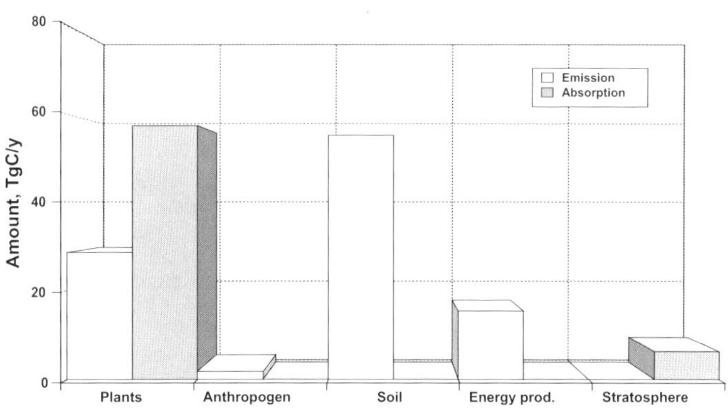

Figure 16. Emission/absorption equilibrium of CO_2 in Hungary.

6. TRADE OF GREENHOUSE GAS EMISSIONS

The Kyoto accord took effect in the last week of January, 2005, with new obligations to reduce the emission of CO_2 and other greenhouse gases (OECD, 2002). The countries signing the accord have a firm commitment to reduce emissions by 5.3%, on average, below those of 1990. Unfortunately, the accord had not been signed by a number of countries, including the USA, Australia and the majority of Middle Eastern and African countries. The principle of the accord is that the total carbon emissions in the ratifying countries are limited (capped) at a fixed level and those nations assign a number of CO_2 and greenhouse gas allowances (1 allowance = 10^6 t equivalent carbon emission) to their major carbon emitting sectors. The participating countries usually offered to reduce emissions by more than 5.3% (8% on average) until 2012. The present situation, however, is a little controversial. In 2002 some countries seem to have significantly exceeded the target, while others have not decreased rates, but rather have produced much higher emissions than in 1990 (Fig. 17). Therefore, the accord makes it possible to trade emissions between countries and companies. In other words, the countries that have unused allowances may sell them, while those that have excess carbon emissions must buy allowances.

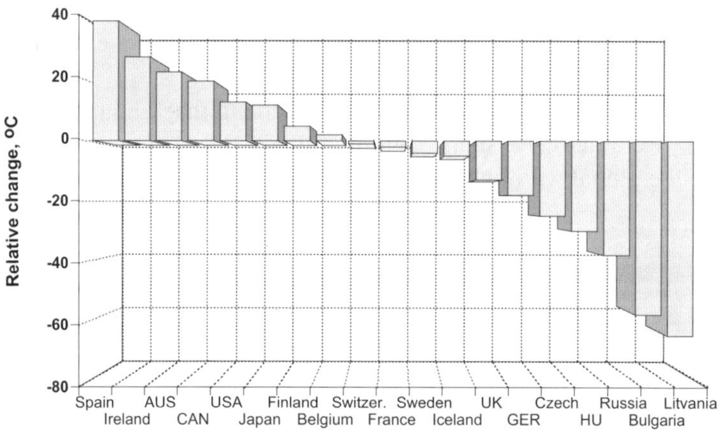

Figure 17. Present greenhouse gas emissions of different countries compared to the level in 1990.

Emission trading is just getting under way and it might be very significant in the future (Bird, 2005). Although only a short time has passed, the financial conditions of emission trading have taken shape and the following average data characterize the present situation:

- Price : ≈ 7 €/ allowance
- Selling unit : 5,000 allowances
- Transaction : 10,000-20,000 allowances
- Weekly volume : 10 million $

The recent trade volumes and the market are still small but they are growing rapidly , and so is the price of an allowance, as the accord took effect. Society has high hopes for emission trading and the Kyoto accord, but some scepticism can also be met. Summarizing the pros and cons, the following arguments can be listed:

Benefits:
1. the accord and emission trading imposes sanctions against the major CO_2 emitting sectors,
2. it encourages the reduction of greenhouse gases emission and
3. it urges the building of clean energy facilities.

Drawbacks:
1. the sanctions do not stop greenhouse emissions,
2. positive discrimination does not encourage technological improvements in those sectors that benefit from emission trading and
3. the new actions do not stop or decrease the concentration of CO_2 and greenhouse gas in the atmosphere.

Thus, parallel with the proposals of the Kyoto accord and emission trading it is fundamental to search for efficient actions and technologies which will provide economic and environmentally-friendly solutions to increase CO_2 and greenhouse-gas absorption and removal from the atmosphere (Orr, 2004). Such options might include:
1. prohibition of clear-cutting and burning of rainforests,
2. planting of more coniferous and deciduous forests world-wide, and
3. sequestration of CO_2 and other greenhouse gases in geological formations, including injection into oil reservoirs, gas-tight underground storage facilities, coal seams and permafrost formations as gas hydrates.

The present decision-makers and stake-holders (scientists, engineers, politicians etc.) cannot shift the responsibility of safeguarding our blue planet and the life of future generations onto somebody else. *The apparent anthropogenic effect on climate clearly proves that the earth is extremely vulnerable to even marginal changes in the atmosphere, hydrosphere and lithosphere. Therefore, all efforts to diminish man-made influence are of vital importance for future generations.* The time is short, the task is tremendous, but integrating efforts will allow the main goals to be fulfilled.

7. CONCLUSIONS

1. Fossil fuels will remain important elements of the energy scenario in the 21st century;
2. Emission of CO_2 and other greenhouse gases will increase until the middle of the century;
3. Diversification of energy sources using renewable energy is fundamental;
4. Replace fossil fuels with biomaterials, fusion and hydrogen energy;
5. Geological evidence shows that levels of CO_2 and other greenhouse gases have remained relatively stable on Earth for the past thousand years;
6. The natural and unnatural increase of greenhouse gases may cause global warming on earth;
7. Climate fluctuates naturally between warm and cool periods. Some researchers say that the 21st century has seen the greatest warming in at least a thousand years, and natural forces can't account for it all;
8. Solid evidence for greenhouse warming has not yet been found, but there is an agreement that CO_2 concentration in the atmosphere is increasing. However, there is no real consensus yet that this is responsible alone for global warming;
9. Despite different arguments it seems evident that anthropogenic CO_2 emissions are relatively small, but their decrease may beneficially influence global warming;
10. The apparent anthropogenic effect on climate clearly proves that the earth is extremely vulnerable to even marginal changes in the atmosphere, hydrosphere and lithosphere. Therefore, all efforts to diminish man-made influences are of vital importance for future generations;
11. Emission trading and sequestration of CO_2 and greenhouse gases have priority in the coming decades.
12. International cooperation is necessary to develop efficient, economic and environmentally-friendly methods, technologies and solutions to reduce emissions of CO_2 and other greenhouse gases and to diminish their concentration in the atmosphere.

ACKNOWLEDGEMENTS

The authors wish to express their appreciation to the Hungarian Science Foundation for their financial support within the framework of the OTKA T034108 and OTKA T047342 projects.

REFERENCES

Al-Fattah, S. M., and Starzman, R. A., 2000 Forecasting World Natural Gas Supply, *J. Pet. Eng.*, pp. 62, May (2000).

Arscott, L., 2003, Sustainable Development in the Oil and Gas Industry, *J. Pet. Eng.*, pp. 60, August (2003).

Bird, M., 2005, Emission Impossible?, *Time, 165*:8, pp. 40 (2005).

BP, 2005, *BP Statistical Review of Word Energy, 2005*, British Petroleum, June (2005).

ExxonMobil, 2004, *The Outlook for Energy: A 2003 View*, ExxonMobil Report (2004).

Falkowski, P. G., 2002 Climate Regulators, *Scientific American*, **8**:56 (2002).

Gerhard, L. C., Harrison, W. E., and Hanson, B. M., 2001, Geological perspectives of global climate change, *AAPG Studies in Geology*, Report 48 (2001).

Glick, D., Montaigne, F., and Morell, V. 2004, Signs from Earth, *National Geographic Magazine*, pp. 15, September (2004).

Holditch, S. A., 2003, The Increasing Role of Unconventional Reservoirs in the Future of the Oil and Gas Business, pp. 34, November (2003).

IPCC, 1995, Temperatures 1856-199. Climatic Research Unit, IPCC Report 95 (1995).

OECD, 2002, *World Energy Outlook*, International Energy Agency, OECD, Paris.

Ónodi, T., 2003, Doubts towards the effect of greenhouse gases", (in Hungarian), *Kőolaj és Földgáz (Hungarian Oil and Gas Journal)*, **136**:10, 119 (2003).

Orr, F. M., 2004, Storage of Carbon Dioxide in Geologic Formations, *J. Pet. Eng.*, pp. 90, September (2004).

Petit, J. R., 1999, Climate and atmospheric history of the past 420,00 years from the vostok ice cCore in Antartica, *Nature*, June (1999).

Shell, 2001, *Energy Needs, Choices and Possibilities: Scenario to 2005*, Shell International Report (2001).

Skov, A. M., World Energy Beyond 2050, *J. Pet. Eng.*, pp. 34, January (2003).

SPE Review, 2004, *Rising to the Environmental Challenge*, 176:6-8 (2004).

UN-ESC, 1996, *Energy and Sustainable Developments: Developments of Energy Resources in the Developing Countries*, Economic and Social Council, United Nations, Report E/C.13/1996/3, January (1996).

UN-ESC, 1998, *Concise Report on World Population Monitoring, 1999: Population Growth, Structure and Distribution*, Economic and Social Council, United Nations, Report E/CN.9/1999/2, December (1998).

UN-ESC, 2003, *Concise Report on World Population Monitoring, 2003: Population, Education and Distribution*, Economic and Social Council, United Nations, Report E/CN.9/2003/2, January (2003).

US-DOE, 2004, *International Energy Outlook, 2004*, Energy Information Administration, US Department of Energy, Report: DOE/EIA-0484 (2004).

US-DOE, 2005, *International Total Primary Energy and Related Information – Primary Energy Production (Supply) and Carbon Dioxide Emissions*, Energy Information Administration, US Department of Energy, June (2005).

WEC, 1995, *Global Energy Perspectives to 2050 and Beyond*, World Energy Council (WEC), International Institute of Applied System Analysis, London (1995).

STAKEHOLDER ACCEPTANCE AND UNDERSTANDING OF CO_2 GEOLOGICAL STORAGE

S. Vercelli and R. Tambelli
Dipt. di Psicologia dinamica e clinica; Università di Roma "La Sapienza", Rome, Italy

Abstract: CO_2 geological storage constitutes a relatively recent scientific technology which could play a major role for the solution of global warming. More research is needed, however at the same time already-available scientific knowledge has scarcely been disseminated. Thus researchers who work in this area are faced with two main problems: 1) get their work to be known and its relevance understood also outside research and academic circles 2) avoid possible misconceptions in the understanding of their research, which could result in negative reactions in public opinion and consequently to the refusal of geological storage. This contribution addresses the need to spread knowledge and make this technology better known for the exploitation of its potentialities. Methodological questions about effective dissemination are dealt with. A psycho-sociological approach is presented and discussed, focusing on: 1) how to develop in the public a positive attitude and interest for learning about geological storage; and 2) how to develop in the public a correct understanding of what geological storage is. This kind of approach works on both the cognitive and emotional levels, beginning with the scientists' representations of their own work and followed by studying the interactions people can develop when coming in contact with a new topic, what kind of reactions are stimulated and how they can be understood depending on specific social contexts, etc. Language study and image issues are explored with specific reference to geological research. Based on Italian experiences for geological research dissemination the role of cultural contexts and psycho-social representations are illustrated. The importance of the analysis of emotional processes is explained. The role of information, communication, storage sites' image and social context in relation to stakeholders decisions is analysed. A possible proactive image strategy is outlined.

Key words: geological storage, climate change, carbon dioxide, scientific dissemination, stakeholders, psychosocial representations, cultural models

S. Lombardi et al. (eds.),
Advances in the Geological Storage of Carbon Dioxide, 345–357.

1. INTRODUCTION

The environment we live in keeps changing all the time. As human beings we participate in this change, trying to understand it, sometimes trying to influence it. We try to understand what happens in the air, on the earth, inside the earth, in the universe. We try to create better environmental conditions, modifying natural features, using our imagination to make the earth a better place to live on. We have to adapt to environment changes, but we are also agents of change.

Change, though, can take unpredictable directions, which sometimes force us to recognise that our influence on natural environment, instead of being beneficial, can become dangerous for our own species.

In 1963 there was a terrible disaster up in the mountains of Lavarone, in the north of Italy. A large piece of the mountain detached itself, fell into the waters of the Vajont dam reservoir, which was a piece of high engineering skill. The water was displaced and flooded the villages in the valley below, causing the death of thousands of people. It was a shock, and we didn't know that it was just one of many disasters still to come due to an unhappy interaction between natural phenomena and man's work. After so many years the dam is still there, empty, useless, perfect in its technological beauty, reminding us how delicate the environmental balance is and how dependent we are, for our life and well being, on the maintenance of its equilibrium.

But the impact of man's work is not always so evident. It can be much more difficult and complex to explain the controlling factors, in human-nature interaction, that can produce negative effects on the environmental balance. A lot of work is being done to precisely evaluate the consequences of our decisions regarding development and the growing need for energy production, but how much of this work comes to be known, how much of this work is really understood outside academic circles?

When the issues are so complex, as is the case for climate change, scientific communication and dissemination is a particularly delicate matter. CO_2 geological storage (CO_2GC) is part of this conceptual context, a context in which it is becoming more and more difficult for common people to understand what's going on and what we should do to avoid being ourselves the cause of major natural catastrophes.

The present paper aims to contribute to the understanding of the issues that have to be considered when communicating such complex subjects.

2. CO_2GS: THEMES AND QUESTIONS

When introducing the subject of CO_2 geological storage numerous other complex and socially-important issues immediately come up.

First of all, CO_2 excess is a world-wide problem and thus it needs to be addressed globally. This is not easy, not only because it is a long and difficult task to find an agreement among all countries, but also because we are just beginning to learn how to face problems that involve a global context. The number of factors which need to be taken into account is great and local policies are no longer local, but rather they get their meaning in relation to many other local policies. If we are dealing with a world problem, how can a "world – scale way of thinking" be developed?

Another major and extremely complex theme is energy production: fossil fuels, nuclear power, renewables, etc. (each with its different advantages and disadvantages) have social, environmental, economic features and costs that form the context within which CO_2GS is going to be evaluated (Shackley et al., 2004). How is CO_2GS to be considered in relation to the general energy situation? What kind of solutions can it offer? How can it help? From the economic point of view, what are the costs to implement this technology and how do they compare to those of other technologies?

People, and especially stakeholders, need to learn about CO_2GS, to be able to decide on its potential and because public collaboration is needed for its implementation. In this respect there are a number of concepts that are crucial for a correct understanding. They need to be studied not only to find scientific information, but also to outline the kind of relationship we feel with them. For example, what is CO_2, what has it come to mean, what is geological storage, how does one imagine it, and how is the effect on the earth (or with a particular territory) felt and characterized?

With some differences among the various countries, some of which are moving forward more quickly than others, the present situation is a preliminary phase to the possible adoption of CO_2GS as a mean for the reduction of excess CO_2 in the atmosphere.

Any decision regarding energy issues will have an enormous impact on society in terms of quality of life and well – being, economic costs and profits, environmental exploitation and protection, social safety and risks, etc.. Thus it is socially very relevant that complete and correct information is available on the various technological opportunities in order to help take decisions. At the same time all the different social groups, with their specific requests, need to enter into the learning process to be able to participate in the decision-making process (Shackley et al., 2004).

The type of information requested on CO_2GS will vary in relation to the needs of different social subjects. For example environmentalists and citizen associations will probably be more interested in safety and health-related issues, politicians to costs and implementation times and plans, industries to costs and advantages such as incentives, and local administrators to control and monitoring of sites, etc.

3. WHY A COMMUNICATION STRATEGY IS NEEDED

CO_2 geological storage is a relatively new technology which is still very little known by the public at large. In the United States just 4% of the population seems to have heard about CO_2GS (Curry et al., 2004). We don't have data for Europe but probably the situation is similar. At the same time, however, CO_2GS is believed to be quite a crucial technology for the reduction of excess CO_2 in the atmosphere.

CO_2 production is expected to increase globally in the future. Newly industrialised countries, such as China, are expected to further contribute to a situation that many observers already consider very dangerous for the earth's health and for our well – being. There is a lot of debate on how to deal with this very serious health and environmental issue, but one point seems to be clear: CO_2 emissions need to be controlled and reduced.

To reduce emissions we must find, improve or increase new ways of energy production that do not result in massive CO_2 production. In the intervening time, during which the new technologies will be developed, we must do all we can to deal with CO_2 that is still being produced through the use of fossil fuels. It is here that CO_2GS comes to be regarded as a safe technology that can potentially have a major role in avoiding the release of large quantities of CO_2 from present technologies into the atmosphere.

We have a problem, CO_2 excess, and we have a potential partial solution to that problem in the form of CO_2GS. What is needed to implement this solution and thus improve the overall situation of CO_2 in the atmosphere? Certainly more research is needed, more studies need to be done on how this technology can best be implemented, on where and how storage sites must be constructed, how they must be monitored and how local populations are to be involved in site selection and implementation. We need technical, legal, economic, and socio – psychological guidelines to be able to facilitate a wide exploitation of this technology (de Conink and Huijts, 2004). But what is probably needed more at this stage is that this technology is better known and understood, both by the public at large and by decision-makers dealing with the problem of excess CO_2.

A correct dissemination of information on CO_2GS issues is fundamental for the promotion of social support and funding, which are both essential for further development and exploitation of this technology. Making CO_2GS better known means:

- **positioning it in the minds of the public as one available option for reducing CO_2.** At the moment most people don't even know about the existence of CO_2GS, therefore it can hardly be considered an option. Furthermore, one has to take care that people get acquainted with the new technology through positive and correct information.
- **facilitating communication with stakeholders.** Stakeholders would find it very difficult to make decisions concerning a technology that is very little known, and therefore implies the risk of a lack of consensus. Clear and tested guidelines can greatly facilitate becoming familiar with the technology and taking appropriate decisions.
- **facilitating communication with the public living near potential sites.** Public consensus is a decisive factor for the construction of new sites. The exploration of the demands and needs of people living near a potential site is required. It is of no use to develop sophisticated technologies if one cannot find a place to put them because people don't know how to evaluate them, and therefore are afraid and don't want them. Getting people to know what CO_2GS is all about must then be a major goal of any policy regarding CO_2GS (Bradbury and Dooley, 2004).

4. STAKEHOLDERS

When we consider the problem of social acceptance we must be careful to take into account any social agency that could be concerned with our problem. In this respect "stakeholders" is a key concept.

What do we mean by stakeholders? In general they can be thought of as those people who have an interest in a particular decision, either as individuals or as representatives of a group of people. This includes people who influence a decision, or can influence it, as well as those who are affected by it. In other words any social group which is directly or indirectly involved with the matter can be viewed as a stakeholder, such as all people living in a site area, all people concerned with environmental issues, all people working in the energy field, all people living in large polluted cities, etc. This way to define stakeholders permits us to take into consideration the various components that contribute to the formation of public opinion and, even more importantly, to avoid neglect of such factors which could arise should our subject strike any sensitive topics in the public. What we mean is that strong reactions can sometimes be elicited and action be taken by previously unknown and unconsidered social groups. It is important to be able to anticipate whether a certain subject could cause particular reactions in the public, in people who are not usually active in taking decisions. Social decisional processes are quite complex and being able to understand the factors which could have a determining influence is greatly facilitating.

Let us consider an example. Recently in Italy it was possible to observe what can happen when stakeholders (in its broadest sense) are not taken into account and thus are not consulted, informed or prepared. In Scanzano, Basilicata, the government had decided to create a dangerous-waste disposal landfill in a site which was considered very safe from a scientific point of view. Unfortunately the local public had not been consulted and no one suspected that a very strong feeling against this waste disposal could arise. In fact, the reaction was so strong that the entire population in the region was actively involved, and for weeks the main roads and railway lines were blocked (with many people sleeping under the stars) until the government had to finally change its decision. What had happened? The area is still a rather well-preserved, uncontaminated region with beautiful landscape, where tourism is expected to increase, and has recently seen the substantial development of organic agriculture. The development potential of this area therefore relies on it being a clean environment. Radioactive waste disposal, even the disposal of low level radioactive waste (such as medical and radiological waste), did not fit with the idea of the local population as how to preserve a clean environment. It was felt that the central government was just interested in discharging at this site something "they didn't know where to put". One can easily understand why this waste disposal site had then come to be considered as such a life threatening issue.

A more common definition for stakeholders is all those social agencies that, in one way or another, have a recognised power to decide or to influence decisions regarding a certain subject, such as local or central administrators, politicians, environmental associations, citizens' associations, scientific centres, etc.. In an even more restricted meaning we can also consider the so called "opinion makers" to be stakeholders, such as specialised journalists, eminent scientists or writers, singers or actors etc.

5. CONSENSUS

Social consensus is a crucial issue for research development and application. It is easily understandable that the greater the interest and the approval a population holds for a certain research domain, the higher the possibilities that that research will be supported. This support will manifest itself not only through research funding, but also in the processes that concern the research being carried out and applied.

Social consensus for research can very often not be an explicit issue. To get an idea of its importance, one can think about television programs organised to raise funds for medical research into particular, and sometimes very rare, illnesses. Looking at one of this programs one can easily understand that the fact that most people feel involved with the researcher's goals makes it possible to raise quite large sums of money, while at the same time ever increasing the interest of people for this kind of research.

It can be quite interesting to reflect on this striking observation: a population's health depends greatly on a safe and clean environment, having clean air to breathe, clean water to drink, good shelter and so on; we also know that the incidence of many illnesses is reduced by living in a healthy environment. Nevertheless it is much easier for the general public to get concerned and involved over certain specific illnesses rather than over issues that can have a much greater influence for the health of entire populations, not to say for the entire global population.

Climate changes can adversely affect the life of millions of people. But to date we haven't seen any television shows raising funds for research on this very subject. How can we understand the scarce involvement and activation level towards issues that can have a massive impact on the public's well – being? Is there a correct distribution of resources with respect to the relative importance of the various problems? From this simple example one can realize the fundamental role of social consensus and its relevance for the development of different scientific domains.

Social consensus can be regarded as an expression of the level at which research goals are shared and understood. That is, there are both an emotional and a cognitive side to the building of consensus.

6. STUDY OF THE IMAGE OF CO_2GS

CO_2GS is presently very little known by the public at large, and thus we have the opportunity to work on its potential image to facilitate its understanding and acceptance. Working on the image implies the identification of the core aspects that qualify a certain subject and the communication goals to be pursued. (Vercelli, 1989).

The image of a given subject is formed both on available information and on the attitude that the object stimulates in the receiver. The first response of the receiver to the image subject can be positive or negative, depending on the feeling and the affective encoding that will be activated. It is very important to get a positive reaction when first contact with the subject takes place. A positive reaction opens the way to further communication, to the demand for more information, to the gradual construction of a relation with the subject that can evolve in time. To summarise we could describe the positive attitude as “tell me more”.

On the other hand, if the first reaction is negative, it will be very difficult to change it, as a negative reaction is bound to activate defences toward something which is regarded as inimical. If the first feeling when coming in contact with the given subject is unpleasant, then any further information might not be taken into account. Thus we could describe the negative attitude as “I’m not interested”.

So, the first and fundamental goal is to find a formula to present our subject in such a way that the majority of people will react wanting to know more.

Our message should be:

- Easy: anybody can feel involved
- Simple: anybody can understand (language).

As we said before the matter we are working on is particularly complex and relatively unknown by the majority of people. At present we can study the image of CO_2GS in the scientific community, looking at some stakeholders which have already been involved (like some environmentalist associations) in those areas where it has already been implemented and in the natural analogues areas . This kind of work will enable us to identify the core aspects that characterise CO_2GS with respect to the way people think and feel about it.

At the same time, the fact itself of conducting this kind of study provides a new context within which the feelings and thoughts related to CO_2GS enter into a process of psycho-social elaboration.

7. PSYCHO-SOCIAL REPRESENTATIONS

7.1 Thinking

When first contact with a new concept or area of knowledge takes place, a process begins which involves the integration of the new information in a pre–existing cognitive network. Previous knowledge, its type and complexity, interacts with the new concepts or data. Depending on the already-assimilated information it can be more or less difficult to understand the new concepts. For example, if I have some knowledge about the earth, how it is constituted, what are the physics of the subsurface, I have a cognitive basis within which geological storage can fit. In the same way people who are used to the scientific approach will have a different comprehension from people who have another cultural background, etc..

Language is also a central issue. It plays an important role in making new knowledge more or less understandable. Terms that are new or particularly relevant require a clear and simple definition. In particular when non–technical stakeholders need to be involved it can be quite crucial that correct information is expressed in simple and everyday language.

7.2 Feeling

On a different level, the new information is categorized on the basis of its affective value. Do I feel this could be meaningful for me, do I feel it can be advantageous or do I feel it might possibly be dangerous? etc... Whereas the basic categorization is friendly/inimical, it is subsequently elaborated depending on specific personal and context variables. This elaboration takes the form of representations that express the kind of relationship with the given subject (Carli and Paniccia 1981; Carli and Paniccia, 2002).

Let us consider an example. In a study for the dissemination of data from the EC-funded Geochemical Seismic Zonation project (Lombardi and Vercelli, 2000), two different groups were studied: one living in Spoleto, a high seismic risk area of central Italy; and one living in Rome, a low seismic risk area.

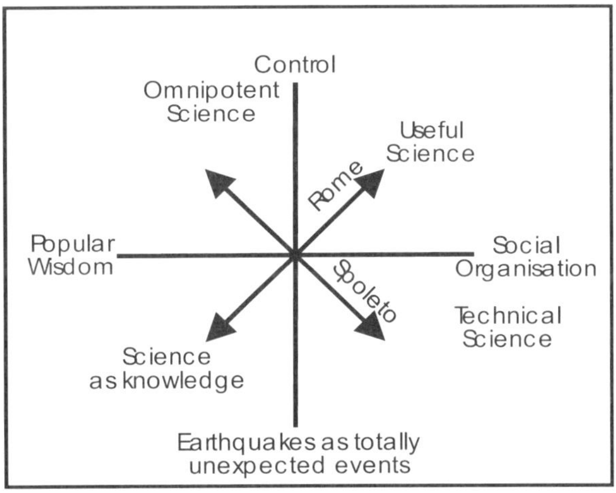

Figure 1. Symbolic dimensions of earthquake culture and science image.

The two groups expressed several earthquake representations, which were linked differently to the image of science and what they expected from science. As can be seen in fig.1, earthquake culture was organized on two main axes.

The first one regarded earthquakes as being, at one extreme, totally unexpected events, that is something we can have no control over whatsoever. This idea was linked to an image of science as something without practical consequences, just knowledge for its own sake, or alternatively to a technical image of science, for example being able to tell how to construct buildings but with no potential for prediction.

At the other end there was a cultural dimension of getting total control over the natural phenomena of earthquakes. This could give place to two opposed conceptions of science, depending on how the community was represented. In a situation in which it was felt that everybody was left to themselves, that the frame of reference was constituted by popular sayings and stories, lived experiences and historical data, science could become a sort of omnipotent ideal. On the opposite side science was felt to be able to supply a certain degree of control, of mastery, when the phenomena was considered something that had to be faced through social organization. The four images of science that emerged help us understand the socio - cultural context and how new information is going to be processed.

In Rome, the low risk area, the "science as knowledge" and "useful science" images were prevalent. Here we have important elements derived from the city's history that "tell" the Romans that usually earthquakes are not too destructive for the city and that there are hundreds of ancient buildings still standing (even if the ground level might have sunk a bit). In this context the scientific contribution could be considered as being scarcely relevant (science as knowledge). What facilitates learning and stimulates an interest is the feeling that society as a whole is taking care of the problem and organizing itself to face it, going beyond the boundaries of one's own area and getting involved in the problem even if one is not personally touched by it (useful science).

In Spoleto, the high seismic risk area, the social representations highlighted quite a different cultural climate. Science images were mostly those of "omnipotent science" and "technical science". In this case the risk was that scientific information could be easily devaluated if it did not meet the expectation of having absolute and certain answers. In this community we would have expected to find more of the "useful science" image. Given the very real risk of earthquakes a good social organization based on updated scientific data could make a significant difference for the community. But this dimension seems at the moment lacking, whereas a more technical approach is more reassuring: we might have no control whatsoever but we can perhaps become more organized towards earthquake consequences.

Here ("technical science") we find more realistic expectations, which positively evaluate the contribution of social organization. It is then possible that an approach that couples scientific data with more technical information will facilitate learning and gradually help develop a more functional cultural representation (could be "useful science").

The way it "feels" can make a great difference in the degree of interest for learning about a new subject. And the study of psycho–social representations can help us understand which is the better way to stimulate positive feelings and thus create a connection with people and stakeholders that will facilitate communication, learning, exchange of ideas and the discussion of problems (Carli, 1987; Carli et al., 1988).

The study of the cultural dimension we enter into, as we start talking about CO_2GS, allows us to identify the possible obstacles to good communication. In this way we get to know the context factors that will influence the way people will look at what we are presenting. There is, though, an even more important outcome. As we start the study of psycho–social representations we give attention to something people live but haven't yet shared, expressed or thought about. We start to make connections and bring into the open issues that constitute implicit ties that need to be discussed. In this way we open a space where scientific data and research can become, and be felt as, the community's property. In other words we actively stimulate the process that allows research to be assimilated into the general culture. At the same time scientific research development is promoted through an open flux between the academic community and the rest of society (Gregory and Miller, 1998).

8. CONCLUSIONS

Geological storage has the potential to be a very valuable technology to help us face the problems that arise from excess CO_2 in the atmosphere. At the moment, however, not only the great majority of the general public but also many relevant stakeholders know very little or nothing about it.

The scientific concepts and data that support this technology are not yet integrated into everyday culture. The way people are going to understand and react when coming in contact with this technology needs to be studied. This is particularly true for people who might be directly involved (because they are living in a potential site area) and for all stakeholders (that is the decision-makers who have the power to implement CO_2GS).

The moment is favourable. Since the matter is for most totally new, there is the opportunity to reach the public with correct information without it already being prejudiced. This same fact, though, makes it particularly difficult to find the right communication channel that can prevent negative reactions and enhance learning and understanding.

CO_2GS is just one aspect of the efforts that are being made to face climate change and that part of it that depends on human activities. It is a potential technology that people need to learn more about, as part of a wider range of measures that can be adopted (Shackley et al., 2004). An organised image strategy is desirable both to meet the demands of stakeholders and the public, and to help the scientific community in the effort of communicating.

Stakeholders acceptance and understanding is the basis for a social decisional process that correctly evaluates advantages and disadvantages of each possible solution. It is essential to understand that cultural dimensions play a fundamental role in social decisional processes. When the gap between new concepts and the current way of thinking is large, it may be very difficult to recognise the social relevance of new technologies and solutions.

The implementation of CO_2GS requires the construction of a context where all different social parties, all stakeholders, can meet and elaborate together a new cultural dimension that takes into account all available information and "reads" it in the light of shared feelings and impressions. This kind of process forms the basis for the understanding of what we, as society, really consider valuable and therefore worth achieving.

The study of psycho–social representations can provide a starting point for such new context, in that it brings to light at least some aspects of the network of thoughts and feelings that are socially shared on a certain subject in a certain social group. In so doing, new symbolic representations and new cognitive networks are stimulated and a new culture can develop where climate change, geological storage and CO_2 issues are presented in such a way that everybody is enabled to understand what is being discussed and which solutions are wise to choose.

REFERENCES

Bradbury, J.A., and Dooley, J.J., 2004, Who's talking? What are the issues? The media's portrayal of carbon dioxide capture and sequestration in the United States, *GHGT7 - 7th International Conference on Greenhouse Gas Control Technologies*, Vancouver, Canada, 5-9 September 2004, 4 p.

Carli, R., 1987, The analysis of the question, *Rivista di Psicologia Clinica* **1**:38-53. (in Italian).

Carli, R., and Paniccia, R.M., 1981, *Psycho-sociology of organisations and institutions*, Bologna: Il Mulino. (in Italian).

Carli, R., and Paniccia, R.M., 2002, *The emotional analysis of text*, , Franco Angeli, Milano, 2002.

Carli, R., Paniccia, R.M., and Lancia, F., 1988, *The group in clinical psychology*. Roma: La Nuova Italia Scientifica (in Italian).

Curry, T., Reiner D.M., Ansolabehere, S., Herzog, H.J., 2004, How aware is the public of carbon capture and storage?, *GHGT7 - 7th International Conference on Greenhouse Gas Control Technologies*, Vancouver, Canada, 5-9 September 2004, 9 p.

De Conink, H.C., and Huijts, N.M.A., Carbon Dioxide Capture and Storage: Public Perception, Policy and Regulatory Issues in the Netherlands, *GHGT7 - 7th International Conference on Greenhouse Gas Control Technologies*, Vancouver, Canada, 5-9 September 2004, 4 p.

Gregory, J., and Miller, S., 1998, *Science in public*, Communication, culture and credibility, Perseus Publishing, Cambridge Massachussetts.

Lombardi, S., and .Vercelli, S.,2000, How to help people enjoy the advantages of scientific knowledge: a project to disseminate the scientific results of the geochemical seismic zonation project, Workshop proceedings of *Mitigation of Seismic Risk Support to Recently Affected European Countries*, November 2000, European Commission-Joint Research Centre.

Lombardi, S., and Vercelli, S., 2000, "Understand" the earthquake: study for the dissemination of scientific data obtained durino the EC-funded project "Geochemical Seismic Zonation", Proceedings of the conference: *Scientific museums and scholarly formation: problems, resources and instruments*, October 2000, Rome Italy. (in Italian).

Shackley, S., McLachlan, C., Gough, C., 2004, The public perceptions of carbon cpture and storage, Tyndall Centre for Climate Change Research, Working Paper 44, January 2004.

Vercelli S., L'immagine dello psicologo: ipotesi sul rapporto tra ricerca d'identità e opinione sociale, Tesi di laurea, Roma 1989.

CO_2GEONET – AN EC-FUNDED "NETWORK OF EXCELLENCE" TO STUDY THE GEOLOGICAL STORAGE OF CO_2

N. Riley
British Geological Survey, Keyworth, Nottingham NG12 5GG, UK

Abstract: The Sixth Framework Programme (FP6)- 2002-2006 - is the European Unions' main instrument for research funding in Europe. FP6 serves two main strategic objectives: to strengthen the scientific and technological bases of industry and to improve competitiveness and innovation in Europe through the promotion of increased co-operation and improved coordination between relevant actors at all levels. Within FP6, the Network of Excellence (NoE) instrument was developed to strengthen excellence on a particular research topic by tackling the fragmentation of European research. One such NoE, formed in 2004, is CO_2GeoNet, which is made up of 13 research organizations from across Europe who have extensive experience and expertise in the study of the geological storage of CO_2. The principle goal of this consortium is to contribute to and form a durable integration of research into CO_2 geological storage, in the European Research Area, as a means of mitigating anthropogenic mass loading of this greenhouse gas to the atmosphere and ocean, while there is continued use of fossil fuels.

Key words: European Network of Excellence, geological storage of CO_2, research consortium, CO_2GeoNet

The EC Framework 6 Network of Excellence, "CO_2GeoNet", a 13 partner research network, contains a critical mass of research activity in the area of underground carbon dioxide (CO_2) storage. World projections of energy use show that fossil fuel dependency will continue to 2030 and beyond; but sustainability will need CO_2 global emissions reductions by at least 60% by 2050. This will be difficult, especially with the rise of the emerging economies such as China, India, Russia and Brazil, and the huge growth in surface transport. By 2020 China and the USA are likely to be

S. Lombardi et al. (eds.),
Advances in the Geological Storage of Carbon Dioxide, 359–362.

responsible for 50% of global emissions. It will require various strategies by Europe, both with its internal energy systems and in international negotiations/policy implementation to set the world on the 60% reduction path within the next decade. The associated rise in global CO_2 emissions, without abatement, will be at an average rate of 1.8% per annum (from the current value of 25Gt p.a., to 38Gt by 2030) - a rise of over 50%. This would be catastrophic for the planet's sustainability, leading to dangerous levels of global warming (+2°C above pre-industrial), sea level rise and ocean acidification (+0.7°C has already been achieved and ocean pH has fallen by 0.1). Urgent action is needed. Europe's CO_2 emissions will rise by an average of 0.6% p.a. up to 2020, from a 2000 level of 3.1Gt to 3.5Gt by 2020.

The rocks under the N. Sea have a theoretical capacity for storing over 800Gt of CO_2. Capturing CO_2 from industrial point sources and storing it underground (a process that mimics nature) is a very attractive route to make cuts in CO_2 emissions. CO_2 capture and storage allows diverse fuel inputs/outputs, enhances security of energy supply and is well aligned with hydrogen production from fossil fuels. About 33% of CO_2 emissions arise from power generation. If hydrogen or electricity based motive power takes over surface transport, then a major proportion of the 60% emission cut could be achieved using CCS. Through the European Commission's Joule 2, FP4 & 5 projects Europe has led the world in R&D in this area, with rapid growth this decade. National programmes are also emerging. This success has a downside, by creating fragmentation through diversification. N. America despite its rejection of Kyoto (except Canada), has recently embraced CO_2 capture and geological storage and is allocating huge resources (over $4bn) over the next 10 years, and similar developments are happening in Australian research programmes. Europe, as a result, risks losing its head start. We therefore must work more effectively and restructure accordingly. The main aim of CO_2GeoNet will be to integrate, strengthen, and build upon the momentum of previous and existing European R&D. In addition it will project European excellence internationally, so as to ensure that Europe remains at the forefront of CO_2 underground storage research and that research leading to implementing large scale and quick deployment of underground CO_2 storage in Europe is delivered as effectively as possible.

The Network focus is on the geological storage of CO_2 as a greenhouse gas mitigation option (not capture). It has several objectives over the 5-year period of EC funding for integration;

- To form a durable and complimentary partnership comprising of a critical mass of key European research centres whose expertise and capability becomes increasingly mutually interdependent. The initial partnership will be between 13 institutes, most of whom have a long and established history of research in geological sequestration. Some new players are also included, either because they are expected to have significant national strategic profile in future CO_2 storage projects, or have capabilities which can be realigned to strengthen the network, or even bring uniqueness. For the first time in an EC FP project marine biologists will be drawn into this research topic.

- To maintain and build upon the momentum and world lead that Europe has on geological CO_2 sequestration and project that lead into the international arena.
- To improve efficiency through realignment of national research programmes, prevention of duplication of research effort, sharing of existing and newly acquired infrastructure and IPR.
- To identify knowledge gaps and formulate new research projects and tools to fill these gaps.
- Seek external funding from national and industrial programmes in order to diversify, build and strengthen the portfolio of shared research activities.
- To provide the authoritative body for technical, impartial, high quality information on geological storage of CO_2, and in so doing enable public confidence in the technology, participate in policy, regulatory formulation and common standards.
- Provide training to strengthen the partners, bring in new network members and sustain a replacement supply of researchers for the future.
- To exploit network IPR, both as a revenue earner to sustain the network and to equip European industry to be competitive in the emerging global low carbon energy markets.

In order to achieve these goals the following five areas of research have been defined, with each partner participating in the areas where they have the most expertise:

- Predictive numerical tools, including geochemical, geophysical, fluid flow and geomechanical tools
- Rock and fluid dynamic experimental facilities, including geochemical, geophysical, fluid flow and geomechanical facilities, data and information.
- Enhanced hydrocarbon recovery methods and tools, including enhanced oil, gas, coal bed methane. Monitoring techniques, including geochemical, geophysical, hydrological, biological and remote sensing techniques

- Risk and uncertainty methods and tools, including ecosystem, health and safety and long term impacts, as well as data and risk scenario uncertainty and mitigation strategy

A total of 7 European nations are represented within the Network, with participation by the following institutions:

1. Denmark
   Geological Survey of Denmark and Greenland - GEUS
2. France
   Bureau de Recherches Geologiques et Minieres- BRGM
   Institute Francais du Petrole - IFP
3. Germany
   Federal Institute for Geosciences and Natural Resources - BGR
4. Italy
   Istituto Nazionale di Oceanografia e di Geofisica Sperimentale- OGS
   Università di Roma "La Sapienza" - URS
5. Netherlands
   Netherlands Organisation for Applied Scientific Research - TNO
6. Norway
   Norwegian Institute for Water Research - NIVA
   Stiftelsen Rogalandsforskning- RF
   SINTEF Petroleumsforskning AS - SPR
7. UK
   British Geological Survey (BGS- NERC)
   Heriot-Watt University - HWU
   Imperial College of Science, Technology and Medicine- IMPERIAL

Further information can be obtained at www.CO_2geonet.com, or by contacting the co-ordinator, Dr. Nick Riley at njr@bgs.ac.uk, tel +44 115 9363312